INTRODUCTION TO
HERPETOLOGY

INTRODUCTION TO
HERPETOLOGY

THIRD EDITION

Coleman J. Goin
MUSEUM OF NORTHERN ARIZONA

Olive B. Goin
MUSEUM OF NORTHERN ARIZONA

George R. Zug
NATIONAL MUSEUM OF NATURAL HISTORY
SMITHSONIAN INSTITUTION

W. H. FREEMAN AND COMPANY
New York

Library of Congress Cataloging in Publication Data
Goin, Coleman Jett, 1911–
    Introduction to herpetology.

    Includes bibliographies and index.
    1.   Herpetology.   I.   Goin, Olive Bown, joint author.
II.   Zug, George R., 1938–   joint author.
III.   Title.
QL641.G6   1978   598.1   77-13554
ISBN 0-7167-0020-4

Printed in the United States of America

3 4 5 6 7 8 9 0   VB   5 4 3 2 1 0 8 9 8 7

CONTENTS

PREFACE
TO THE THIRD EDITION

This edition of the *Introduction to Herpetology* follows the same organization and philosophy as the first two editions. Our intent is not to be encyclopedic, but rather to cover the wide spectrum of amphibian and reptilian biology. We believe that through this approach the reader will obtain a solid foundation in herpetology and a sense of the flavor of current herpetological research.

With the exception of the two chapters on structure, all chapters have been rewritten and reorganized. Speciation and geographic distribution have been combined into a single chapter because of the close relationship between the two subjects. Maps of the distribution of the major amphibian and reptilian families have been added.

With each revision, our indebtedness increases. In addition to those acknowledged in the first and second editions, we wish to thank Charles J. Cole for the excellent photographs of karyotypes in Chapter 11 and Bart Kaveruck for help with the artwork. Edwin H. Colbert has been unfailingly kind and helpful. Rather than single out a few colleagues for special acknowledgment, we wish to thank all who have contributed ideas and corrections through conversations, correspondence, and publications. Finally, special thanks to Pat Zug for her patient typing and retyping of this revision.

As always, we hope our readers will continue to point out errors, inaccuracies, and omissions.

Coleman J. Goin
Olive B. Goin
October 1977   *George R. Zug*

PREFACE
TO THE FIRST EDITION

This book is planned for use as a text in a one-semester course in herpetology. It is designed for students who have had one year of college biology, but who may have had no more than one year.

Courses in herpetology usually consist of two parts—a series of lectures or discussions in which basic principles are presented, and a series of laboratory and field exercises. We believe that laboratory work should be based primarily on local faunas and on the specimens available in local institutions. These differ from region to region. Techniques useful for collecting animals in one climate may be of little use in another; the season of the year at which animals are abundant and active varies from place to place; the characters used in identifying specimens from one faunal region may not apply to those from another. Moreover, excellent field guides, keys, and local lists are available for most places where courses of herpetology are presented today. We have therefore left the organization of the laboratory and field work to the individual instructor, and have concentrated instead on the major aspects of herpetology that apply throughout the world.

In these days when so much of biology is concerned with happenings at the molecular level within the individual cell, we believe there is a definite need for the student to appreciate that these processes have biological meaning only as they help us to understand the living, functioning animal. We agree with Professor Romer that:

"It is not enough to name an animal; we want to know everything about him: what sort of a life he leads, his habits and instincts, how he gains his food and escapes enemies, his relations to other animals and his physical environment, his courtship and reproduction, care of his young, home life (if any). Some aspects of these inquiries are dignified by such names as *ecology* and *ethology;* for the most part they come broadly under the term *natural history.* Many workers who may study deeply—but narrowly—the physiological processes or anatomical structure of animals are liable to phrase this, somewhat scornfully, as '*mere* natural history.' But, on reflection, this attitude is the exact opposite of the proper one. No anatomical structure, however beautifully designed, no physiological or biochemical process, however interesting to the technical worker, is of importance except insofar as it contributes to the survival and welfare of the animal. The study of the functioning of an animal in nature—to put it crudely, how he goes about his business of being an animal—is in many regards the highest possible level of biological investigation." *(The Vertebrate Story,* A. S. Romer, Univ. Chicago Press, Copyright 1959 by the University of Chicago.)

In preparing this volume, our task has been twofold: we have had first to decide on the basic organization that we felt a text in this field should have, and second to synthesize a mountain of original literature into a volume of modest size.

In the former task, we have worked under the firm conviction that the proper approach was to discuss basic biological principles as exemplified by amphibians and reptiles. This we have tried to do. In the latter task, we have of necessity shown a great deal of personal bias in deciding just what should and what should not be included. Because of the wealth of original literature, there are many fascinating facts of herpetology that we have had to leave unmentioned. We know that some of our professional colleagues will feel that, like good Anglicans, "we have left out those things which we ought to have put in, and we have put in those things which we ought to have left out."

In the first chapter we indicate the position of the amphibians and reptiles in the animal kingdom, discuss briefly certain basic principles of classification, and give a resume of the rise of herpetology as a science. The next four chapters deal with the structure and evolutionary history of the two classes. Chapters 6 through 11 are concerned with natural history and with the mechanisms of speciation and geographic distribution. The last six chapters give a summary of the living amphibians and reptiles to the family or subfamily level, with notes on life history and geographic distribution.

Since herpetology is a worldwide subject, we have tried, in our choice of examples and illustrative material, to strike a balance between native and exotic forms.

The references given at the end of each chapter are intended simply as suggestions to the interested student who may wish to pursue a particular topic further than is possible in an introductory text. Most are compendiums. Occasionally, we have included original papers that are of exceptional interest and importance and that give information not generally available elsewhere.

It is a pleasure to acknowledge the help we have received from so many people in the preparation of this text. We are indebted to M. Graham Netting for reading Chapters 8, 9, 11, 13, and 17, and to Archie Carr for reading Chapters 8, 9, and 14. We wish especially to thank Kenneth W. Cooper, not only for the critical reading of Chapter 10, but also for the continuing interest he has shown in this work, and for the many pertinent references he has sent us. Henryk Szarski read Chapter 4 and Carl Gans read portions of an earlier draft of the manuscript. Walter Auffenberg assisted us materially in dealing with the classification of the snakes and crocodilians.

Mr. and Mrs. J. C. Battersby, Charles M. Bogert, Robert F. Inger, and Alfred S. Romer all responded most kindly to appeals for special information and material.

Not least have been the intangible benefits we have received from discussions with these and other colleagues, among whom we should like to mention in particular Doris M. Cochran and Ivor Griffiths.

Except where otherwise indicated, the excellent photographs were made by Isabelle Hunt Conant, many of them especially for this volume. The line drawings are from the gifted pen of Evan Gillespie, many of them from sketches made originally by Ester Coogle.

To James E. Bohlke, Alice G. C. Grandison, W. S. Pitt, Oswaldo Reig, and Charles K. Weichert we are indebted for special illustrative material.

We wish to thank Dean George T. Harrell for the use of a dictaphone during the preparation of the first draft, and Mrs. Sue P. Johnson for her careful typing of the final draft.

Since everything we have ever done has had imperfections, we feel sure that this book will have its share. We would like to request that our friends be kind enough to point out to us our errors, both of omission and of commission, so that in the future we may mend our ways.

Coleman J. Goin
March 1962   *Olive B. Goin*

INTRODUCTION TO
HERPETOLOGY

CHAPTER ONE

INTRODUCTION

THE SCIENCE OF BIOLOGY has become enormously complex. No longer is it possible for the work of one person to encompass all of its ramifications. To keep up with advances, biologists have been increasingly obliged to limit themselves to various subsciences, such as anatomy, genetics, embryology, or ecology. These subsciences can be visualized as extending vertically through the parent science. But this is not the only way biology can be approached; it can be subdivided according to the kinds of organisms studied. These subdivisions include such disciplines as ornithology, entomology, and herpetology. They extend across and interweave with the primary divisions mentioned above.

Biology might thus be visualized as a vast tapestry, with the threads of the warp formed by the many subsciences and those of the woof formed by the groups of organisms under study. Whether one wishes to follow a thread of the warp and study something like the anatomy of one or more structures in many different kinds of animals (comparative anatomy) or to follow a thread of the woof and study the anatomy, behavior, and distribution of a particular group of organisms (such as the snakes) is a matter of personal inclination. When we study herpetology (*herpeton* = crawling thing, *logos* = reason or knowledge), we follow those threads of the woof that consist of the amphibians and the reptiles. Any aspect of the biology of these animals is legitimately part of the subject of herpetology.

ZOOLOGICAL POSITION OF AMPHIBIANS AND REPTILES

We cannot profitably study any of the broader aspects of zoology until we know just what animals we are dealing with and where they fit in the whole pattern. No one knows exactly how many different kinds of animals there are, but one careful estimate gives 1,120,000, and this is within reason. To bring order and meaning into this bewildering array of forms, we must classify and divide them into groups and categories. There are a number of different schemes of classification that we might adopt. We could, for example, divide animals according to their habitat, such as forest, desert, lake, or ocean. Some studies require this kind of classification, but the standard, universally accepted classification today, the one we mean when we speak of animal classification, is based on the degree of relationship through descent from a common ancestor. Closeness of relationship is usually judged by similarity of morphological characters.

The animal kingdom is divided into a number of large groups called phyla (e.g., phylum Mollusca: shellfishes, like the oysters, clams, snails, and squids; phylum Arthropoda: joint-footed animals, such as insects, spiders, centipedes, and lobsters). The phylum Chordata comprises animals that at some stage in their life history have pharyngeal pouches, a hollow dorsal nerve cord, and a notochord (a stiffening rod running along the back). The phylum Chordata is divided into several small subphyla and one large one, the subphylum Vertebrata, to which belong the animals most familiar to us—the fishes, amphibians, reptiles, birds, and mammals. Vertebrates are animals in which the notochord, though still present in the early embryonic stages, has been largely replaced in the adult by a jointed vertebral column composed of a number of separate structures, the vertebrae. The anterior end of the nerve cord is expanded to form a brain, which is enclosed in a protective box, the cranium.

Members of the subphylum Vertebrata are divided into a number of classes. Included are several classes of fishlike vertebrates plus the classes Amphibia, Reptilia, Aves, and Mammalia. These comprise about 38,000 kinds of living animals:

Fishlike vertebrates	17,000 ± species
Amphibians	3,000 ± species

Reptiles	6,000 ± species
Birds	8,600 ± species
Mammals	3,500 ± species

Amphibians

Amphibians are vertebrate animals whose body temperature is dependent upon the external environment; that is, they are ectothermic. They have soft glandular skins that are for the most part without scales. They lack the paired fins of the fishes: instead, most have limbs with digits, as have the higher animal forms, the reptiles, birds, and mammals. These four classes of animals with limbs are sometimes linked in the superclass Tetrapoda (four-footed). The amphibian egg lacks a shell and, to prevent the developing embryo from desiccating, must be laid in water or in humid surroundings.

Ancestral amphibians were derived from primitive fishes. They were the first vertebrates to move onto land, and they gave rise to all the other terrestrial vertebrates: the reptiles, birds, and mammals. Two hundred fifty million years ago, they were a prominent element in the world's fauna; today, they are the smallest tetrapod stock with only about three thousand living members. These are divided into four orders:

Order Gymnophiona (Apoda), caecilians	150 ± species
Order Meantes (Trachystomata), sirens	3 species
Order Caudata (Urodela), salamanders	310 ± species
Order Salienta (Anura), frogs and toads	2,510 ± species

Reptiles

Reptiles, like amphibians, are ectothermal vertebrates that do not have paired fins. They differ from amphibians, however, in that they all have scales. The reptilian egg usually has either a parchmentlike or a calcareous shell and is laid on land. In addition to a yolk sac, the embryo has three extraembryonic membranes—the *amnion, chorion,* and *allantois*—that are not present in amphibian and fish embryos. Since these membranes are also present in birds and mammals, these three classes are sometimes called amniotes, to distinguish them from the anamniotes, the fishes and amphibians.

Reptiles are descendants of the early amphibians and were the dominant animals on the earth during the Mesozoic era; they gave rise to the two classes of endothermic vertebrates with internal temperature control, the birds and mammals. Reptiles have lost the dominant position they held during the Mesozoic, although they are still far more numerous than amphibians. Living reptiles are divided into the following orders:

Order Testudines (Testudinata), turtles	230 ± species
Order Rhynchocephalia, tuatara	1 species
Order Squamata	
Suborder Amphisbaenia, worm lizards	140 ± species
Suborder Serpentes (Ophidia), snakes	2,700 ± species
Suborder Sauria (Lacertilia), lizards	3,000 ± species
Order Crocodylia, alligators and crocodiles	21 species

Amphibians and reptiles have traditionally been studied together. This is partly for historical reasons—at one time the differences between the two groups were not recognized as being important enough to justify their placement in separate classes. It is also partly a matter of convenience—amphibians are a small group, and the methods of collecting and preserving them are similar to those used for reptiles. To avoid repeating the rather cumbersome phrase "amphibians and reptiles," we will hereafter use "herps" as a general term for the members of both classes.

With our knowledge of biology becoming more and more detailed and the literature more and more voluminous, it is difficult for a worker even in the restricted field of herpetology to keep abreast of current developments. Some modern herpetologists restrict themselves almost entirely to the study of amphibians, or of reptiles, or perhaps even of a single order.

SYSTEMATICS AND TAXONOMY

Historically, the task of naming, describing, and classifying amphibians and reptiles has necessarily had priority, and even at the present time a portion of the literature of herpetology is concerned with such studies. Here herpetology interweaves with taxonomy and systematics. These two terms are often used interchangeably, but it seems better to restrict *taxonomy* to the frequently very complicated task of assigning names to groups of animals, and *systematics* to the formulation of a classification that will describe the relation of the animals to one another. The two do overlap, of course. Current taxonomic practice requires that the taxonomist describe the form he is naming and include in the description an indication of the animal's relationships. The systematist frequently finds that he must name one or more new forms or resolve a nomenclatural problem before he can discuss intelligibly the relationships of the animals he is classifying.

Our present system of classification comprises a series of categories, each less inclusive than the preceding one. Phyla are divided into classes, classes into orders, orders into families, families into genera (singular genus), and genera into species (singular species). A Box Turtle is thus classified:

Phylum Chordata
Class Reptilia
Order Testudines
Family Emydidae
Genus *Terrapene*
Species *Terrapene carolina*

These are the standard categories. Each category is intended to represent a distinct level of evolutionary relationship. We can visualize the evolutionary history of a group (its phylogeny) in the form of a branching tree (the phylogenetic tree). The species are the twigs of the phylogenetic tree. As the categories become higher, the branches they represent extend further back in time and represent older points of divergence. Ideally, all forms placed in a given category should represent a single, monophyletic lineage (a clade), all of whose members are descended from the same ancestral population. All the species (twigs) placed in one genus should arise from a single small branch, and all the genera placed in one family should arise from a single larger branch. Since the fossil record is woefully incomplete, our classification is necessarily based largely on similarities among extant species. As a result, convergence may trick us into linking together species or higher groups that share many features because of a similar way of life, but do not have a common ancestor. Such a group is polyphyletic and is called a grade. Usually, when a systematist decides he is dealing with a grade rather than a clade, he reclassifies the group. For example, the Australian tree frogs closely resemble some South American tree frogs of the genus *Hyla,* and were long placed in that genus. But recent work indicates that the former evolved from a different branch of the phylogenetic tree, so they have been removed from *Hyla* and are placed in the genus *Litoria.*

The superclass, infraorder, subfamily, and other intermediate categories reflect a more detailed knowledge of relationship. They are categories of convenience. Every animal is a member of a species, genus, family, order, class, and phylum; not every animal belongs to a superclass or subfamily. The latinized name applied to a category designates a particular group of organisms, called a taxon (plural taxa). For example, the taxon *Rana pipiens* refers to all populations of frogs that belong to that species; the taxon Serpentes includes all snakes.

Of all the taxonomic categories, the species is the only one that can be objectively defined, at least for bisexual vertebrates. *A species is a population or group of populations of similar animals that interbreed, or are potentially able to interbreed, and produce fertile offspring.* This definition formalizes our intuitive impression that like tends to breed with like and produce more of the same. Species—or, rather, their component populations—are the units of evolutionary change. Evolution is change in the genetic composition of a population through natural selection. As such, evolution operates on the population, although natural selection operates on individuals. Natural selection must not be viewed as "nature red in tooth and claw," but as a process of differential reproduction, by which one set of parents are able to place more of their offspring into the breeding pool for succeeding generations than can another set of parents.

Because of evolution, species are not static units of life. They possess both spatial and temporal dimensions, and owing to these extensions in space and time, they are often difficult to delimit. Since most species are spread over large geographic areas, they are divided into small, semi-isolated populations by geographic barriers, such as rivers or mountains, and by the discontinuous nature of their preferred habitats. Each population is subjected to a different physical and biological environment and becomes adapted to that specific environment. The differences between populations may range from small variations in gene frequency to striking differences in morphology or physiology. The extent of the differences is largely determined by the amount of gene exchange between adjacent populations, the intensity of selection, the length of time since the founding of a population, and the genetic composition of its founding members. When the differences between populations of a given species are evident and occur more or less abruptly, these populations are considered to be distinct geographic races or subspecies. For example, the Spring Peepers in most of the eastern United States are uniform pale pink below, while those of northern Florida are heavily spotted below. The two populations are therefore regarded as different races.

No matter how different two populations may be, if they are capable of freely interbreeding, they are members of the same species. This criterion is very precise in principle, but often extremely difficult to apply in practice. Too often we simply do not know whether two given forms can or do interbreed in nature. The systematist must use his judgment, must decide whether specimens collected from two populations perhaps many kilometers apart are similar enough to represent a single species. But mere morphological similarity is not always a reliable guide. The Leopard Frog of the southeastern United States does not differ markedly from that of the northeastern United States, and the two were long thought to be separate geographic races of a single species. Then it was found that the two forms, though they may inhabit the same pond, do not interbreed. It is now known that the Leopard Frogs, once considered a single, wide-ranging species, *Rana pipiens*, in reality comprise at least four, and probably more, separate species.

Such failure or inability to interbreed results from the evolution of barriers to the production of fertile offspring. These reproductive barriers, or isolating mechanisms, may operate either before or after mating. Typical premating mechanisms are seasonal or habitat isolation, behavioral isolation, and mechanical isolation. Males and females of different species may not interbreed, even though they live in the same area, because they occupy different habitats or because their reproductive periods occur at different times. Two species may also fail to interbreed because the reproductive signal given by members of one sex of one species is not recognized by members of the opposite sex of the other species. If these isolation mechanisms fail, mating may still be impossible owing to mechanical incompatibilities such as differences in size.

Postmating mechanisms may cause failure to fertilize the egg, death of the embryo, or sterility of the offspring. Selection favors premating mechanisms, for postmating mechanisms waste an individual's gametes (sperm or eggs) and energy, which could have been spent more productively mating with one of its own kind. The selection for premating isolation mechanisms is sometimes overtly apparent in areas where two species are sympatric (occur together). Reproduction-associated characters of two species may become strikingly different in the sympatric area, yet remain similar in allopatric (where the two species do not occur together) areas. This kind of character displacement occurs frequently in the calls of frogs.

Since many pairs of closely related species (sibling species) cannot be recognized by differences in morphology, systematists have been enlarging their repertoire of technical tools. Experimental cross-fertilization was one of the first such tools to be used in anuran systematics. Using the techniques of embryology, the eggs of different species were fertilized with testicular extracts. Cross-matings could then be compared by the degree of successful development of the embryos. It was discovered that the offspring from crosses between closely related species are frequently as viable as those from intraspecific crosses. Thus this technique may fail to distinguish species, although the survivorship of the embryos does demonstrate the degree of genetic compatibility of different species, which in turn indicates the degree of relationship. Recording and analysis of frog voices has become an important technique in the recognition of sibling species. The breeding calls are species-specific, and a female will only go to a calling male of her own species. Different calls identify different species. *Hyla versicolor* and *Hyla chrysoscelis* can be distinguished only by their calls and their karyotypes (chromosome complements); in external appearance they are identical.

Techniques for measuring protein similarity are rapidly gaining prominence in the analysis of the genetic composition of species and the degree of genetic relationship between species. Since immediate products of the genes are being compared, these techniques provide a more direct measure of the genotype (genetic constitution) of the animal than do morphological or behavioral traits. They have the additional advantage over cross-fertilization and vocal analysis that they can be compared at all life stages and for all types of animals, not just frogs. These techniques, combined with morphological and natural-history observation, enable herpetologists to identify species more clearly.

NOMENCLATURE

Our system of nomenclature is based on the one first universally applied to all animals by Linnaeus in the tenth edition of his *Systema Naturae,* published in 1758. Linnaeus gave each species known to him a name consisting of two parts, the name of the genus to which it belonged, plus a specific epithet, the trivial name (e.g., *Rana esculenta*). Over the years, other biologists adopted the Linnaean system of designating species. But as more and more new species were found and named, and more and more papers were published using these names, confusion inevitably crept in. Sometimes a biologist, either deliberately or through ignorance, gave a name to a species that had already been named something else, creating a *synonym.* Sometimes two investigators, working on different groups, happened to give the same name to entirely different forms. Such a name is called a *homonym.* But if we, as biologists, are to understand one another, we must be sure that each species has only one name and that each name applies to only one species.

The need for a set of rules for taxonomists to follow in proposing new names, or in deciding which of several names already in use should be accepted, soon became acute. A number of codes were drawn up in different countries and for different groups. The English followed one code, the French another, the Germans still a third; ornithologists had a code of their own, and so did paleontologists. Finally, in 1895, the Third International Zoological Congress appointed a committee that drew up the *International Code of Zoological Nomenclature,* commonly known as the Code. This Code was accepted by the Fifth International Zoological Congress in 1901 and is now universally followed. It has been revised from time to time, most recently in 1961. A permanent International Commission of Zoological Nomenclature serves as a "supreme court" to resolve the knotty problems of interpretation that seem to be inevitable under any code of laws.

Most of the rules and recommendations of the Code deal with technicalities of interest only to professional taxonomists. A few of them, however, should be familiar to every zoologist, if only because an understanding of them will clarify his reading and facilitate his writing on zoological subjects.

No genus of animals may have the same name as any other genus, but the same trivial name may be used in different genera (though not more than once within a single genus). The specific name of an animal consists of the generic name plus the trivial name. Thus the specific name of the Box Turtle classified in the preceding section is not *carolina*, but *Terrapene carolina*. When a specific name is written, the generic name always begins with a capital letter, the trivial name with a small letter, and both are italicized (indicated in a manuscript by underlining the name). If a species is found to comprise two or more geographic races, each is given a third name, which follows the trivial name and, like it, is never capitalized. The subspecies that most nearly represents the material on which the trivial name of the species was based is always given a third name that is the same as the trivial name. Thus the race of the Box Turtle in the eastern United States is *Terrapene carolina carolina,* while the Floridian race is *T. c. bauri.** Sometimes a scientific name is followed by the name of the scientist who first proposed it, e.g., *T. c. carolina* Linnaeus. The name of a higher category (family, order, class, etc.) begins with a capital letter but is not italicized. Family names always end in "-idae" (Emydidae) and subfamily names in "-inae" (Emydinae).

DEVELOPMENT OF HERPETOLOGY

Among the early writers on natural history, Aristotle and Pliny stand out. Aristotle (384–322 B.C.) has been called the originator of biological classification, not because his system bears much resemblance to the one we use today, but because he first felt the need to establish categories based on characteristics, particularly anatomical ones, of the animals. He says, "Four-footed beasts that produce their young alive have hair; four-footed beasts that lay eggs have scales." Compare this with the biblical classification of animals as "clean" or "unclean."

The Roman Pliny (A.D. 23–79) wrote a *Naturalis Historiae* in which he listed all the animals of which he knew or had heard. Pliny was not a systematist, and there is little order to his arrangement. He was also very credulous. Many of his animals are fabulous, and many of the tales he tells of real animals are equally fabulous. But his accounts are lively and, where

---

*Note that when the full name has been written out once in a paragraph, we may, if we use the name again in the same paragraph, abbreviate the generic name and, when a trinomial is used, the trivial name.

he deals with animals he could study himself, contain some accurate information.

The works of these two men were handed down in manuscript form during the Middle Ages, and were among the first books made widely available after the invention of printing in the mid-fifteenth century. Pliny's "Natural History" was published in 1469, and some of the works of Aristotle, including his "On the History of Animals," were printed in the period 1495–1498.

The reawakened interest in the world of nature that came with the Renaissance, and the enormous widening of the horizons of that world following the voyages of the explorers, resulted in the discovery of many more kinds of animals. The works of Pliny and Aristotle no longer sufficed. Several new compendia appeared, largely based on the classification of Aristotle and the style of Pliny. Notable is the *Historia Animalium,* published between 1559 and 1583 by Conradus von Gesner, a talented Swiss naturalist. Gesner's greatest contribution was the introduction of scientific illustrations (Figure 1–1). His woodcuts of a frog, a toad, a ringed snake,

RANA PERFECTA.   FOETVS RANÆ CAVDATVS.

FIGURE 1–1
"Rana perfecta" as illustrated by Gesner in his *Historia Animalium*, published in 1586, one of the earliest herpetological illustrations. The name is not a specific name in the Linnaean sense, but means the adult or "perfect" form of the frog. [Courtesy of the Smithsonian Institution.]

and a viper are probably the first published figures of amphibians and reptiles.

The first English-language book on natural history is by the Reverend Edward Topsell, who, in 1608, published *Historie of Foure-footed Beasts, describing the true and lively figures of every Beast . . . collected out of all the volumes of C. Gesner and other writers.* The next year, Topsell brought out his *Historie of Serpents or the Seconde Book of Living Creatures.* Topsell's conception of serpents included practically everything that creeps and crawls: spiders, scorpions, and many fabulous creatures in addition to the amphibians and reptiles.

The first truly critical and systematic approach to herpetology was made by John Ray, an Englishman, son of a village blacksmith. He was primarily a botanist, but his *Synopsis Methodica Animalium Quadrupedum et Serpentini Generis,* published in 1693, is concerned with the herps. Ray was the first to group these animals together, for the reason that their hearts have a single ventricle, in contrast to the two-chambered ventricle of birds and mammals. He further distinguished between the harmless and poisonous snakes by the character of their teeth. His work was excellent for its time.

It is unfortunate that Linnaeus (1707–1788) did not adopt Ray's classification. He named the herps he knew and gave them a place in his classification, but he had little interest in them or liking for them, as is shown by his summary in *Systema Naturae:*

> These foul and loathsome animals are distinguished by a heart with a single ventricle and a single auricle, doubtful lungs, and a double penis.
>
> Most Amphibia are abhorrent because of their cold body, pale colour, cartilaginous skeleton, filthy skin, fierce aspect, calculating eye, offensive smell, harsh voice, squalid habitation, and terrible venom; and so their Creator has not exerted his powers (to make) many of them.

The classification of Linnaeus is faulty and inadequate, but the principles he laid down and the system of nomenclature he established allowed other workers to place systematic herpetology on a firm foundation.

In post-Linnaean times, herpetology has made rapid advances. We shall here mention only a few of the more important early workers. Among these was André Marie Constant Duméril, a French naturalist and herpetologist. With his colleague, Gabriel Bibron, Duméril wrote an important ten-volume work, *Erpétologie Générale,* which was published from 1835 to 1854. Some of these volumes were coauthored by his son, Auguste Henri André Duméril.

Albert C. D. G. Günther, who served as keeper of reptiles in the British Museum from 1856 to 1895, gave tremendous impetus to herpetology with his series of catalogues of amphibians and reptiles in the museum. He also did spadework for neotropical herpetology in his volume on amphibians and reptiles in the series *Biologia Centrale Americana.* George Albert Boulenger, who worked in the British Museum from 1882 to 1920, laid the groundwork for the worldwide study of herpetology with a nine-volume catalogue of the batrachians (amphibians) and reptiles in the British Museum (Figure 1–2).

FIGURE 1–2
George Albert Boulenger. [Courtesy of
the British Museum (Natural History).]

FIGURE 1–3
Edward Drinker Cope. [Courtesy
of the Academy of Natural Sciences
of Philadelphia.]

In the New World, Edward Drinker Cope (1840–1897) was publishing paper after paper on fundamental herpetology (Figure 1–3). Cope not only made significant contributions in his long series of descriptive papers on amphibians and reptiles, but also, through his great knowledge of comparative anatomy and paleontology, provided a foundation for much of our modern classification of these animals. Two of his most important works, *The Batrachia of North America* (1889), and *The Crocodilians, Lizards and Snakes of North America* (1900), are still fundamental books in the library of anyone concerned with North American herpetology. The latter book was published posthumously.

The work of Leonard Stejneger (1851–1943) is perhaps not appreciated as much as it should be by many younger North American herpetologists. From the time he was appointed curator of reptiles in the United States National Museum in 1889, until his death in 1943, he published voluminously on the herpetology not only of America but of the world. Among his important books are *Herpetology of Japan* and *The Poisonous Reptiles of North America*. But perhaps his outstanding contribution was in the work he undertook with his younger friend and associate, Thomas Barbour, director of the Museum of Comparative Zoology. In 1917, the two brought out the first edition of the *Check List of North American Amphibians and Reptiles,* a volume that went through five editions under their joint authorship. It is hard indeed for many young herpetologists to appreciate fully the difficulties faced by researchers in the days before a checklist and handbooks for every group were readily available. Stejneger and Barbour, by synthesizing North American herpetology into a readily digestible form, had an incalculable effect in stimulating research in the field.

Finally, we mention the name of Raymond L. Ditmars, late curator of reptiles at the New York Zoological Society. At a time when the publication of popular books was frowned upon as unworthy of a true scientist, Ditmars persisted and turned out a series of popular volumes on reptiles, particularly snakes, that awakened many a youngster to an interest in herpetology. It was only after this interest was developed that these young workers could begin to digest the more fundamental work of such men as Frank N. Blanchard, Emmett R. Dunn, G. Kingsley Noble, Karl P. Schmidt, and John Van Denburgh. Ditmars was thus performing one of the most important functions of a scientist, and one that is all too often overlooked. A scientist has the responsibility of interpreting his specialty to the general public and of answering the questions of amateurs and interested laymen. Since Ditmars' time, several other herpetologists have recognized this responsibility and have written excellent popular books on the herps.

These books are helping to diminish the widespread, long-standing, and unwarranted public antipathy toward the herps. Nevertheless, too many people still share the opinion of Linnaeus and wonder why anyone should waste his time on such unpleasant creatures. One answer is that a thorough knowledge of the herps will elucidate most of the biological principles that apply throughout the animal kingdom, and may provide a clue to some of the unanswered problems of basic biology. But in the last analysis, the herpetologist studies reptiles and amphibians because, in contrast to Linnaeus, he likes them and finds them interesting. *De gustibus non disputandum.*

READINGS AND REFERENCES

Bellairs, A. *The Life of Reptiles,* vol. 1 & 2. London: Weidenfeld and Nicolson, 1969.

Carr, A. *The Reptiles.* New York: Life Nature Library, 1963.

Cochran, D. M. *Living Amphibians of the World.* New York: Doubleday, 1961.

Frazer, J. F. *Amphibians.* London: Wykeham, 1973.

Gans, C., and T. S. Parsons (eds.). *The Biology of the Reptilia,* vol. 1–5. London: Academic Press, 1969–76.

International Commission on Zoological Nomenclature. *International Code of Zoological Nomenclature.* London: International Trust for Zoological Nomenclature, 1961.

Mayr, Ernst. *Principles of Systematic Zoology.* New York: McGraw-Hill, 1969.

Minton, S., and M. Minton. *Giant Reptiles.* New York: Scribner's, 1973.

Noble, G. K. *The Biology of the Amphibia.* Garden City: Dover, 1954, reprint.

Nordenskiold, Erik. *History of Biology.* New York: Tudor, 1928.

Oliver, J. A. *The Natural History of North American Amphibians and Reptiles.* Princeton: Van Nostrand, 1955.

Peters, J. A. *Dictionary of Herpetology.* New York and London: Hafner, 1964.

Porter, K. R. *Herpetology.* Philadelphia: Saunders, 1972.

Schmidt, K. P., and R. F. Inger. *Living Reptiles of the World.* New York: Doubleday, 1957.

Smith, M. A. *The British Amphibians and Reptiles,* rev. ed. London: Collins, 1954.

CHAPTER TWO

STRUCTURE
OF AMPHIBIANS

IF AMPHIBIANS HAD NOT MADE the transition from aquatic to semiterrestrial existence, the evolution of the higher vertebrates could not have taken place. The cause and course of this shift of habitat have been subjects of much speculation. The individual modern amphibian accomplishes the same transition in weeks or months, instead of thousands of years, and because we are so familiar with the change in life style, we accept it as one of the commonplaces of nature. However, for neither the individual nor the group is the transition complete; most adult amphibians are bound to humid environments, and those that lay eggs must do so either in water or in moist surroundings. Their transitional position imposes structural and functional complications. Some of their characteristics are adaptations to life in the water, others to life on land. Some reflect the necessitites of an individual life cycle that goes from an aquatic larva to a terrestrial adult, which nevertheless must return to the water to breed. The larva, moreover, does not resemble the fish ancestor of the amphibians, and the steps by which it metamorphoses into an adult are not the same as those by which the fish evolved into an amphibian. The tadpole, for instance, does not start with fins that change into legs.

The amphibians have no unique structures, such as the feathers of birds or the mammary glands of mammals, that set them off from the other groups of tetrapods. There is no structure of which we can say, "This is found in all amphibians and only in amphibians." Even the characteristic life history that gives the group its name (*amphi* = both, *bios* = life) does not hold for all members, since some amphibians lay their eggs or bear their young on land and have no aquatic larval stage, yet no one doubts that they are true amphibians. Therefore, any definition of the class Amphibia must be based on a combination of characteristics, rather than on definitive, unique ones.

DEFINITION OF AMPHIBIA

Most members of the class Amphibia—frogs, salamanders, sirens, and caecilians—are small animals that have smooth, moist skins without scales. They lay their eggs in water or in moist surroundings. The egg is covered by several gelatinous envelopes, rather than a shell, and usually hatches into a larva. The larva differs structurally from the adult and metamorphoses (that is, changes abruptly rather than gradually) into the adult body form.

The larva has gills, which in some species may be enclosed within a gill chamber. In the adult, respiration takes place either through lungs, through gills, through the skin or mucous membrane lining the mouth and pharynx, or through some combination of these. The lungs are simple in structure and usually appear before metamorphosis.

Amphibians are ectothermic. Multicellular mucous and poison glands occur in the skin. There are no true nails or claws, although in some forms horny epidermal structures may be present at the tips of the toes. The heart is three-chambered. The skull is flattened and composed of fewer bones than is the fish skull; it articulates with the vertebral column by two occipital condyles (a condition characteristic of mammals but not of reptiles or birds). As in the fish, there are only ten pairs of cranial nerves, rather than the twelve pairs present in the higher tetrapods.

ANATOMY OF AMPHIBIANS

Detailed anatomical descriptions will not be presented in this book. We are here concerned chiefly with those structural advances from the fishes that, in combination, make an animal an amphibian.

Integument

We usually think of the evolution from aquatic fish to more or less terrestrial amphibian as consisting mainly in the shift from aquatic to aerial respiration and the transformation of fins into legs. Actually, the process entailed basic structural changes throughout the body. Some of the most important of these took place in the integument, since the skin is the part of the animal directly in contact with the external environment, and, in amphibians, an important respiratory organ as well.

The amphibian skin, like that of all vertebrates, consists of two parts: an outer layer (the epidermis) and an inner layer (the dermis) (Figure 2–1).

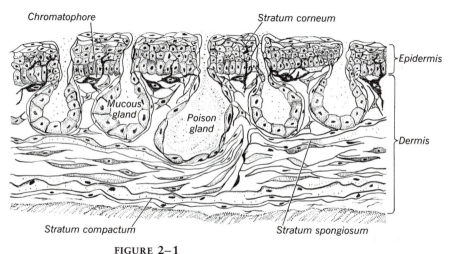

FIGURE 2–1
Cross section through the skin of *Rana pipiens*.

Epidermis. The epidermis comprises several layers of cells, and already consists of two or more such layers before the larva hatches. One of these is the stratum corneum, an outer layer of horny, dead cells that protects the deeper living cells and helps reduce moisture loss. This is an adaptation to life on land; it is not present in fishes and is lacking among the amphibians only in a few aquatic salamanders. During ecdysis (the shedding of the skin), it is the stratum corneum that is sloughed. In some forms (e.g., the Leopard Frog, *Rana pipiens*), the skin comes off in bits and pieces. In others, (e.g., the toads, *Bufo*), it is pulled off as a whole. We never find a shed toad skin, as we do a snake skin, because the toad eats it immediately.

Dermis. The dermis is relatively thin. It is composed of two layers, an outer, loosely organized stratum spongiosum and an inner, more compact stratum compactum. Since it serves as a respiratory organ, it is usually well supplied with blood vessels. Pigment in amphibians is found in special cells called chromatophores, which are usually located in the uppermost layer of the dermis or between the dermis and the epidermis.

Glands. Epidermal glands may be either unicellular or multicellular. The unicellular glands that are so common in fishes have almost disappeared in amphibians. Patches of them develop in the head region of the embryo and secrete an enzyme that digests the gelatinous envelopes of the egg, and so aids in hatching. Unicellular glands known as the glands of Leydig are present in the epidermis of some larval salamanders; their function is unknown. Other unicellular glands are found in sirenids. By contrast, multicellular mucous glands are numerous and well developed in amphibians. The mucus secretion helps keep the skin moist and provides a medium for the exchange of gases, which is very important for animals depending wholly or partly on integumentary respiration.

Multicellular poison glands are well developed in many species, especially in the more terrestrial anurans. The so-called warts of toads and the parotoid glands on the back of the neck are actually masses of poison glands. *Hyla vasta,* a giant tree frog found in Haiti, gives off a poisonous secretion so strong that the unwary collector who picks one up with bare hands may suffer inflammation of the skin. Many herpetologists have found all the frogs in their collecting bags dead at the end of the day except one species. Apparently the poison of one species may be lethal to members of other species, but not to individuals of its own kind. Of course, the story of getting warts from handling toads is fictitious.

Some specially modified integumentary glands are present in frogs and toads. Many of the tree frogs have glandular discs at the tips of the fingers and toes that apparently aid in climbing, and the glandular thumb pads of breeding male anurans help them to clasp the female. Male salamanders may have hedonic (pleasure-giving) glands under the chin, on the face, or on the tail. The secretions of these glands stimulate the female during courtship.

Scales. As a general rule, the dermal scales characteristic of the fishes have been lost in modern amphibians, and the epidermal scales of the reptiles have not yet developed. Rudimentary dermal scales are found buried in the skin of some caecilians, and a few toads have bony plates embedded in the skin on the back. The spadefoot toads *(Scaphiopus* and *Pelobates)* have highly cornified areas on the feet that might be considered epidermal scales. The African Clawed Frog, *Xenopus laevis,* has dark, cornified, epidermal structures—the so-called claws—on the first three digits of the hind foot, and some salamanders, such as *Onychodactylus,* have cornified, epidermal, clawlike structures at the ends of the digits. These are not true claws, for they lack the underpart, the so-called sole horn, or subunguis, of the true claw; they are modified epidermal scales.

Digestive System

The amphibians have a digestive tract little changed from that of their fish ancestors. The most striking modification is the development of a manipulative tongue.

Tongue.   A well-developed, definitive tongue is apparently an adaptation to life on land. It first appeared in the amphibians and is characteristic of all the higher land vertebrates. A fish's food is already wet when captured and can be swallowed whole; it does not need to be moved around in the mouth for moistening or chewing before it is swallowed. The tongue of a fish is a primary tongue: a fleshy fold on the floor of the mouth that lacks intrinsic muscles and can be moved only within narrow limits. It is useful, perhaps, in pushing food farther back in the mouth for swallowing, but it cannot be used to capture prey. Some aquatic salamanders (e.g., the Mud Puppy, *Necturus)* have only a primary tongue, and the tongue of some aquatic toads has degenerated and is almost or entirely absent. Other frogs and salamanders have a definitive tongue, which has, in addition to the part representing the primary tongue, an expanded glandular portion supplied with intrinsic musculature. The tongue of most frogs is attached by its base near the anterior margin of the jaw, and, at rest, is folded back on the floor of the mouth with the tip pointing toward the throat. Anyone who has seen a toad snap at a fly knows how quickly the tongue can be flipped out and how accurately it can be aimed. The tongue of some salamanders has a more extensive attachment, is mushroom-shaped (boletoid tongue), and can be shot forward to trap unwary insects. In both groups, mucous glands in the mouth provide a sticky secretion that clings to the tongue and aids in the capture of prey. Oral glands, like the definitive tongue, are a characteristic of the amphibians that is not present in lower forms.

Teeth.   With the exception of the epidermal, toothlike structures that are found in certain larval forms, the teeth of amphibians, as of most other vertebrates, are true teeth—that is, they have a hard layer of enamel surrounding the softer dentine and central pulp cavity (Figure 2–2). Toads of the genera *Bufo* and *Pipa* are toothless, but most other anurans and salamanders have simple, cone-shaped teeth. Except for a few genera (the South American hylid *Amphignathodon,* the African ranids *Dimorphognathus* and *Phrynodon,* and the male *Petropedetes natator*), frogs have no

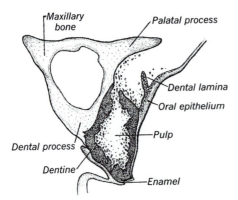

FIGURE 2-2
Section through a maxillary tooth of *Hyla cinerea*, showing its relation to the maxillary bone, oral epithelium, and dental lamina from which the new tooth is formed. [After Goin and Hester, *J. Morphol.*, 1961 (1962).]

teeth on the lower jaw. For the most part, amphibian teeth are located on the jaws, but they may also occur on bones of the roof of the mouth—the palatines and vomers. In a few genera of frogs, they are also attached to the parasphenoid bones. Amphibian teeth are polyphyodont (may be replaced an indefinite number of times) and basically homodont (all the teeth along the jaw are similar). Most frogs, salamanders, and caecilians have the crown of the tooth separated from the pedicel by a zone of weakness.

Gut. The amphibian esophagus is very short, being little more than a constricted area of the alimentary tract. Usually, both the esophagus and the mouth are lined with cilia that sweep small food particles into the stomach. Secreting cells in the esophageal epithelium of some frogs produce an enzyme, pepsin, which does not begin to function until it reaches the stomach. The stomach of some salamanders is simply a straight portion of the gut, whereas in the frogs it has differentiated into a cardiac end leading from the esophagus and a short, narrow pyloric end leading into the intestine. Digestive glands are absent from the stomachs of some fishes, but are present in all amphibians. The intestine of the caecilians has not yet differentiated into large and small intestines and shows only a slight degree of coiling. Intestinal coiling is more evident in the salamanders, and there is differentiation into a large and small intestine. In the anurans, the coiling tendency is even more marked, and the large intestine is plainly set off from the small intestine. The amphibian intestine opens into a cloaca, a common chamber into which the urinary, reproductive, and digestive systems empty. A ventral outfolding of the amphibian cloaca gives rise to the urinary bladder.

Respiratory System

By means of the respiratory system, oxygen passes from some outside medium (air or water) into the bloodstream of an animal, and carbon dioxide passes from the bloodstream into the outside medium. (In a sense, all respiration takes place through water, since respiratory surfaces must be kept constantly moist.) This requires some structure or structures to bring a rich supply of blood into close contact with the medium. Nowhere is the transitional position of the amphibians more clearly shown than in the variety of their respiratory processes. They have three types of highly vascular structures (ones well supplied with blood vessels) that may be used for respiration: gills, lungs, and the surface of the skin and lining of the mouth and pharynx. As a general rule, larval amphibians use gills and adults use lungs, but there are many exceptions. Cutaneous respiration (through the skin) is important in aquatic larvae and in those adults that have moist skins.

Gills.   Most fishes have internal gills, which develop from tissue on the inner surfaces of the gill arches; amphibians have external gills, which originate from the integument and are borne on the outer surfaces of the gill arches. During embryonic development, several paired pouches form in the wall of the pharynx. In most amphibians, three of these pairs break through to the outside to form gill slits. Cartilaginous supporting rods, the visceral arches, are formed in the septa between the gill slits. The gills, which consist of filaments covered with ciliated epithelium, are borne by the visceral arches anterior to the gill slits. Gills of tadpoles are usually smaller and simpler than those of salamander larvae. In one family of caecilians, Typhlonectidae, the embryos in the uterus develop two large, baglike gills from the dorsal part of the pharynx. It has been suggested that these gills function in the absorption of oxygen from the intrauterine fluid. Other caecilian embryos have three pairs of more normal, branching gills.

Among the frogs, a gill cover, the operculum, forms shortly after the external gills appear. A sheet of tissue grows backward from the hyoid region to cover the gill slits, the gills, and the region from which the forelimbs will eventually develop. It then fuses with the body behind and below the gill region to enclose the gills in a branchial chamber. This chamber has an opening to the outside—the spiracle—which may be either ventral or lateral; tadpoles of a few species have paired spiracles. Shortly after the operculum forms, the original gills degenerate and new gills develop from the walls of the gill clefts. Although they are enclosed in a branchial chamber, these are external gills derived from the integument. An opercular fold also forms in the salamanders and caecilians but it is small, consisting simply of a crease anterior to the gill region, so that no branchial chamber is formed.

During metamorphosis, the gills of anurans are resorbed, the gill slits close, and the lungs take over. Salamanders show more variety. They seem to be experimenting with different solutions to the problem of breathing on land. Most terrestrial salamanders lose gills and gill slits (see Table 2–1) and acquire lungs just as frogs do. Members of the family Plethodontidae never

TABLE 2–1
Distribution of Gills, Gill Slits, and Lungs
among Adult Salamanders and Sirens

Family	Number of pairs of gills	Number of pairs of slits	Lungs
Hynobiidae	0	0	yes[a]
Cryptobranchidae			
Cryptobranchus	0	1	yes
Andrias	0	0	yes
Ambystomatidae[b]	0	0	yes[c]
Salamandridae	0	0	yes[d]
Amphiumidae	0	1	yes
Plethodontidae[b]	0	0	no
Proteidae	3	2	yes
Sirenidae			
Siren	3	3	yes
Pseudobranchus	3	1	yes

[a]Except *Onychodactylus.*
[b]Certain members of these families fail to metamorphose, and retain the larval gills and gill slits when sexually mature.
[c]Reduced in *Rhyacotriton.*
[d]Except *Chioglossa* and *Salamandrina.*

develop lungs; the vast majority are terrestrial and depend entirely on respiration through the skin and lining of the mouth and pharynx. The aquatic *Amphiuma* and *Cryptobranchus* develop lungs and lose their gills, but retain the openings of one pair of gill slits. Adult Proteidae possess both gills and lungs, and two gill slits remain open.

Adult sirens have lungs, gills, and gill slits. If the gills of a Dwarf Siren *(Pseudobranchus)* are removed, it can survive by coming frequently to the surface to gulp air. If a Dwarf Siren is put in an aquarium with a screen to prevent it from coming to the surface, its gills expand and it continues to survive. The importance of the role of cutaneous respiration in sirens and aquatic salamanders has not been determined. The Proteidae and Sirenidae are sometimes grouped together as perennibranchs because both retain gills as adults, but they are not closely related.

Gills of caecilians are usually resorbed before the young hatch or are born.

Lungs. The lungs of amphibians develop before metamorphosis and are relatively simple in structure. The left lung of caecilians is usually very short, with alveoli (singular alveolus), little pockets at the ends of the bronchioles, only in the right lung. Salamander lungs are paired elongated sacs, the left one usually longer. In some salamanders, the lining is smooth; in others, alveoli are present in the basal part. Lung linings are more complex in the frogs, which is what we should expect, since they have been more successful, on the whole, in making the transition from gills to lungs than have the salamanders. The walls of their lungs are made up of many folds lined with alveoli. In general, the more terrestrial the frog or toad, the larger are the alveolar respiratory surfaces in the lungs.

Respiratory Passages.   The nasal passages of amphibians lead from the external nares (nostrils) to the internal nares or choanae, openings in the roof of the mouth just inside the upper jaw. Terrestrial amphibians have mechanisms for controlling the size of the external aperture, whereas many larval salamanders, the adult sirens, and frog tadpoles have valves around the internal nares to control the direction of water flow. The mouth leads into the pharynx, a gateway chamber that passes into the esophagus of the digestive system and is connected to the lungs by the trachea. The trachea is short in most salamanders, but in *Amphiuma* and *Siren* it may be four or five centimeters long. Frogs generally have such a short trachea that it can hardly be said to exist, except for the modified anterior portion that forms the larynx. A trachea is definitely present in the aquatic Pipidae, in which the lungs act as hydrostatic organs. The trachea divides at the lower end into two bronchial tubes, which lead to the lungs.

During aquatic respiration, water passes from the mouth to the pharynx and out the gill slits. In salamanders and sirens, this sets up a current of water that flows over the gills. In anuran tadpoles, water enters the branchial chamber, flows over the gills, and passes to the outside through the spiracle. Amphibians breathing air keep the mouth tightly closed and draw air in or push it out through the nasal passages by lowering or raising the floor of the mouth.

Larynx.   Since amphibians are the lowest form of vertebrate with a true ear, it is not strange that they are the first ones to have the anterior end of the trachea modified to form a voice box, or larynx. The larynx opens into the pharynx via a slitlike glottis, bounded on each side by a bar of cartilage derived from the last visceral arch. These cartilages are usually divided into upper (arytenoid) and lower (cricoid) ones. Sometimes the cricoid cartilages fuse to form a ring. Frogs and toads have two muscular bands, the vocal cords, stretching across the larynx parallel to the glottis. Air passing over the vocal cords makes them vibrate to produce sounds. Tightening or relaxing the vocal cords causes variations in pitch.

Voice is well developed in the anurans. A female frog may grunt when held in the hand or give a "mercy scream" when caught by a snake. The calling of a chorus of breeding males is one of the most familiar of all animal sounds. Some frog species may lack voices, but not all of those reported to be silent are indeed so. *Eleutherodactylus cundalli* of Jamaica, which was for many years considered a voiceless species, has been found to call only from caves and crevices.

Salamanders lack vocal cords, and most of them are voiceless. Plethodontids also lack trachea and larynx. A few salamanders have "voices", but the sounds they produce seldom amount to more than faint squeaks or grunts. The Pacific Giant Salamander, *Dicamptodon ensatus,* has a low-pitched, rattling note, and makes a screaming sound when in danger. The genus *Siren* was so named because it was originally reported to "climb trees and quack like young ducklings." It does neither, but does emit a train of audible clicks and a distress yelp.

Vocal Sacs.   The males of many frogs have vocal sacs, outpouchings from the mouth cavity that extend ventrally and laterally under the skin and muscles of the throat. Sometimes there is a single vocal sac, sometimes a pair. In some species they are very large, and extend posteriorly into the large lymphatic spaces that underlie the skin. When a frog calls, its vocal pouch is filled with air, sometimes swelling outward as a glistening bubble that collapses at the end of the call. How effective these pouches are as resonators is clear to anyone who has heard the bellowing of a Bullfrog *(Rana catesbeiana)* or the ear-splitting blast of a Great Plains Toad *(Bufo cognatus)*. The calls of different species of frog are just as characteristic as the songs of birds. Furthermore, individual frogs are apparently able to vary the pitch of the call. Some species of frog have a "rain call" that can be easily distinguished from the mating call given at breeding ponds.

Circulatory System

In conjunction with the development of lungs, the amphibians also evolved a double circulatory system comprising a pulmonary and a systemic circulation.

Heart.   Instead of the simple two-chambered heart that is characteristic of most fishes, amphibians have a three-chambered heart with a right atrium, a left atrium, and a single ventricle. Blood returning from the systemic circulation empties into the right atrium; pulmonary veins empty into the left atrium. The lungless plethodontids lack pulmonary veins, and their left atrium is small. The amphibian ventricle is not divided, but nevertheless the two bloodstreams, pulmonary and systemic, mix only to a slight degree. Some amphibians have a rather complicated system of valves and partitions that helps keep the two streams separate. In sirenids the ventricle is nearly divided by a septum, so that the heart is almost four-chambered.

Arteries.   In the primitive ancestor of the vertebrates, blood apparently left the heart through a large artery extending forward, the ventral aorta, which gave off six pairs of branches, the aortic arches. These arches curved around the gut to unite on the dorsal side and form the dorsal aorta, the main systemic artery distributing blood to the body. The evolution of the vertebrate arterial system has largely been a trend toward the loss of the aortic arches. Most fishes have lost the first two arches. The last four are divided into afferent and efferent parts joined by the capillaries passing through the gills. In larval amphibians and the gilled salamanders, the aortic arches are not interrupted by gill capillaries. Instead, the arches give off vascular loops, which pass into the gills and branch into capillaries; these are gathered together again to rejoin the aortic arches dorsally. This simplifies the change from gill to lung respiration at metamorphosis: the vascular loops degenerate, the arches expand, and the flow of blood from heart to body continues without interruption.

Salamanders are more primitive than frogs in the evolutionary development of their aortic arches (Figure 2–3). The first two pairs of arches are

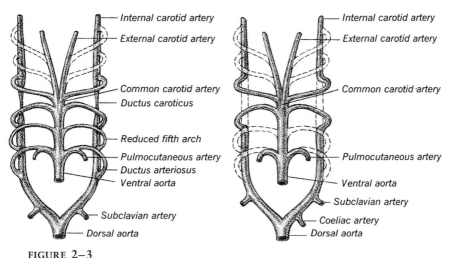

Internal carotid artery

External carotid artery

Common carotid artery
Ductus caroticus

Reduced fifth arch

Pulmocutaneous artery
Ductus arteriosus
Ventral aorta

Subclavian artery

Dorsal aorta

Internal carotid artery

External carotid artery

Common carotid artery

Pulmocutaneous artery

Ventral aorta

Subclavian artery

Coeliac artery
Dorsal aorta

FIGURE 2–3
The aortic arches as found in most salamanders (*left*) and frogs (*right*), ventral view.

lacking; the forward extensions of the ventral aorta from which they originally arose form the external carotid arteries, which carry blood to the jaws. The third pair of arches form the internal carotid arteries, which extend forward to supply the face and brain. Frequently a connection remains between the third and fourth arches. The fourth pair are the systemic arches, which form the main part of the dorsal aorta. The fifth pair of arches may persist, although, if so, they are much reduced in size. The sixth pair of arches give rise to the pulmocutaneous arteries, which carry blood mainly from the right side of the heart to the lungs and skin to be oxygenated; each arch retains its connection to one of the arms forming the dorsal aorta through a ductus arteriosus. Thus some of the unoxygenated blood passing through the sixth pair of arches may go to the body, rather than to the lungs or skin. In frogs, there is never any connection between the third and fourth arches, the fifth pair have disappeared, and the sixth have lost their connections with the dorsal aorta, so that the unoxygenated blood they carry must all go to the lungs or skin.

Veins. Like the other vertebrates, amphibians have an hepatic portal system. The veins carrying blood from the intestine join to form the hepatic portal vein, which branches into a network of small sinuses in the liver. These vessels are then gathered together again into the hepatic veins to return blood toward the heart. The liver removes and transforms substances that have been absorbed by the blood from the intestine, storing some in the form of glycogen, returning others to the blood in a form that can be used by the cells. It also removes the harmful nitrogenous waste products of cell metabolism, changes them into harmless urea (in most adult amphibians), and empties the urea into the blood to be removed by the kidneys. Obviously, it is to the advantage of an animal to have as much blood as possible pass through the liver to be cleansed and restocked with food before its return to the heart. In this the amphibians show an advance over the fishes. Part of the blood returning from the hind legs and posterior part of the body empties directly into the hepatic portal vein through an anterior abdominal vein. The rest of the blood from the hind legs passes through the renal portal system of the kidneys. Amphibians also have pulmonary veins, which are usually lacking in fishes.

Excretory System

Getting rid of the waste products of metabolism and regulating the salt content and water balance of the body are the functions of the excretory system, which consists of the kidneys and the tubes leading from them outside the body. The excretory and reproductive systems are closely connected, especially in males, so that it is difficult to discuss one without referring to the other.

Kidney. Three successive types of kidney appear in the evolution of the vertebrates. The kidney of the adult amphibian, as in most fishes, is an opisthonephros, midway in the evolutionary series between the primitive, anterior pronephros of the hagfish and the compact, posterior metanephros of the higher tetrapods.

The kidneys of salamanders and sirens are long and are divided into a narrow anterior part and an expanded posterior part (Figure 2–4). Those of the sirens fuse posteriorly to form a knoblike "tail kidney." Caecilian kidneys are also long, extending most of the length of the body cavity, and are uniform in width throughout. Frogs have more compact kidneys that are

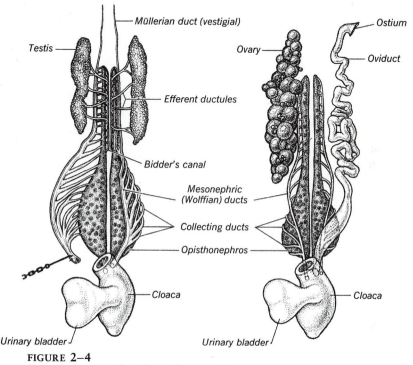

FIGURE 2–4
Urogenital organs of a salamander, ventral view: (*left*) male; (*right*) female.

posterior, flattened dorsoventrally, and not markedly divided into anterior and posterior parts (Figure 2–5). Caecilians and sirens have well-developed glomeruli (capillary nets) in the anterior part of the kidney, but in salamanders and frogs the anterior glomeruli are reduced or absent. Presumably, when these glomeruli are lacking, the anterior kidney has lost its excretory function. In male amphibians, ducts from the testes enter the anterior part of the kidney and empty into the nephric ducts. These in turn join to form collecting ducts that lead into the mesonephric or Wolffian duct and then into the cloaca. Collecting ducts from the posterior part of the kidney also drain into the Wolffian duct in caecilians and sirens. The posterior kidney ducts of salamanders and frogs tend to fuse to form a ureterlike structure that may empty into the posterior part of the Wolffian duct or directly into

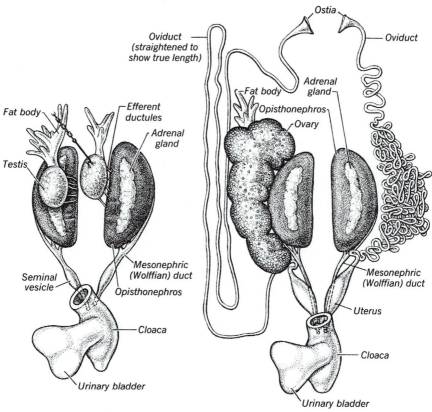

FIGURE 2–5
Urogenital organs of a frog, ventral view: (*left*) male; (*right*) female.

the cloaca. This fusion is most marked in males. The Wolffian ducts runs along the outside of the kidney in caecilians and salamanders but is embedded in the kidney tissue in frogs and sirens.

Bladder. Amphibians have a thin-walled urinary bladder that opens into the cloaca and is not connected with the Wolffian ducts from the kidneys. Urine passes down these ducts directly into the cloaca and then backs into the bladder for storage. The liquid discharged by a captured toad on the hands of the collector is not urine, but reserve water stored in the cloaca.

Reproductive System

As is typical in vertebrates, the amphibian reproductive system is composed of primary sex organs (gonads) and accessory organs, which include ducts and other structures that help bring the germ cells together.

Female Reproductive Organs. The amphibian ovary is saccular (sac-shaped), with an enclosed lymphatic cavity. In vertebrates, the size of an individual ovum is largely determined by the amount of contained yolk. Amphibian eggs contain a moderate amount of yolk: more than the microscopic eggs of most mammals, in which the embryo receives nourishment from the maternal bloodstream, but less than the large-yolked eggs of reptiles, in which the young reach an advanced state before hatching. The eggs of many amphibians are about two millimeters in diameter and may be very numerous. Frogs, in particular, may produce thousands of mature ova at one time. Ovaries filled with ripe eggs are irregular lumpy structures that fill the greater part of the body cavity. The eggs escape into the coelom through the external wall of the ovary.

Ovaries vary in shape with variation in body form. Salamander ovaries are elongated, but are not as long and narrow as those of caecilians (see Figure 2–4). Frogs, on the other hand, have decidedly shorter, more compact ovaries (see Figure 2–5). The cavity of the salamander ovary is single and continuous, but in the frog ovary it is divided into a number of pockets.

Associated with the ovaries of amphibians are fat bodies, or corpora adiposa. The fat bodies of salamanders are long, slender structures running parallel to the ovaries along their median edges. Those of frogs lie just anterior to the gonads and consist of several yellowish, fingerlike processes. Caecilians have leaflike fat bodies lying along the lateral edges of the ovaries. That they apparently serve as storage places for nutritive material for the developing ova is indicated by changes in the size of the fat bodies during the year: they shrink as the eggs enlarge, being smallest at the close of the breeding season, after which they gradually increase in size, until the maximum is reached just before the ova begin rapid growth.

The tubes that transport the products of the ovaries to the outside are the paired Müllerian ducts. These have the same general pattern for all amphibians. Anteriorly, each tube begins as an expanded, funnellike opening, the ostium, which is situated far forward in the body cavity. Eggs that have escaped from the ovaries into the body cavity are directed toward the ostium by cilia located on the peritoneum of the body wall, on the liver, and on adjacent structures. For most of its length, the Müllerian tube has a thickened wall with a glandular lining. During the breeding season, the duct becomes greatly enlarged and markedly coiled, and the glandular lining epithelium secretes a clear, gelatinous substance (Figure 2–6). After entering the ostium, the egg is forced along by peristaltic waves. As it passes down the oviduct with a twisting, spiral motion, the glands deposit several layers of jellylike material around it. Posteriorly, each oviduct is enlarged to form a uteruslike structure, which serves only as a temporary storage place for eggs that are soon to be laid. The uteri of most amphibians enter the cloaca separately, but in *Bufo*, the two uteri unite and have a common opening into the cloaca. Most female salamanders possess a dorsal diverticulum of the

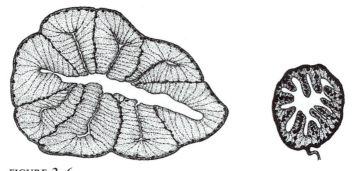

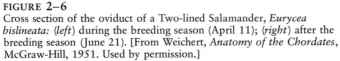

FIGURE 2–6
Cross section of the oviduct of a Two-lined Salamander, *Eurycea
bislineata:* (*left*) during the breeding season (April 11); (*right*) after the
breeding season (June 21). [From Weichert, *Anatomy of the Chordates*,
McGraw-Hill, 1951. Used by permission.]

cloaca, the spermatheca, which serves to store sperm until the eggs are ready
to be fertilized. Male amphibians usually have rudimentary Müllerian ducts.

Male Reproductive Organs.   The primary male reproductive organ is the
testis, a compact structure made up of a mass of seminiferous (sperm-
bearing) tubules that connect, by means of ducts, to the outside. Each testis
is really a compound tubular gland. Its shape, like that of the ovary, corre-
sponds generally to the animal's shape. The testes of primitive caecilians are
elongated structures that look like strings of beads. The beadlike swellings,
consisting of masses of seminiferous tubules, are connected by a longitudinal collecting duct. More advanced caecilians tend to display lobular fusion.
Salamander testes are shorter and more irregular in outline. The testes of
frogs are compact and either oval or nearly rounded. There is a pronounced
difference in the size of the testes during the breeding and nonbreeding
seasons. Like the ovaries, the male gonads have fat bodies associated with
them. These resemble the fat bodies of the ovaries in position and appearance and, like them, fluctuate in size with the onset and passing of the
breeding season.

Male toads of the family Bufonidae have peculiar structures—Bidder's
organs—that lie between the testes and the fat bodies. If the testes are
removed, these lobelike structures will develop into functional ovaries in
about two years, bringing about a complete reversal of sex. When this
happens, the otherwise rudimentary oviducts enlarge, seemingly in response
to female sex hormones produced by the transformed Bidder's organs. This
is a striking illustration that each sex in the vertebrates has rudimentary
organs of the other.

The efferent ductules of the salamander carry sperm from the testis to
Bidder's canal, a fine longitudinal canal that runs along the median edge of

the kidney. Bidder's canal connects with the nephric tubules in the anterior portion of the kidney by means of a number of short ducts. (Bidder's canal is present, though in a rudimentary form, in the female as well.) The tubules emerge from the lateral edge of the kidney to join the Wolffian duct. In those species in which the anterior kidney has lost its excretory function and the posterior kidney ducts enter the cloaca directly, the male Wolffian duct is wholly reproductive in function, and is thus a true ductus deferens.

The longitudinal collecting duct of the caecilian testis gives off small transverse canals, which pass into the kidney and join the nephric tubules that enter the Wolffian duct. Some species show traces of a Bidder's canal.

The urogenital ducts of frogs are similar to those of salamanders, but there are minor variations. In some forms, the efferent ductules connect directly with the nephric tubules; but in others they join a Bidder's canal that is close to the median border, but within the kidney. Spermatozoa are then transferred from Bidder's canal through the nephric tubules to the Wolffian duct. A Bidder's canal is present in the female, but its function, if any, is unknown. The Wolffian duct emerges from a point near the posterior end of the kidney and passes into the cloaca. Males of many species of frog have a dilation in the Wolffian duct close to the junction with the cloaca. This forms a seminal vesicle for the temporary storage of sperm. Seminal vesicles are poorly developed in *Rana pipiens* and *Rana catesbeiana,* the two species most commonly studied in introductory zoology courses. Seminal vesicles, when present, are largest during the breeding season and shrink after breeding is over. Such seasonal changes indicate that their development and function are under hormonal control.

The Wolffian duct, particularly that of salamanders, also shows seasonal variation in size, becoming largest at the height of reproductive activity. In the nonbreeding season, the duct may be reduced to a mere thread, particularly anteriorly, where it functions solely in reproduction.

Males of all known salamanders except the Hynobiidae and Cryptobranchidae undergo an enlargement of the glandular lining of the cloaca as the breeding season approaches. The glands secrete a jellylike material around a cluster of sperm to form a packet, which rests on a jelly base that is also secreted by glands of the cloaca. Together they form a toadstool-shaped structure, the spermatophore.

Skeleton

Bone may develop either in the skin (the dermal skeleton or exoskeleton) or deep within the body (the endoskeleton). The endoskeleton is laid down first as cartilage, which during development is replaced more or less completely by bone. Dermal bone forms directly, with no cartilaginous precursor. Generally, when we speak of the skeleton of an animal, we mean the endoskeleton and those parts of the exoskeleton, particularly in the skull region, that have dissociated themselves from the skin and moved inward to join the endoskeleton.

The main function of the exoskeleton is to protect the body. The endoskeleton, on the other hand, provides a place of attachment for the muscles, a framework against which they can pull when moving the body. When the vertebrates moved to land, a third function of the skeleton became very important. The bodies of aquatic animals are supported by the buoyancy of water. Air provides no similar support for terrestrial animals; their skeletons assume the supportive role. The appendages shift from steering fins to supportive and propulsive legs; the vertebral column becomes an arch on which to suspend the viscera. The skeletons of land vertebrates are more completely ossified, more firmly linked together, and have more complex articulations than those of fishes, because of the increased need for support and locomotion. The beginnings of these changes are evident in amphibians.

Except for a few bones of the skull, the functional exoskeleton has disappeared in most amphibians. Certain vestiges (scales, claws) have already been mentioned in the section on the integument.

Skull. The skulls of modern amphibians show considerable variation in shape and amount of ossification, depending on the habits and habitats of the species. In general, however, in contrast to the fish skull, the amphibian skull is flattened, contains far fewer dermal roofing bones, particularly the posterior ones, and has an ossified brain case (chondrocranium). The jaw is attached directly to the skull (autostylic), whereas in fishes, the jaw links to the skull through a modified gill arch bone (hyostylic).

Vertebral Column. The vertebral column of fishes consists of a series of lozenge-shaped centra joined by a persistent notochord, which constricts to a thin thread as it passes through each centrum and expands into a ball between centra. Each centrum bears a dorsal neural arch, which protects the spinal cord, and, in the tail region, a ventral haemal arch for caudal blood vessels. There is no regional differentiation other than that between the trunk vertebrae without haemal arches and the caudal vertebrae with them. In the amphibians, the demands of weight support and appendicular locomotion have resulted in additional regional differentiation and more complex vertebral structure. The first (neck or cervical) vertebra has two concave facets for more mobile articulation with the skull; the trunk vertebrae have paired facets fore and aft for firmer support of the trunk; the last trunk (sacral) vertebra bears enlarged transverse processes (diapophyses) to articulate with the pelvic girdle, and the tail (caudal) vertebrae also bear paired intervertebral facets. In frogs, the caudal vertebrae have fused into a long, rodlike bone, the urostyle. There is considerable variation in the mode

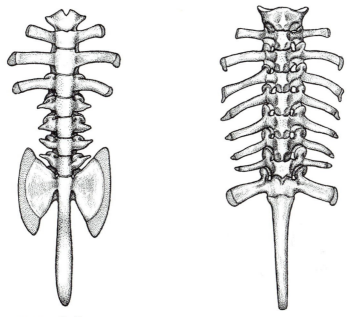

FIGURE 2–7
Vertebral columns of (*left*) a frog with expanded sacral diapophyses, *Scaphiopus couchi*; (*right*) one with rounded sacral diapophyses, *Rana virgatipes*. [After Noble, *The Biology of the Amphibia*, McGraw-Hill, 1931 (reprinted Dover, 1954).]

of ossification and the ultimate shape of the vertebrae between and even within the major amphibian groups (Figure 2–7).

Ribs. Ribs are present, though poorly developed, in modern amphibians. Usually they attach to the vertebrae by two heads. They are longest in caecilians, shorter in salamanders, and in most frogs are lacking entirely or are only present as cartilaginous tips on the transverse processes of the vertebrae.

Sternum. The sternum, or breastbone, which is characteristic of the higher tetrapods, appears for the first time in the amphibians, although not in a well-developed state. The ribs do not attach to it to form a definite thoracic basket. It is absent in caecilians and in some of the elongated salamanders; other salamanders have a small cartilaginous plate lying between the halves of the pectoral girdle. The sternum in primitive frogs resembles the salamander sternum, but in more advanced frogs it may be rod-shaped and partly ossified.

Girdles and Limbs.   In fishes, locomotion is accomplished by undulatory waves of the body. This axial locomotor mechanism is not entirely lost in the amphibians, but the limbs begin to dominate forward movement (appendicular locomotion). The shift from axial to appendicular locomotion has caused significant modifications of the appendicular skeleton. The pectoral girdle has lost its connection to the skull and is anchored to the trunk by a muscular sling. In addition, its dorsal segments are somewhat reduced in comparison with those of fishes, whereas the ventral parts form a complex articulation mechanism, particularly in the girdles of frogs. The pelvic girdle is large and articulates firmly with the vertebral column through the sacral vertebra.

The limb skeleton in tetrapods is a linear column of articulating bones, rather than a row of bracing bones as in the fins of fishes. From body to tip, the tetrapod limb consists of a single long bone, a pair of long bones, a series of small sliding bones, and a series of elongated digit bones. It provides a terrestrial animal with an effective support, steering, and propulsive mechanism. This basic organization was established in the early amphibians and has been variously modified in the recent ones. The basic pentadactyl (five-digit) plan has been modified to no more than four fingers in all amphibians, and often only four toes. Salamanders commonly show fusion of some carpal bones and loss of one or more digits. Sirens have lost the hindlimbs, and caecilians the entire appendicular skeleton. Frogs have fused the paired bones of both fore- and hindlimbs, and elongated a pair of ankle bones, to produce a skeleton specialized for jumping.

Endocrine Glands

There is little to be said here about the structure of the endocrine glands. Most of them are found in all vertebrates, and the morphological variations from one group to another seem, in general, to have little evolutionary significance.

Parathyroids appear for the first time as definitive structures in amphibians, and seem to be concerned with controlling the level of calcium salts in the blood. They are present in all higher vertebrates, and their removal results in death.

The two component parts of the adrenal gland are closely associated in amphibians, as in reptiles and birds, instead of being separated, as in the lower vertebrates. The reason for this difference is unknown.

Although morphologically the endocrines are of little interest, functionally they are among the most fascinating of organs. The thyroid of amphibians is the gland controlling metamorphosis. If it is removed from a tadpole, the tadpole grows into a large, fat supertadpole, but never into a frog although it does develop lungs and reproductive organs. If thyroid extract is fed to such a tadpole, it metamorphoses. On the other hand, if a very small tadpole is fed thyroid extract, it stops growing and quickly metamorphoses into a midget frog. The thyroid gland also controls ecdysis. An amphibian from which this gland has been removed does not shed its skin, and the dead layers pile up until the animal appears much darker than normal. When fed thyroid extract, it soon sheds the dead epidermal layers.

In amphibians, as in other tetrapods, the pituitary produces a number of trophic hormones that control the activity of other endocrine glands (Figure 2–8). These include thyrotrophin (TSH), which controls the thyroid, gonadotrophins (FSH and LH), which stimulate the production of gametes and sex hormones by the gonads. Additional hormones of the pituitary include prolactin (LTH), which has been shown to stimulate newts to migrate to water. Increase in the production of prolactin in the spring may thus be the factor that triggers the return of amphibians to their breeding sites. Melanocyte-stimulating hormone (MSH) causes dispersal of pigment granules in pigment cells in the skin, resulting in a general darkening in

FIGURE 2–8
Small-mouthed Salamanders, *Ambystoma texanum: (lower)* normal control animal; *(center)* a few hours after removal of pituitary gland, pigment cells contracted; *(upper)* several weeks after pituitary removal—the dark color is due to the animal's failure to shed the corneal layer of epidermis, which has become very thick. The last effect is undoubtedly caused by the failure of the thyroid gland to function in the absence of the pituitary gland. [From Weichert, *Anatomy of the Chordates,* McGraw-Hill, 1951. Used by permission.]

those amphibians that are able to change color. Vasotocin, which is antidiuretic and increases the permeability of frog skin to water, helps regulate water balance.

Nervous System

Animals on land face a far more complicated environment than do those in water. The chief evolutionary change in the vertebrate brain is the enlargement of the cerebral hemispheres, the lateral lobes of the forepart of the brain. Primitively, the roof and sides of these hemispheres are composed of a nonnervous epithelial layer, the pallium. The amphibians have made few important advances over the fishes in the structure of the nervous system, but in them for the first time scattered nerve cells appear in the walls of the pallium, foreshadowing the enormous development of the cerebrum in mammals.

Like fishes, amphibians have ten cranial nerves; higher vertebrates have twelve. Primitive fossil amphibians apparently had twelve, and the reduction to ten in the modern forms seems to have resulted from a shortening of the cranium, so that the eleventh and twelfth now appear as spinal nerves. The movements of tetrapod limbs are more complicated than are those of fish fins and require a more complex innervation. Enlargements of the spinal cord in the cervical and lumbar regions result; these appear first in the amphibians. The spinal cord of the salamanders extends to the tip of the tail, but that of frogs is much shortened, and the spinal nerves continue through the neural canal as a brushlike structure, the "horse's tail," or cauda equina.

Sense Organs

As the amphibians adapted to a terrestrial existence, their sense organs were modified to perceive through an aerial medium instead of the aqueous medium of their fish ancestors.

Eye.   The cornea, the clear window in the outer layer of the vertebrate eyeball through which light rays enter, becomes opaque when dried. Animals that live on land must have some mechanism to keep the cornea moist and to wash off specks of dirt. Terrestrial amphibians have developed a series of glands in the tissue around the eye, and movable eyelids to wash the secretions of these glands across the eyeball. The eyeball can be protruded or withdrawn into the eye socket. When it is drawn in, the lower lid moves up to close over it. Permanently aquatic salamanders and sirens lack eyelids, as do all amphibian larvae. Eyes are very poorly developed in caecilians and in many cave-dwelling salamanders.

Ear. The amphibian ear shows several advances over the ear of fishes, for the amphibians have two additional sensory patches (basilar papilla and amphibian papilla) in the membranous labyrinth of the inner ear. The basilar papilla responds to high-frequency sounds and appears to be the forerunner of the sensory structures of the cochlear duct in the higher tetrapods; the amphibian papilla responds to low-frequency sounds and has no successor in the other tetrapods.

Low-frequency—environmental—vibrations are received by the forelimbs and pectoral girdle, transmitted to the inner ear by an opercular bone, and received by the amphibian papilla. High-frequency sounds—courtship calls—are received by the ear drum (tympanum), transmitted by a bone, the stapes (columella) to the membranous labyrinth of the inner ear, and received by the basilar papilla. All modern amphibians appear to use the opercular–amphibian papilla mechanism, whereas, so far as is known, only the frogs use the stapes–basilar papilla complex, and then only for territorial and courtship behavior. Since all modern amphibians have basilar papillae, did their common ancestor have a voice?

Lateral Line. Fishes perceive low-frequency vibrations by means of the lateral line, a series of sense organs arranged in rows on the head and sides of the body. There is evidence that its chief function is the detection of nearby objects, stationary or moving, by wave reflection. Lateral line systems are still present in all aquatic amphibians—for example, free-living larval stages, aquatic salamanders, and pipid frogs—but such systems function only in water and are absent in terrestrial amphibians.

Jacobson's Organ. The sense organ known as Jacobson's organ appears for the first time in the amphibians. It is also well developed in lizards and snakes, but is vestigial in most other tetrapods. In amphibians, it consists of a pair of blind sacs connected by ducts to the nasal cavities. Since it is innervated in part by a branch of the olfactory nerve, it probably plays a part in the recognition of food.

READINGS AND REFERENCES

Ecker, A., R. Wiedersheim, and E. Gaupp. *Anatomie des Frosches,* 2nd ed., 3 vols. Braunschweig: 1888–1904.

Florkin, M., and B. T. Scheer (eds.). *Chemical Zoology,* vol. IX, *Amphibia and Reptilia.* New York: Academic Press, 1974.

Francis, E. T. B. *The Anatomy of the Salamander.* London: Oxford University Press, 1934.

Gans, C. "Sound production in the Salienta: mechanism and evolution of the emitter." *American Zoologist,* vol. 13, 1973.

————. *Biomechanics.* Philadelphia: Lippincott, 1974.

Holmes, S. J. *The Biology of the Frog,* 4th ed. New York: Macmillan, 1927.

Lombard, R. E., and I. R. Straughn. "Functional aspects of anuran middle ear structures." *Journal of Experimental Biology,* vol. 61, 1974.

Moore, J. A. (ed.) *Physiology of the Amphibia,* vol. 1. New York: Academic Press, 1964.

Parsons, T. S. "Evolution of the nasal structure in lower tetrapods." *American Zoologist,* vol. 7, 1967.

Trueb, L. "Bones, frogs, and evolution." *In* J. L. Vial (ed.), *Evolutionary Biology of the Anurans.* Columbia: University of Missouri Press, 1973.

Wake, D. B. "Aspects of vertebral evolution in the modern Amphibia." *Forma et Functio,* vol. 3, 1970.

————, and R. Lawson. "Developmental and adult morphology of the vertebral column in plethodontid salamander *Eurycea bislineata,* with comments on vertebral evolution in Amphibia." *Journal of Morphology,* vol. 139, 1973.

Whitear, M. "Flask cells and epidermal dynamics in frog skin." *Journal of Zoology,* vol. 175, 1975.

CHAPTER THREE

STRUCTURE
OF REPTILES

Two great evolutionary advances make the reptiles more successful as invaders of the land than their more primitive amphibian relatives. The first advance is the amniote egg. As the embryo develops, folds of tissue grow around it and fuse above it to form two closed sacs. The outer sac—the chorion—surrounds the whole egg with a protective membrane; the inner sac—the amnion—is filled with fluid to provide the embryo with an aquatic environment in which to develop without danger of desiccation (Figure 3-1). A third extraembryonic membrane—the allantois—is a saclike structure that grows from the hindgut of the developing embryo and presses against the chorion; it acts as a respiratory organ and also as a storage place for metabolic waste products. A fourth membrane—the yolk sac—develops around the yolk. The entire complex of developing embryo and extraembryonic membranes is surrounded by a watery albumen (except in snake and lizard eggs) and is covered with a tough shell. Since the reptilian eggshell is permeable, some environmental moisture is necessary to prevent desiccation. But the egg is much better adapted to deposition on land than the unprotected egg of even the most terrestrial amphibian.

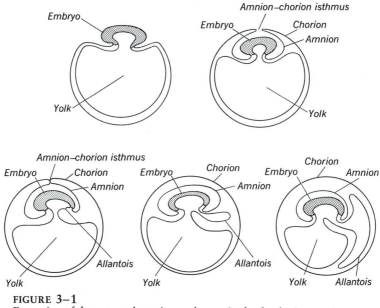

FIGURE 3-1
Formation of the extraembryonic membranes in the developing amniote egg.

The second advance that makes reptiles more successful on land than amphibians is a tougher and less permeable epidermis. The amphibian epidermis is a thin, secreting tissue that offers little protection against desiccation, abrasion, solar radiation, and other hazards of terrestrial life. The reptilian epidermis has lost most of its secretory function and forms a thick protective barrier against abrasion and other physical injury, while acting as a shield against desiccation. Scales, thickened keratinous plates, are an additional development to improve the protective function of the epidermis.

DEFINITION OF REPTILIA

The members of the class Reptilia—turtles, lizards, snakes, worm lizards, crocodilians, and tuatara—are amniotes that have epidermal scales but lack feathers, hair, or mammary glands. Development is direct, since the need for a larval stage ended with the evolution of a shelled egg that could be laid on land. The stratum corneum is much better developed than that of amphibians, and the skin is dry. In addition to heavy epidermal scales, many reptiles have bony dermal plates lying under the epidermis. Reptiles have fewer skin glands than amphibians. Respiration, as would be expected, is entirely by means of lungs, except some through vascular tissue in the pharynx of aquatic turtles. A secondary palate, which in higher amniotes separates the nasal cavity from the oral cavity, is present but incomplete in most reptiles, so that a partial separation occurs. In crocodilians this palate becomes complete. The regional differentiation of the vertebral column characteristic of mammals makes its appearance in the reptiles, though only the crocodilians have five clear-cut regions. There is a single occipital condyle. The legs usually have five digits, which end in true claws. The kidney is metanephric, like that of birds and mammals. The heart is either three-chambered or, in the crocodilians, four-chambered. There are twelve pairs of cranial nerves. Like the amphibians, reptiles are ectothermic.

ANATOMY OF REPTILES

Present-day reptiles do not match the birds or fishes in number of species or individuals. They have, however, successfully invaded many habitats, so that they now occupy the seas, fresh water, land, trees, and soil; the Flying Dragons *(Draco)* have even become gliders. For so small a group, the reptiles show a wonderful variety of structural modifications.

Because reptiles are so diversified, it is difficult to give a concise structural account that is applicable to all. The following generalizations may serve as a basis for understanding the groups discussed in later chapters.

Integument

The reptilian skin has lost its respiratory function, and thus need not be moist. In fact, one of its main functions is to prevent water loss. This is accomplished primarily by means of a thicker epidermis with surface layers composed of dead, compacted cells. These modifications do not establish complete impermeability, but greatly improve it while also improving the skin's protective function.

Epidermis. The reptilian epidermis contains a minimum of six to eight cell layers, and often many more. The stratum germinativum produces all these layers, and as they are pushed outward by succeeding generations of cells, they show regional differentiation, eventually forming a thick stratum corneum.

All reptiles periodically shed the superficial epidermal layers. In lizards and snakes, the entire superficial skin is shed as a single unit. Their epidermis possesses an outer and an inner tissue layer above the stratum germinativum. Each tissue layer consists of four specialized layers of cells. When the shedding cycle begins, the two tissue layers separate simultaneously over the entire body, and the outer layer is removed. The inner layer is now an outer layer, and no inner layer exists. After a week or so, the stratum germinativum begins to divide, and soon a new inner tissue layer is produced; once it is fully differentiated, the cycle can repeat itself.

Dermis. The dermis is a thick connective tissue layer supporting and nourishing the epidermis. The chromatophores (color cells) lie predominantly in the dermis at its junction with the epidermis. Chromatophores are named for the pigments they bear. The melanophores contain melanin (black pigment granules) and are the most abundant. Each melanophore is a many-branched, amoeboid cell; neither the cell nor its arm move. Color changes result from the movement of melanin within the cells; lighter color results from melanin aggregating in the central cell body, darker color from melanin dispersing into the arms.

Scales. The scales and scutes of reptiles are not homologous with the dermal scales of fishes, but are different structures derived from the epidermis. In some cases, these epidermal scales are supported internally by bony plates (dermal scales or osteoderms). The osteoderms of lizards and crocodilians remain largely as discrete bony nodules, usually in a one-to-one ratio with the epidermal scales. In turtles, however, the osteoderms have fused with one another and with parts of the endoskeleton to form the carapace and plastron (upper and lower shell). *Sphenodon* and crocodilians have fused osteoderms (gastralia) in the ventral body wall, which form an abdominal, rib-cage-like structure to support the viscera.

Reptilian scales may not be strictly homologous among living reptile groups. In all of them, the scales derive from the stratum germinativum, but the developmental pattern, final organization, and manner of shedding are different. Lepidosaurians (*Sphenodon* and squamates) are completely encased in scales, usually overlapping, and the base of each scale is joined to those of the surrounding scales; the scales are shed as a single unit. Crocodilian scales cover the entire body in a tilelike arrangement, and as the surface of each scale wears away, it is replaced, apparently continually, by the cell layers below. Turtles have large, thin epidermal scutes covering the shell, and occasionally clusters of scales on the legs and head. The scales and associated skin of the legs and head are usually shed in small ragged pieces. Underlying the scutes is a thin Malpighian layer (the basal layer of the epidermis). From time to time, a center of this layer beneath each scale cornifies to form a new scute, slightly larger than the one above it. The old scute either falls off or adheres to the new one, creating a pyramid of scutes at each such center.

Only for the squamates (snakes, lizards, and related forms) are the histology and developmental sequence of scales well known. The shedding cycle has been described briefly above. The histology of the scales is more complex than is usually described. The scale consists of two tissue layers, an outer and an inner epidermal generation (Figure 3–2). These tissues are essentially identical except for the maturity and associated cornification of

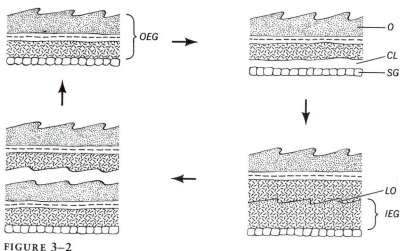

FIGURE 3–2
Cycle of epidermal scale development in squamates. *CL*, clear layer; *IEG*, inner epidermal generation; *LO*, lower oberhautchen; *O*, oberhautchen; *OEG*, outer epidermal generation; *SG*, stratum germinativum. [After Maderson, *Forma et Functio*, 1970.]

the outer tissue. Each generation consists of four layers of specialized cells: the alpha or lacunar cells, the mesos, the beta cells, and the oberhautchen. The two generations are separated by a clear cell layer that acts as the cleavage surface during shedding. The two superficial layers of the outer tissue form the heavy cornified or keratinous covering of the scale, and it is this surface that forms the primary epidermal barrier. The rattles of a rattlesnake are modified epidermal scales (Figure 3-3).

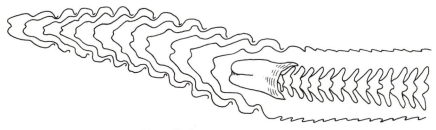

FIGURE 3–3
Structure of the rattle of a rattlesnake.

Digestive System

The reptilian digestive apparatus is not strikingly different from that of the amphibians. The main differences occur in the structures of the mouth. New oral glands appear and become further specialized; the teeth also become specialized and begin to show regional differentiation.

Glands.   The oral glands, which help moisten the prey to prepare it for swallowing, are better developed in reptiles than in amphibians. A palatine gland is present on the roof of the mouth. Reptiles also have lingual (tongue), sublingual (below the tongue), and labial (lip) glands. The poison glands of the poisonous snakes are modifications of the labial glands in the upper jaw. There is one on each side, which opens by a duct into the groove or cavity of the poison fang. The poisonous lizard *Heloderma* has a sublingual gland on each side that is modified to produce poison. Four ducts lead from each gland through the bone of the lower jaw to empty the poison into the vestibule in front of the grooved teeth. Thus, whereas a rattlesnake can inject its poison at a single strike, as with a hypodermic needle, *Heloderma* must hang on and chew to force the poisonous saliva into the wound. Marine turtles and crocodilians, which take their prey in water, have poorly developed oral glands.

Tongue.   The tongue of crocodilians and most turtles is not protrusible, and simply lies on the floor of the mouth as an aid to swallowing. The squamates have well-developed, protrusible tongues. The tongue of most lizards is divided into a protrusible foretongue to aid in food gathering and a stationary hindtongue for swallowing. In all lizards, the tip of the foretongue serves a sensory role by carrying odor particles to the openings of Jacobson's organ in the roof of the mouth. This role becomes predominant in the anguinomorph lizards; their foretongue is narrow with a deep median cleft, and is retractible into the hindtongue; the hindtongue is reduced or has lost its role in swallowing. The narrow, deeply forked snake tongue is the end point of this modification, serving only as a bearer of odor particles.

Teeth.   Throughout the evolution of the vertebrates, teeth have decreased in number and location. Fish may have teeth almost anywhere in the mouth and pharynx. Amphibian teeth occur on the jawbones and on the bones (palatines, vomers, parasphenoids) of the roof of the mouth. Snakes and lizards have teeth on the palatines, pterygoids, and jawbones. Crocodilian teeth are restricted to the jawbones, as in mammals. Turtles have no teeth, and the jaws bear a horny covering with sharp cutting edges (tomia).

Like amphibians, reptiles replace their teeth an indefinite number of times (polyphyodont). Tooth replacement occurs in a wavelike sequence from back to front. Adjacent teeth are in different waves, so that functional teeth are always present throughout the length of the jaw. Reptilian teeth commonly show some regional specialization in size or shape (heterodont). The teeth of squamates attach either to the upper jaw margin (acrodont) or to the inside jaw margin (pleurodont); the crocodilians have tooth sockets (thecodont) like mammals (Figure 3–4).

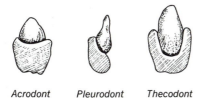

FIGURE 3–4
Types of tooth attachment found in reptiles.

Acrodont   Pleurodont   Thecodont

Gut.   The esophagus of reptiles is generally longer than that of amphibians, and its walls are gathered into longitudinal folds to permit expansion during the swallowing of large prey. The stomach is clearly set off from the esophagus. Snake and lizard stomachs are long and spindle-shaped. Crocodilians have part of the stomach modified to form a muscular, gizzardlike region. The small intestine of reptiles is long and coiled. The large intestine usually has a colic cecum, a blind pouch at the juncture with the small intestine. Colic ceca are absent in crocodilians, but are present in most birds and almost all mammals. The large intestine empties into a cloaca, which opens to the outside through the vent.

Respiratory System

Since reptiles as a group are the first completely terrestrial vertebrates, they are the first to depend almost entirely on the lungs to aerate the blood. Pharyngeal respiration supplements lung breathing in some aquatic turtles, enabling them to stay under water for long periods, but in general reptiles are air breathers. Gills have disappeared completely, and the reptilian lungs are better developed than those of amphibians.

Among the higher tetrapods, the anterior part of the respiratory tract is separated from the anterior part of the digestive tract. The separation is initiated in the reptiles, in which a hard palate begins to form. The openings of the internal nares are pushed to the back part of the mouth, and the nasal passages between external and internal nares are elongated.

The reptilian larynx is generally no better developed than that of the amphibians. Most reptiles are voiceless, but some lizards produce harsh sounds, and alligators bellow vociferously during the mating season. A few turtles also have voices.

Circulatory System

When they became terrestrial, the reptiles were released from the necessity of circulating blood to the gills, but the resulting increased dependence on lungs called for improvements in the pulmonary circulation and the heart.

Heart. The heart of most reptiles is three-chambered, but crocodiles and alligators have achieved a completely four-chambered heart. The right atrium, which receives unoxygenated blood from the body, is completely separated from the left, which receives oxygenated blood from the lungs. Both atria of reptiles with three-chambered hearts empty into a single ventricle, but an incomplete septum growing from the apex toward the center of the ventricle essentially separates the two bloodstreams. The septum of crocodilians, however, is complete.

Arteries. The third, fourth, and sixth aortic arches are present in reptiles, but their connections have been considerably modified. The truncus arteriosus, the large blood vessel that extends forward from the heart to give rise to the arches, splits at its base to form three main trunks. One of these, the pulmonary aorta, leaves the right side of the ventricle and divides into two pulmonary arteries, which represent the sixth aortic arches. The left aorta (left fourth aortic arch) has become connected with the right side of the ventricle. The rest of the truncus arteriosus (right aorta) connects with the left side of the ventricle and carries oxygenated blood. In the crocodilians there is an opening—the foramen of Panizzae—between the right and left aortae as they leave the heart. The right aorta crosses over the left and runs forward on the right side. First it gives off the right (fourth) aortic arch, then it passes forward and divides into the common carotid arteries. These split into internal carotids, representing the third aortic arches, and external carotids, the forward extensions of the old ventral aorta (Figure 3–5). The left and right fourth aortic arches fuse to form the dorsal aorta. Tests of the oxygen content of the blood in the various arches show that the left aortic arch, even though it comes from the right side of the ventricle, carries oxygenated blood. It has been suggested that because pressure is normally lower in the pulmonary circuit than in the systemic circuit, the unoxygenated blood in the right ventricle follows the path of least resistance during ventricular contraction and passes into the pulmonary aorta. The left aorta receives its blood, not from the right ventricle, but via an interventricular canal or, in the crocodilians, via the foramen of Panizzae from the left ventricle. Thus all the blood entering the systemic circulation is oxygenated, and the reptiles have essentially a double circulation.

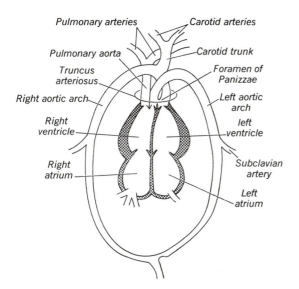

FIGURE 3–5
Relation of the aortic trunks
to the heart in the Alligator.

Veins. The venous system of the reptiles shows little change from that of the amphibians. The large veins bringing blood back from the body are shifted over to empty into the right atrium.

Excretory System

The reptiles are the first vertebrates to have a true metanephros as a functioning kidney. The mesonephros appears only temporarily during development. During embryonic development, a ureter grows forward from the posterior end of the Wolffian duct and fuses with the collecting tubules of the metanephric kidney. The Wolffian duct then carries only reproductive products and gives rise to such male structures as the epididymis, ductus deferens, and seminal vesicles.

The bladder is an outgrowth of the cloaca. Snakes and crocodilians lack a bladder, but most lizards and turtles have a well-developed, bilobed structure. As in amphibians, urine passes into the cloaca and then backs up into the bladder. Some turtles have accessory bladders that are used as reservoirs for water storage.

Reproductive System

The change that, above all others, made it possible for the reptiles to become truly terrestrial was the development of the shelled amniote egg, which could be laid away from water. Surprisingly, the structural modifications in the reproductive system that accompanied this revolutionary shift in habit were relatively minor. Perhaps the most obvious evolutionary modification is the development of the male copulatory organs, and even these apparently are not really necessary, since *Sphenodon* and most birds achieve internal fertilization without the aid of such structures.

Female Reproductive System.   Reptilian ovaries, like those of amphibians, are paired structures lying within the body cavity. Squamates have saccular ovaries like those of amphibians, but the other reptiles have solid ovaries, as do the higher tetrapods. Reptile eggs are polylecithal (large-yolked), whereas amphibian eggs are mesolecithal (moderate-yolked). This is because the young reptile hatches at a far more advanced state of development than does the larval amphibian, and hence needs more nutriment to carry it along until it is able to get food for itself. The size of the ova depends largely on the size of the species, but all reptilian ova are considerably larger than amphibian ova. However, because reptiles have a much smaller number (usually less than 100) of ova that mature simultaneously, the ripe ovaries, though sometimes very large, do not seem to dominate the body cavity as completely as they do in frogs. Ovaries of *Sphenodon,* turtles, and crocodilians are rather broad and symmetrically placed, but those of the squamates are elongated. This is particularly true of snakes, in which the right ovary lies somewhat in advance of the left, an accommodation to the snake's long, narrow body form.

Like those of the amphibians, reptile eggs break through the wall of the ovary into the coelom. The oviducts (Müllerian ducts) open into the coelom by narrow, slitlike ostia. Ova entering the ostium of the oviduct are forced along the tube by ciliary action and muscular contractions of the wall. Each oviduct is differentiated into regions that perform different functions in depositing the envelopes that surround the egg when it is laid. Fertilization must take place before these envelopes are formed. Because the cilia on the walls of the oviduct beat so that the egg moves from the ostium to the uterus, there must be some mechanism to allow sperm to travel against the main cilia-produced current, in order for them to reach the ovum soon after it enters the oviduct. Turtles have a narrow band of cilia along the side of the oviduct that beat toward, instead of away from, the ostium. This is presumably the path followed by the sperm to reach the ovum. Whether a similar path exists in other reptiles is not known.

Sphenodon, turtles, and crocodilians possess glands in the upper part of each oviduct for secreting albumen about the ovum. These glands are lacking in the squamates, and consequently their eggs lack albumen. The lower part of the oviduct, the so-called uterus, is specialized as a shell gland to secrete the shell around the egg. The two uteri enter the cloaca independently. The oviducts vary in size with the seasons, being largest at the height of breeding activity. The right oviduct of many reptiles, particularly of snakes, is longer than the left.

Since reptiles have a metanephric kidney with a new duct, the old Wolffian duct of the opisthonephros no longer functions in females. It persists, however, as a vestigial structure in close association with the ovary in snakes, turtles, and, to a lesser extent, other reptiles.

Male Reproductive System. Like amphibians, reptiles have paired testes suspended within the body cavity. Usually these lie at about the same level, but in snakes and lizards one is frequently farther forward than the other. The seminiferous tubules are long and coiled. The testes may be oval, round, or pyriform; they fluctuate in size like those of amphibians, growing larger with the approach of the breeding season.

With the degeneration of the embryonic mesonephros, the Wolffian duct no longer functions in excretion, but persists as a reproductive duct to transport sperm from the testis to the outside. The end of the Wolffian duct closest to the testis is much coiled and forms part of a tubular mass, the epididymis. Persistent mesonephric tubules are modified into efferent ductules that connect the seminiferous tubules of the testis to the Wolffian duct. These ductules form the remainder of the epididymis, which may be even larger than the testis. The Wolffian duct continues posteriorly as the ductus deferens, which is sometimes straight but more often convoluted. In most reptiles, the deferent duct on each side joins the metanephric ureter and with it enters the cloaca through a common opening at the tip of a urogenital papilla. The Müllerian duct commonly persists, though generally it is reduced in size. The Müllerian ducts of the male European Green Lizard, *Lacerta viridis,* are as well developed as those of the female. Like the gonads, the epididymes and deferent ducts of reptiles studied so far show a seasonal modification in size and are apparently under endocrine control. Many lizards undergo a periodic enlargement of some of the posterior urinary tubules of the metanephros. These enlarged tubules produce an albuminous substance that, presumably, forms part of the seminal fluid in which the spermatozoa are suspended.

Snakes and lizards have glandular structures in the cloacal walls that contribute to the seminal fluid. No accessory glands are known in turtles.

With the development of the amniote shelled egg, internal fertilization became mandatory. This is usually accomplished by copulatory organs. Two different types of copulatory organ are found in reptiles; each apparently represents a separate evolutionary development.

A single (occasionally multilobed), protrusible copulatory organ—the penis—is present in turtles. Only the distal end is free. There is a groove along the surface, providing a passageway for the sperm. The turtle penis is apparently derived from thickened portions of the cloaca wall and is made of both connective and erectile tissue. The mass of erectile tissue is the corpus cavernosum. During mating, the corpus cavernosum becomes filled with blood, enlarging and stiffening the penis so that it can be extruded through the cloacal opening. This erection enables the penis to serve as an intromittent copulatory organ.

The protrusible penis of crocodilians is longer and more slender than that of turtles, and the groove is deeper. A spongy structure—the glans penis—at the outer end of the groove resembles the mammalian glans.

Snakes and lizards have paired copulatory organs—the hemipenes (singular hemipenis). The word means "half a penis," but this is not really correct, since each organ is separate and complete (Figure 3–6). The hemipenes are not homologous with the penis of turtles, crocodilians, or mammals.

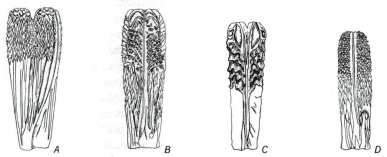

FIGURE 3–6
Representative hemipenes of snakes: (A) Common King Snake, *Lampropeltis getulus*, southeastern United States; (B) Horseshoe Snake, *Coluber hippocrepis*, Italy; (C) Rosy Boa, *Lichanura roseofusca*, Baja California; (D) Puff Adder, *Bitis arietans*, Africa.

The hemipenes lie on either side of the base of the tail and form distinct thickenings, so that it is frequently possible to determine the sex of a lizard or snake without dissection. Each hemipenis is a tubular structure that can be turned inside out like the finger of a glove. It bears a groove—the sulcus spermaticus—to transport the semen from the cloaca to the tip of the hemipenis. The distal end of the organ is drawn back after copulation by means of a long retractor muscle. In a preserved specimen, the length of the hemipenis depends largely on the state of contraction of this muscle at the time of death. The external openings for the hemipenes can be seen on a snake by lifting the scale over the vent. The hemipenis of a lizard is usually short and broad, and the inner surface (outer when the organ is everted) is typically pleated and folded; that of a snake is longer, and the surface may be covered with spines and fingerlike projections arranged in rosettes called calyces (singular calyx). These apparently hold the hemipenis in the female cloaca during copulation. The hemipenis may be bilobed and the sulcus bifurcated.

In mating, usually only one hemipenis is inserted into the female cloaca. Which hemipenis is inserted is determined simply by which side the male happens to be on during copulation.

Skeleton

The skeletal modifications for terrestrial life that originated in the amphibians are further developed in the reptiles.

Skull. Reptilian skulls show even more diversity than those of amphibians. They range from the light, kinetic skulls of snakes, with their loosely articulated parts, to the robust, solid skulls of crocodilians. In general, the bones are heavier and more completely ossified than in amphibians. Only one occipital condyle is present. The various modifications of the temporal region are described in Chapter 5.

Vertebral Column. A typical reptilian vertebra consists of a ventral, spool-shaped centrum supporting dorsally a solid neural arch. Small, crescentic intercentra may be ventrally wedged between the centra of successive vertebrae. To add rigidity, yet allow flexibility to the vertebral column, each vertebra bears a pair of sliding articular surfaces (zygapophyses) on the front and rear of its neural arch. The prezygapophyses of one vertebra slide beneath the postzygapophyses of the preceding vertebra, forming with the centra a firm link between successive vertebrae. This is the typical tetrapod pattern. Snakes and some lizards (strangely enough not the legless, snakelike ones) have two more bearers of sliding articular facets on the neural arch. The anterior zygosphenes of one vertebra fit into grooves of the zygantra (singular zygantrum) of the preceding vertebra (Figure 3–7).

The reptilian vertebral column shows regionalization, but not to the extent of the mammalian column. Four regions are usually recognizable: cervical, trunk, sacral, and caudal. The first two vertebrae of the cervical region are modified to support the head, yet allow rotation of the head and neck. The first vertebra is called the atlas, because it bears the head, as Atlas of mythology bore the world on his shoulders. The second vertebra—the axis—permits the head to be rotated from side to side. Other cervical vertebrae differ little from the succeeding trunk vertebrae, but in most limbed reptiles they lack articular facets for the free ribs. The trunk vertebrae extend from the pectoral girdle to the sacral region; most support free ribs. The sacral vertebrae (two or three) bear large, robust transverse processes that articulate with the ilium of the pelvic girdle. The caudal or tail vertebrae usually support small haemal arches (chevrons). In *Sphenodon,* many

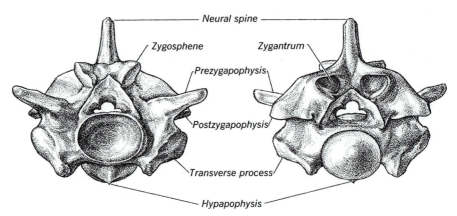

FIGURE 3–7
Vertebra of a snake, *Eunectes murinus*: (*left*) anterior view; (*right*) posterior view.

lizards, and a few snakes, each caudal vertebra has a transverse cleavage zone in its middle to permit the tail to be broken off (autotomy).

In snakes and legless lizards, only precaudal and caudal vertebrae are clearly differentiated. Turtles have eight uniquely articulated cervicals, which permit great flexibility, and trunk vertebrae fused to the dermal bones of the carapace. Only in crocodilians are the five mammalian divisions of the vertebral column easily discernible: in addition to the cervicals, sacrals, and caudals, the trunk vertebrae are divided into thoracics, with large articular facets for the ribs, and lumbars without such articular surfaces.

Ribs. Along with the variation in the vertebral columns of reptiles goes considerable variation in the number, structure, and attachment of the ribs. They may be attached to the vertebrae by two heads or by a single head, and they may articulate with the centra or with the neural arches. Furthermore, in a single animal the points of attachment may shift according to the position of the vertebrae in the spinal column. At their ventral ends, the ribs of the anterior dorsal region usually join a sternum, which is frequently cartilaginous. Trunk ribs of *Sphenodon* and lizards may join a parasternum lying between the sternum and the pelvic girdle. Turtles and snakes lack a sternum. The trunk ribs of turtles are fused to the carapace. In snakes, the ventral ends of the ribs are bound by muscular connections with the abdominal scales. The posterior cervical and anterior dorsal ribs of *Sphenodon* and crocodilians each bear a curved, uncinate process that projects posteriorly to overlap the rib behind, giving strength to the thoracic body wall.

Girdles. Usually, the pectoral girdle on each side is made up of a coracoid, procoracoid, and scapula, with a clavicle and interclavicle frequently present. The pelvic girdle is formed of three bones on each side: a dorsal ilium that is fused to the sacral ribs, a ventral pubis, and an ischium. The two pubic bones join ventrally in a pubic symphysis, and the two ischia also form a symphysis. Between these two symphyses is a large, heart-shaped space—the cordiform foramen. The reptile limb typically has five digits, but there is a marked tendency toward a reduction of the limbs in lizards. Vestiges of the hind limbs remain in some primitive snakes, and of the front limbs in one genus (*Bipes*) of worm lizards.

Nervous System

The spinal cord of reptiles, like that of salamanders, extends the entire length of the vertebral column. Cervical and lumbar enlargements, made up of the cell bodies of the neurons that go to the limbs, are present in all reptiles except snakes and limbless lizards. The cervical enlargements of turtles seem unduly massive, but this is purely relative: the trunk muscles of these reptiles are so reduced that the portion of the cord between the enlargements, from which nerves pass to these muscles, is also reduced and is more slender than in other reptiles.

Some snakes and limbless lizards possess a distinct, albeit poorly developed, lumbosacral plexus, a network of nerves that ordinarily leads to the hind legs. This indicates that these animals must have arisen from ancestors that had legs.

Reptiles have twelve pairs of cranial nerves, as do all the higher tetrapods. The eleventh and twelfth pairs presumably represent the first two spinal nerves of the amphibians.

The cerebral hemispheres of the brain are larger than those of the amphibians, and a new area—the neopallium—appears in their roof. In crocodilians, the migration of nerve cells to the outer wall of the neopallium results in the formation of the first true cerebral cortex.

Sense Organs

On the whole, the broad changes in the sense organs necessitated by terrestrial existence first appeared in the amphibians, and have simply been further developed in the reptiles.

Eye.   Snakes and some lizards lack movable eyelids, but have instead a transparent window—the brille—covering the eye. It is this that gives snakes their glassy, unwinking glare. They truly cannot close their eyes, even though the eyes are always covered. Other reptiles have well-developed, movable lids and a transparent nictitating membrane, or third eyelid.

In lizards, and to a lesser extent in crocodilians, a conical papilla extends into the vitreous chamber from the blind spot, where the optic nerve enters the eye. It consists of numerous fine blood vessels enclosed in a framework of ectodermal neural tissue, and provides nourishment for the inner retinal layer of the eye. A similar structure occurs in snakes, but it is derived from mesoderm and thus apparently is not homologous with that in lizards. Snakes also have a fine network of blood vessels extending across the retina.

Ear. The auditory part of the inner ear is more highly developed in the reptiles than in the amphibians. Most reptiles have a tympanic membrane, which usually lies flush with the head, and a middle ear cavity through which sound waves are transmitted to the inner ear. Snakes lack the middle ear cavity. The bones that carry sound vibrations, instead of abutting on a tympanic membrane, connect with the jawbones, and it is through these that the snake hears. The old saying "deaf as an adder" was probably based on the absence of external ear openings, and is, of course, erroneous.

Jacobson's Organ. The vomeronasal organ known as Jacobson's organ is small in turtles and crocodilians. It is highly developed in snakes and lizards and no longer connects to the nasal passages, but opens directly into the mouth (Figure 3–8). The flickering, two-pronged tongue of the snake picks up chemical particles from the air. When the tongue is retracted, its tips are inserted into the openings of the pockets of Jacobson's organ. The sense of smell plays a significant role in the recognition of prey, enemies, or potential mates.

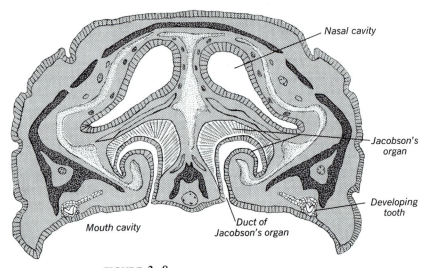

FIGURE 3–8
Transverse section through the head of a lizard, *Lacerta*, showing Jacobson's organ.

READINGS AND REFERENCES

Bojanus, L. H. *Anatome Testudinis Europaeae.* Facsimile Reprint in Herpetology, no. 26, 1970.

Etheridge, R. "Lizard caudal vertebrae." *Copeia,* no. 4, 1967.

Gaffney, E. S. "An illustrated glossary of turtle skull nomenclature." *American Museum Novitates,* no. 2486, 1972.

Gans, C., and T. S. Parsons (eds.). *Biology of the Reptilia,* vol. 1–5. London: Academic Press, 1969–1976.

———, and P. F. A. Maderson. "Sound producing mechanisms in recent reptiles: review and comment." *American Zoologist,* vol. 13, 1973.

Grasse, P. P. *Traité de Zoologie. Reptiles,* vol. 14, pt. 2. Paris: Masson et Cie Editeurs, 1970.

Harris, V. A. *The Anatomy of the Rainbow Lizard.* London: Hutchinson, 1963.

Kardong, K. V. "Kinesis of the jaw apparatus during the strike in the cottonmouth snake, *Agkistrodon piscivorus.*" *Forma et Functio,* vol. 7, 1974.

McDowel, S. M. "The evolution of the tongue of snakes, and its bearing on snake origins." *In* T. Dobzhansky, M. Hecht, and W. C. Steere (eds.), *Evolutionary Biology,* vol. 6. New York: Appleton-Century-Crofts, 1972.

Maderson, P. F. A. "Lizard glands and lizard hands: models for evolutionary study." *Forma et Functio,* vol. 3, 1970.

Oelrich, T. M. "The anatomy of the head of *Ctenosaura pectinata* (Iguanidae)." *Miscellaneous Publications, Museum of Zoology, University of Michigan,* no. 94, 1956.

Oldham, J. C., H. M. Smith, and S. A. Miller. *A Laboratory Perspectus of Snake Anatomy.* Champaign: Stipes, 1970.

Taub, A. M. "Comparative histological studies on Duvernoy's gland of colubrid snakes." *Bulletin of American Museum of Natural History,* vol. 138, 1967.

Waring, H. *Color Change Mechanisms of Cold-blooded Vertebrates.* New York: Academic Press, 1963.

Wever, E. G. "The tectorial membrane of the lizard ear: types of structure." *Journal of Morphology,* vol. 122, 1967.

CHAPTER FOUR

ORIGIN
AND EVOLUTION
OF AMPHIBIANS

DURING THE LONG COURSE of evolutionary history, new groups occasionally arise as a result of fundamental adaptive shifts. They are able to adopt new modes of living and move into environments that were closed to their immediate ancestors. Surely one of the most important such shifts was that made by the amphibians, the first vertebrate animals to leave the water and spread out over the land. Indeed, they were the only vertebrates to do so, and from them have come all the higher types—the reptiles, birds, mammals, and man himself. Unusual interest, then, attaches to the early evolutionary history of these relatively inconspicuous inhabitants of the land.

SOME PALEONTOLOGICAL CONSIDERATIONS

Unfortunately, amphibians as a group do not leave very good fossils. Many of them are small and have delicate skeletons, and their bones are easily scattered or crushed. This is especially true of the more recent forms, most of which are smaller than their early amphibian ancestors. And so there are great gaps in our record of amphibian evolution. Some gaps are being slowly filled, as paleontologists devote more of their attention to the tiny scattered vertebrae that may be all that remains of a once numerous form. Some gaps will never be filled, and thus there will always be a certain amount of guesswork in our attempts to construct amphibian family trees. But the guesses of paleontologists are not blind stabs in the dark. Through studies of forms that have left clear and long-continued records, certain evolutionary principles have been derived to serve as guides. One of these is that there is a strong tendency toward irreversibility in evolution: an animal that has lost its legs does not later give rise to an animal possessing legs, and a bone that has disappeared from the skull of an ancestral form will not reappear in the skulls of its descendants. We may never be able to say definitely just which fossil forms were the forerunners of the modern amphibians, but we can be quite sure that they were not animals that had completely lost their legs. Thus, by a process of elimination, we may be able to narrow the field to one or a few probable ancestors, even for those modern animals that have no fossil history.

It is sometimes assumed that the basic criterion in the classification of animals is how recently two forms have diverged from a common ancestor. Thus if two species are put in the same genus, their common ancestor is thought to be closer to us in time than the common ancestor of these two and a third species of a different genus. But this criterion is really quite limited in its applicability. The length of time two forms have been separated is only roughly correlated with the degree of divergence they show. Organisms evolve at different rates. Of two lines, one may evolve very slowly and remain close to the ancestral type, the other may evolve rapidly into something quite different. Very early in the evolutionary history of the amphibians, perhaps even before they were true amphibians, the stock apparently split into several groups. One gave rise to the amphibian ancestors of the reptiles and higher tetrapods. Whether it also gave rise to any or all of the groups of modern amphibians is still a matter of lively debate among paleontologists. It is almost certain, though, that the line leading to the modern frogs had diverged from that leading to the salamanders before the birds evolved from the reptiles. But because the frogs and salamanders retain a basically similar structural pattern and mode of life, we place them in the same class, whereas the reptiles, birds, and mammals, which have departed widely from the amphibian pattern, are placed in different classes. Degree of divergence, rather than closeness of descent, is thus the basic criterion of classification.

Introducing the factor of time brings another complication into schemes of classification. Suppose at a given period we have two genera that are obviously closely related. Each of these genera gives rise to a line, the later members of which are so distinct as to be placed in different families. Should we classify the two original genera in the families to which they gave rise, a vertical classification, or should we put them together in a third family, a horizontal classification? Either would be technically correct. Which scheme we adopt will probably depend on the completeness of the fossil record that connects the various forms, and this is more or less a matter of chance. Most classifications are actually based on combinations of the two methods.

It should be stressed that any system of classification is not an objective reality, but a man-made system that is necessary because we find it difficult to study a large number of objects unless we can organize them into categories. If we knew all there is to know about all living organisms and had a complete fossil record of every species that has ever lived, there would still be room for differences of opinion on the rank to be assigned to different groups. As it is, we know almost nothing about the detailed structure of most living species, and the fossil record is, and will necessarily remain, fragmentary. As we study more intensively the material already available, and as new forms, both living and extinct, are discovered, our ideas of the composition and relationships of the groups will shift. The classifications given in this and following chapters will almost certainly be revised in the future. They should not be taken as final or as indicating that everything is known about the relationships of the herps.

In studying the evolutionary history of a class, it is possible to use either a horizontal or a vertical approach. With the former, we would consider all the amphibians of the Mississippian period, then all those of the Pennsylvanian, and so on. This would give a clear picture of the fauna of any given period, but would make it difficult to follow the various lines of descent. We shall here adopt very largely a vertical approach, following one line through its evolutionary course, then going back and picking up another line. This method requires a certain amount of mental gymnastics, of jumping backward and forward in time. To do so, it is necessary to keep the geological time scale clearly in mind. Figure 4–1 shows the time scale with the known duration of amphibian groups, for reference during the following discussion.

ORIGIN OF AMPHIBIANS

The first animals that are clearly amphibians are the ichthyostegids, which come from fresh-water deposits of the Upper Devonian. *Ichthyostega* and two closely related genera had a fishlike body form, but strong, short limbs. The limbs and their girdles, the absence of opercular (gill flap) bones, and the presence of otic (ear) notches are the main structures identifying them as amphibians. Most of their other characters are shared with a group of ancient fishes known as rhipidistian crossopterygians.

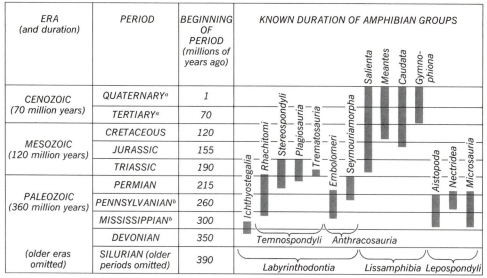

ERA (and duration)	PERIOD	BEGINNING OF PERIOD (millions of years ago)	KNOWN DURATION OF AMPHIBIAN GROUPS
CENOZOIC (70 million years)	QUATERNARY[a]	1	
	TERTIARY[a]	70	
MESOZOIC (120 million years)	CRETACEOUS	120	
	JURASSIC	155	
	TRIASSIC	190	
PALEOZOIC (360 million years)	PERMIAN	215	
	PENNSYLVANIAN[b]	260	
	MISSISSIPPIAN[b]	300	
	DEVONIAN	350	
(older eras omitted)	SILURIAN (older periods omitted)	390	

[a]The Tertiary is frequently divided into epochs. Beginning with the oldest, these are: Paleocene, Eocene, Oligocene, Miocene, and Pliocene. The Quaternary is also divided into Pleistocene and Recent.
[b]The Mississippian and Pennsylvanian are often grouped together as the Carboniferous period.

FIGURE 4–1
The geologic time scale, showing the distribution of the amphibians in time.

The crossopterygians (lobe-finned fishes) were stout-bodied, predatory fishes with heavy, fleshy fins. Their ancestors diverged from the main line of bony fishes (Class Osteichthyes) early in the Devonian and soon split in two lineages, the crossopterygians and the lungfishes (Dipnoi). Both groups have been proposed as the exclusive ancestors of the amphibians. Even now, some scientists believe that the dipnoans gave rise to the salamanders, and the crossopterygians to the anurans, the caecilians, and most of the extinct amphibian groups. However, the bulk of the evidence seems to indicate the crossopterygians as the sole ancestors of all amphibians.

Both groups (classed together as Sarcopterygii, or fleshy-finned fishes) were almost exclusively fresh-water fishes; they are now almost extinct. In contrast, the other main evolutionary line of bony fishes (Actinopterygii, or ray-finned fishes), though they arose in fresh water, soon invaded the sea. Today they are the dominant fishes in both habitats. They possess a fin skeleton of thin, cartilaginous rays—unlikely precursors for the stout limb skeleton required to support a terrestrial animal. Similarly, the habits and fin structure of dipnoans tend to eliminate them as amphibian ancestors. The dipnoan fin has an elongated leaf shape, supported internally by a long median row of thin bony elements and a series of thin bony spines radiating from this median row. Devonian lungfishes appear to have shared the habits of their living representatives, aestivating when their pool of water evaporated and thus reducing the need for stout fins to aid in short-term terrestrial locomotion. The crossopterygian fin skeleton has stouter bones, arranged in a sequence comparable to the bones of a tetrapod limb (Figure 4–2).

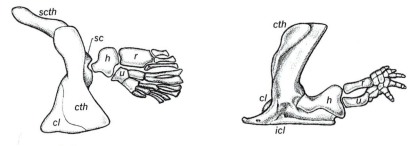

FIGURE 4–2
Shoulder girdle and pectoral fin of (*left*) a crossopterygian and (*right*) an ancient fossil amphibian, placed in a comparable pose to show the basic similarity in limb pattern: *h*, *r*, and *u* are the humerus, radius, and ulna, respectively, of the tetrapod, and their obvious homologues in the fish fin; *cl*, clavicle; *cth*, cleithrum; *icl*, interclavicle; *sc*, scapula; *scth*, supracleithrum. [From Romer, *The Vertebrate Body*, Saunders, 1955.]

Latimeria, the sole surviving crossopterygian, manipulates its mobile fins in a way far more reminiscent of tetrapod limb movement than the simple back and forth waving of the actinopterygian fin.

Other characters support the idea of a crossopterygian–amphibian link. Both groups possess true internal nares, or choanae. Most fishes have blind nasal sacs with no connections to the mouth. In some crossopterygians, a passage leads from each nasal sac to an opening on the roof of the mouth. This passageway permits the nasal sac to retain its olfactory function, yet provides a pathway for air flow to the lungs when only the external nostrils are above water. The brain case of crossopterygians is hinged transversely into anterior and posterior halves; apparently this hinge permits a wider gape as an aid in catching prey. A trace of a transverse hinge remained in *Ichthyostega*. The teeth of early crossopterygians have a peculiar labyrinthine infolding of the enamel. This condition is found in no other fishes, but is characteristic of the earliest amphibians. The vertebral column of early crossopterygians and *Ichthyostega* retained a large notochord and, thus, poorly ossified vertebral centra. Each centrum was either a complete bony ring pierced by the notochord, or a small bone (pleurocentrum) on either side of the notochord supporting the neural arch and a larger U-shaped bone (intercentrum) beneath the notochord.

Whatever environmental factors led to the evolution of the amphibians, it must be emphasized that in spite of their seemingly terrestrial characters, such as limbs, the earliest amphibians were highly aquatic and probably spent little time on land. The terrestrial environment of the early Devonian was quite bare. Only a few plants and invertebrates had become fully terrestrial, and these were confined to a thin border along the water's edge. The landscape had the hostile appearance of a great barren desert broken occasionally by narrow corridors of low, green-edged streams and lakes, while the water teemed with life—thick aquatic vegetation and myriad invertebrates and fishes. By mid-Devonian times, treelike plants and a diversity of spiders had appeared, and by the end of this period true forests stretched back from the ponds and streams. But the earliest amphibians, with their fishlike bodies and tails, were ill adapted to travel far inland. The amphibian adaptations may have evolved as a means to exploit the shallow water for its abundant food supply or, in the young, to avoid being eaten. Mobility in shallow water or over short expanses of dry land would have been enforced by the periodic droughts, which characterized the Devonian climate and would create a selective advantage for those animals that could move from a drying pool to a watered one.

Lungs or gas bladders were present in the earliest bony fishes, and provided an accessory respiratory mechanism when oxygen content in the water was low, whether as a result of stagnation, heavy vegetation, or high temperature. Since all the crossopterygians appear to have been active predators, the well-developed fins may have evolved for slow bottom-walking in order to stalk a prey before the final swimming dash. Later they would take on a supportive function when the fish moved about in shallow water with part of its body exposed, losing the weightlessness provided by water. Each structure would be slowly modified as one line of crossopterygians slowly changed their aquatic habits to amphibious ones.

DIVERSITY OF ANCIENT AMPHIBIANS

Apparently very early in the evolution of amphibians, they diverged into two distinct and very different groups, Labyrinthodontia and Lepospondyli. These two subclasses or superorders contain the extinct amphibians of the late Paleozoic and early Mesozoic eras. The Lissamphibia (the superorder containing the living orders) arose somewhat later, perhaps from the labyrinthodonts. The fossil record does not show any unquestionable evidence of relationships among the three lineages, and thus we can only speculate about the manner and precise time of their divergence from one another (Figure 4–3).

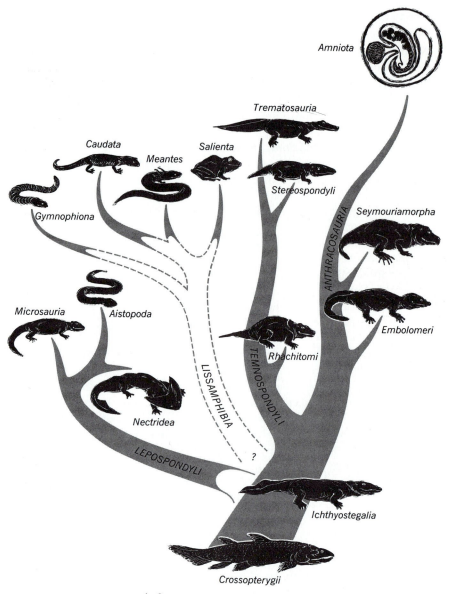

FIGURE 4–3
Phylogenetic arrangement of the orders of amphibians.

Superorder Labyrinthodontia

Judging from the fossil record, the labyrinthodonts were the most abundant and diverse amphibians of the Carboniferous, Permian, and Triassic periods. Perhaps this dominance reflects the line's general tendency to improve their active, predatory role in shallow water and terrestrial habitats. These diverse amphibians shared few characteristics. Most possessed strong limbs and girdles, strong jaws, vertebral centra consisting of intercentrum and pleurocentrum, and, the most widely shared link, labyrinthine enamel folds on the teeth.

The basal group of labyrinthodonts were the Ichthyostegalia. They gave rise to the Anthracosauria and the Temnospondyli. The anthracosaurs were less successful as amphibians than the temnospondyls, but evolutionarily more successful, since they were the ancestors of the reptiles.

These two later labyrinthodont groups and their subgroups are largely recognized by their vertebral structure. The general trend is for the temnospondyls to enlarge the intercentrum and lose or reduce the pleurocentrum; in anthracosaurs the trend is the reverse (Figure 4–4). There are, of course, reversals and deviations from the general pattern. Some evidence also indicates a fusion of inter- and pleurocentra in some of the later temnospondyls.

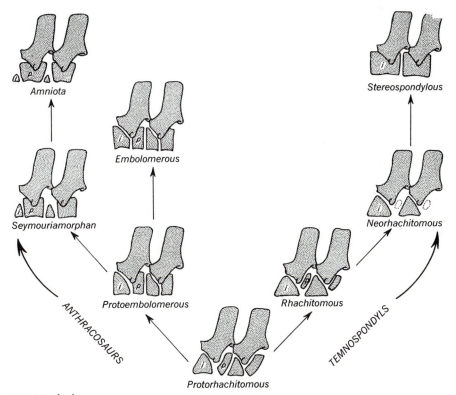

FIGURE 4–4
Phylogeny of vertebral types; *i*, intercentrum; *p*, pleurocentrum (true centrum). Viewed from the left side. [After Romer, 1966.]

Order Ichthyostegalia. The ichthyostegalians were primitive amphibians, in many ways intermediate between fishes and tetrapods in appearance. *Ichthyostega* had short, stubby, pentadactyl limbs instead of fins, and a stout, long tail with a caudal fin supported by fin rays. The head was broad and fishlike, with a short, rounded snout and a long skull roof posterior to the orbits. In later amphibians, the skull became shorter posteriorly and the snout proportionally longer. Lateral line canals were present, indicating persistent aquatic habits. The centrum was like that of some crossopterygians, with a U-shaped intercentrum and a pair of pleurocentra beneath the neural arch—another indication of aquatic habits, since this type of vertebral column would not have been an effective arch for supporting the body on land.

Besides *Ichthyostega,* this order includes several other genera of late Devonian and Carboniferous amphibians. All seem to have been active predators. Some of them, at least, retained an armor of fishlike, bony ventral scales and traces of scales on the back and sides. Although primitive in most respects, all possessed specializations indicating that they were side branches, rather than in the main stem of amphibian evolution. For example, the intertemporal bone of the skull is absent in all ichthyostegalians, but present in other primitive amphibians.

Order Temnospondyli. The earliest temnospondyls were aquatic, but some later forms were well adapted to life on land. In the late Permian and Triassic, the temnospondyls again became aquatic, only to die out at the end of the Triassic.

The temnospondylous rhachitomes were the most abundant of all fossil amphibians in number of genera. Most possessed stout, broad bodies and heads and stout, but short, limbs. The earliest forms were the peculiar loxommids of the late Mississippian, with strangely elongated and enlarged orbits. Throughout the Pennsylvanian, rhachitomes appear to have been moving toward more terrestrial habits, culminating in strongly terrestrial genera, such as *Eryops* and *Cacops,* of the early Permian. Thereafter, the trend seems to have been back to aquatic habits. Some members of one group, the trematosaurs, apparently made a successful transition to the marine environment; they are the only amphibians to have done so. The rhachitomes became extinct early in the Triassic.

Another group of temnospondyls, the stereospondyls, appeared in the late Permian and lasted through the Triassic. They were highly aquatic and, apparently, bottom dwellers. Their bodies were broad and flat, their limbs small and weak. One line, the long-snouted capitosaurs, had a variety of genera ranging from medium-sized forms, with skull widths of 10 to 20 cm, to the giant of all labyrinthodonts, *Mastodonosaurus,* with a skull over 1 m wide. The short-snouted stereospondyls were small to medium-sized animals, whose skulls tended to be greatly elongated and widened posteriorly.

The strangest group of temnospondyls were the plagiosaurs. They had broad and very flat bodies, armored above and below, with reduced limbs and tail. These armored pancakes possessed external gills and were presumably neotenic (retaining larval characters into adulthood), like some modern salamanders.

Order Anthracosauria.   The anthracosaurs were a small, short-lived group. They first appeared in the Mississippian, and became extinct as amphibians at the end of the Permian. But before they disappeared, they gave rise to true terrestrial vertebrates, the reptiles. An early group, the embolomeres, were largely aquatic animals with long bodies, powerful tails, small limbs and usually long-snouted skulls. The aquatic trend was reversed in the seymouriamorphs. These were mostly moderate in size and stockily built, had stout limbs, and were certainly capable of walking on land. Although they share a number of characters with reptiles, such as a pentadactylous forefoot and a true pleurocentrum in combination with a reduced intercentrum, they apparently were not the ancestors of reptiles. The presence of aquatic seymouriamorph larvae indicates that they did not lay shelled, amniote eggs on land.

The ancestor of the reptiles must have diverged from the main anthracosaur line sometime in the late Mississippian or earliest Pennsylvanian, since the captorhinomorph and pelycosaur reptiles were in existence by the middle Pennsylvanian. This ancestor certainly shared a number of terrestrial adaptations with the seymouriamorphs. However, these shared characters probably evolved in parallel, rather than from a common ancestor. The suggested environmental and selective factors causing the evolution of a truly terrestrial vertebrate are as speculative as those for the crossopterygian–amphibian transition. It must be noted, though, that the amphibian transition was to semiaquatic adaptation and principally involved structural modifications, whereas the reptile transition was to terrestrial adaptation and involved developmental and physiological characters as well.

Superorder Lepospondyli

The lepospondyls were small, salamanderlike amphibians that lived in the ponds, streams, and marshes of the Carboniferous and Permian periods. Most appear to have been highly aquatic forms with elongated bodies, small limbs, and a small, often bizarre-shaped head with weak, short jaws. With few exceptions, they retained this body form and aquatic habit throughout their entire existence.

On their first appearance in the fossil record, three distinct lineages of lepospondyls are apparent: Aistopoda, Nectridea, and Microsauria. Exactly how these are related to one another and to the other amphibians remains a puzzle. All three share an unpaired bony centrum perforated by the notochord, and a skull lacking the otic notch, so they probably arose from a common ancestor. However, who the ancestor was and when it diverged from the transitional amphibian line remain unanswered questions.

Order Aistopoda. Next to the labyrinthodont ichthyostegalians, the snakelike aistopodans are the oldest known amphibians, appearing first in the early Mississippian and disappearing in the early Permian. Only three genera are known; all had more than two hundred vertebrae divided clearly into cervical, trunk, and caudal segments, indicating that these limbless amphibians had limbed ancestors. It has been suggested that the aistopodans and the two following lepospondyl groups had direct development, but, as yet, there is no positive evidence for this conjecture.

Order Nectridea. Two groups of aquatic amphibians, largely confined to the Carboniferous, constitute the nectrideans. The newtlike forms, usually represented by the terminal genus *Diplocaulus,* had short trunks, small but fully developed limbs, and a swimming tail with large dorsal and ventral crests. The caudal vertebrae had expanded, fanlike neural and haemal arches. In most, the skull had backward-projecting horns. The other group had slender, eel-shaped bodies, reduced or absent limbs, and long, pointed heads. Both groups appear to have been fully aquatic.

Order Microsauria. The long-bodied and usually weak-limbed microsaurs have aptly been called little bogus reptiles. Their skulls have repeatedly been confused with those of the early captorhinomorph reptiles; however, they appear to be in no way related or ancestral to the reptiles. They varied considerably in structure: some paralleled the aistopodans and nectrideans in reduction of limbs and elongation of body, but most had a more normal tetrapod build. A few may have been terrestrial or, at least, semiaquatic.

Superorder Lissamphibia

The lissamphibians comprise the four orders of living amphibians and several fossil groups clearly related to the living forms. A tremendous temporal and structural gap separates the lissamphibians from the two ancient superorders (labyrinthodonts and lepospondyls). This has led to much confusion and debate about the relationships of the lissamphibians to the ancient superorders and the interrelationships of the lissamphibian orders to one another. Whether the lissamphibians are indeed more closely related to one another than to any of the fossil groups remains uncertain. The fossil record gives no help. Frogs extend back to the Triassic and salamanders to the Jurassic, but they show no signs of converging toward a common ancestor, nor of approaching any of the Paleozoic groups. Their vertebrae resemble those of the lepospondyls, their skulls are more similar to those of the temnospondyls.

A serious difficulty in clarifying the relationships of the modern amphibians is that most of their shared characters are soft anatomical ones, and thus not preserved in fossils. Such characters as the opercular bone of the lissamphibian middle ear are found in no other vertebrate group, living or fossil. But even in living lissamphibians, this bone is frequently cartilaginous or fused to the stapes, and its presence can be determined only by embryological studies. Thus the opercular bone and other supposedly unique characters of the lissamphibians may have been present in some of the fossil groups, though they are undetectable now. Nonetheless, the consensus holds that these unique characters indicate a single common ancestor for the frogs, salamanders, sirens, and caecilians. When they diverged from the ancient amphibians is not known, but it must have been late in the Paleozoic, because the great structural and physiological diversity of the modern groups and the appearance of frogs in the Triassic indicate a long separation from the older groups. Who their ancestor is remains unknown. The best guess may be that it lies among the temnospondyls.

ANCESTORS OF MODERN AMPHIBIANS

The modern amphibians are uniquely linked by a small suite of characters. Most possess pedicellate teeth; each tooth is divided transversely by a zone of weakness into an upper and a lower segment. Their hearing apparatus is composed of two separate mechanisms, an operculum–amphibian-papilla unit for low-frequency sounds and a columella–basilar-papilla unit for high-frequency sounds. Their eyes contain "green rods," a special type of visual cell found in no other animals. Their fat bodies are derived from the gonadal ridge and not from elsewhere, as in other vertebrates. Their skin is specialized for cutaneous respiration and also shares a number of unique glandular structures.

Caecilians

The caecilians (Order Gymnophiona) are tropical, wormlike creatures that are limbless and practically blind. No fossil caecilians have been found to bridge the gap between the modern ones and their remote ancestors. The single fossil known, from the Eocene of Brazil, seems to be related to a modern African form.

Frogs

The first anuran to appear in the early fossil record is *Triadobatrachus* (Order Proanura) from the early Triassic deposits of Madagascar. *Triadobatrachus* is clearly a frog, although it may not be on the main line of frog evolution or may be a metamorphosing individual. It shows no clear relationships with any living frog family, nor does it serve as a link with the ancient amphibians.

The remaining Mesozoic frogs are all referable to the more primitive and aquatic families of modern frogs (Order Salientia). The early and late Jurassic deposits of Argentina have yielded two ascaphid frogs, the small (30 mm) *Vieraella* and the large (12 cm) *Notobatrachus* (Figure 4–5). Both were

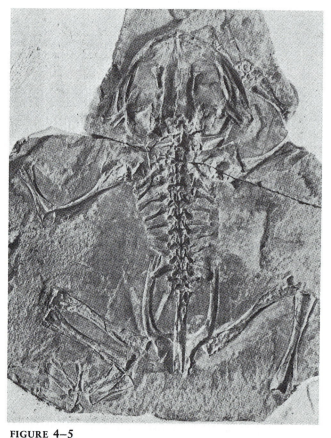

FIGURE 4–5
Notobatrachus degiustoi Reig, a primitive frog from the Jurassic of Patagonia. [Courtesy of Professor Osvaldo Reig.]

presumably aquatic. From the late Jurassic of Europe, the two genera *Montsechobatrachus* and *Eodiscoglossus* are early representatives of the discoglossids. Six genera of pipid frogs have been found in Cretaceous and Paleocene deposits of the Middle East, North America, Africa, and Argentina, indicating a much wider distribution than that of their modern relatives; however, all seem to have been as fully aquatic as present pipids.

In either the late Mesozoic or the very early Cenozoic, a sideline of pipoid frogs developed into what is now recognized as the family Paleobatrachidae. An ill-fated group comprising only two genera and five species, it was extinct by the middle Tertiary.

All remaining fossil frogs have been placed in families that have living representatives. As pointed out above, the discoglossids and probably the pipids go back to the Jurassic. The Pelobatidae probably extend back to the Cretaceous and are definitely known from the Eocene. Both Bufonidae and Leptodactylidae are known from the Paleocene, Rhinophrynidae from the Eocene, Hylidae and Ranidae from the Oligocene, and the Microhylidae from the Miocene. Although the remaining families have not been reported from the Tertiary, future fossil recoveries will probably extend the histories of most frog families back to the late Mesozoic. One point of interest is that most fossil frogs are discovered within the limits of their family's present distribution or in areas adjacent to it. It is also noteworthy that both *Triadobatrachus* and the oldest true frog are from the southern hemisphere, whereas all fossil salamanders are from the northern hemisphere.

Salamanders

The salamanders comprise two main evolutionary lines, the sirens (Order Meantes) and the "true" salamanders (Order Caudata). Biochemical evidence suggests that the two lines diverge as early as the late Permian or early Triassic. As yet, this evidence is not supported by the fossil record. The sirens are now known only from the late and, perhaps, the early Cretaceous. Even then, they were highly specialized aquatic animals with an eellike body.

The "true" salamanders have long, slender bodies and tails, and small but well-developed limbs. They first appear in the fossil record in the late Jurassic. These oldest salamanders were permanent larvae, like the Mud Puppies *(Necturus)*, and may belong to that group; other proteids show up sporadically in the Cretaceous, Eocene, and Miocene. Salamanders—*Scapherpeton* and *Lisserpeton*—next appear in the Cretaceous, and these were also neotenic forms. They are placed in a separate family, Scapherpetonidae; there is some evidence that they may be related to the cryptobranchids or ambystomatids. Several other ambystomatids—*Bargmannia, Geyeriella,* and *Wolterstorffiella*—are known from the early Tertiary.

Salamandrids may also extend back to the early Cretaceous and unquestionably go back to the Paleocene. Two other salamanders, *Opisthotriton* and *Batrachosauroides,* are early Tertiary forms. Like most fossil salamanders, they are neotenic forms (even today, most of the pond and river salamanders are neotenic, and most fossil sites are from such deposits) and seem to have affinities with the modern plethodontids; however, they have been placed in their own family, Batrachosauroididae. Another salamander, *Prodesmodon,* occurred throughout the Cretaceous in North America, only to disappear there and reappear in the Miocene of France. This salamander matches quite closely the characters of an aquatic desmognathine. It also shares a number of characteristics with the early Cretaceous *Prosiren* (which is not a true siren), and the two are placed in the family Prosirenidae.

Of the other living families of salamanders, amphiumids appear in the Cretaceous, cryptobranchids in the Oligocene, and plethodontids in the Pliocene. Hynobiids do not appear as fossils, unless the Paleocene ambystomatids of Europe are actually hynobiids.

READINGS AND REFERENCES

Dessauer, H. C. "Biochemical and immunological evidence of relationships in Amphibia and Reptilia." *In* C. A. Wright (ed.), *Biochemical and Immunological Taxonomy of Animals.* London: Academic Press, 1974.

Estes, R. "Prosirenidae, a new family of fossil salamanders." *Nature,* vol. 224, no. 5214, 1969.

———. "Origin of the Recent North American lower vertebrate fauna: an inquiry into the fossil record." *Forma et Functio,* vol. 3, 1970.

——— and O. V. Reig. "The early fossil record of frogs: a review of the evidence." *In* J. L. Vial (ed.), *Evolutionary Biology of the Anurans.* Columbia: University of Missouri Press, 1973.

Goin, O. B., and C. J. Goin, "DNA and the evolution of the vertebrates." *The American Midland Naturalist,* vol. 80, no. 2, 1968.

Jurgens, J. D. "The morphology of the nasal region of Amphibia and its bearing on the phylogeny of the group." *Annale Universiteit van Stellenbosch,* vol. 46, no. 2, 1971.

Laurent, R. F. "La distribución des amphibiens et les translations continentales." *Memoirs du Muséum National d'Histoire Naturelle,* ser. A, vol. 88, 1975.

Olson, E. C. (ed.). "Evolution and relationships of the Amphibia." *American Zoologist,* vol. 5, no. 2, 1965.

———. *Vertebrate Paleozoology.* New York: Wiley-Interscience, 1971.

Parsons, T. S., and E. E. Williams. "The relationships of the modern Amphibia: a re-examination." *Quarterly Review of Biology,* vol. 38, no. 1, 1963.

Piveteau, Jean (ed.). *Amphibiens, Reptiles, Oiseaux*. Traité de Paléontologie, vol. 5. Paris: Masson, 1955.

Romer, A. S. *Vertebrate Paleontology*, 3rd ed. Chicago: University of Chicago Press, 1966.

_____. *Notes and Comments on Vertebrate Paleontology*. Chicago: University of Chicago Press, 1968.

Salthe, S. N., and N. O. Kaplan. "Immunology and rates of enzyme evolution in the Amphibia in relation to the origins of certain taxa." *Evolution*, vol. 20, no. 4, 1966.

Schaeffer, B. "The evolution of concepts related to the origin of the Amphibia." *Systematic Zoology*, vol. 14, no. 2, 1965.

Schmalhausen, I. I. *The Origin of Terrestrial Vertebrates*. New York: Academic Press, 1968.

Szarski, Henryk. "The origin of the Amphibia." *Quarterly Review of Biology*, vol. 37, no. 3, 1962.

Tihen, J. A. "The fossil record of salamanders." *Journal of Herpetology*, vol. 1, no. 1–4, 1968.

Watson, D. M. S. "The evolution and origin of the Amphibia." *Philosophical Transactions of the Royal Society of London*, ser. B, vol. 215, 1926.

CHAPTER FIVE

ORIGIN
AND EVOLUTION
OF REPTILES

THE AMPHIBIANS WERE LIKE the early explorers of the New World, who still called the Old World home. They invaded and explored the land, but most could not travel far from their aquatic homeland, and necessarily returned to it to reproduce. The reptiles were the true colonizers who settled down to live and reproduce in the New World of dry land.

THE AMNIOTE EGG

The development of an amniote egg marked the dividing line between amphibians and reptiles. The amniote or cleidoic (closed) egg is a self-contained aquatic environment that permits the complete development of an embryo, freeing the parents from the necessity of seeking water when ready to reproduce. Although some living amphibians lay large-yolked, terrestrial eggs and lack free-swimming larval stages, these eggs require moist environments and often parental protection (see Chapter 6). The amphibian ancestors of the reptiles probably possessed a similar egg with similar requirements, and may have evolved their terrestrial egg under similar conditions. The present-day

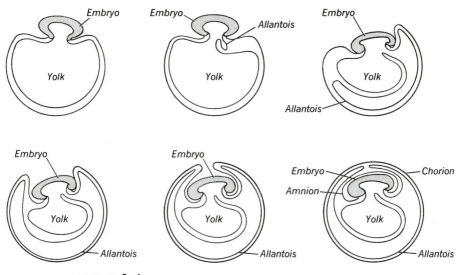

FIGURE 5–1
Hypothetical origin of the extraembryonic membranes in reptiles.

amphibians that reproduce on land seem to have evolved terrestrial eggs in humid mountainous areas, where there were few open bodies of quiet water for breeding, but sufficient environmental moisture to delay desiccation. The Pennsylvanian, when the reptiles first appeared, was a humid period with much local mountain building.

It has been suggested that the first of the amniote extraembryonic membranes to evolve was the allantois, which develops as an outgrowth of the posterior part of the embryonic gut. Reptiles presumably evolved from amphibians that already laid a small number of large-yolked eggs beside the water. The large amount of yolk allowed a prolonged developmental period and eliminated the need for an aquatic, feeding larva. A longer developmental period, though, exposed the embryo to an increased danger of desiccation. Water conservation became a problem. Much of the water loss during development results from the need to dispose of the nitrogenous waste products of metabolic activity. Aquatic larval amphibians excrete ammonia, which is both highly toxic and highly soluble. Embryos developing in eggs laid on land could not afford to expend the large amounts of water needed to remove the ammonia from the body. It is suggested that the embryos of the amphibian ancestors of the reptiles excreted urea, which is much less toxic than ammonia, just as terrestrial adult amphibians do today. Solutions of urea have a high osmotic pressure and, if retained within the egg, might have helped the embryo to absorb water from the environment. This in turn could have led to the development of an enlarged storage sac (allantois) at the hind end of the gut to hold both the accumulated water and dissolved urea (Figure 5–1). The water could then be resorbed by the allantoic blood vessels. As the allantois increased in size it would have spread between the layers of extraembryonic tissue, forcing them into folds around the embryo. When the folds met and fused, they formed the other two extraembryonic membranes, the amnion, which surrounds the embryo proper, and the chorion, which surrounds the embryo, amnion, allantois, and yolk sac. This is not the way the membranes develop in the embryos of modern reptiles, but it may well be the way they first evolved.

ORIGIN OF REPTILES

Reptiles first appear in the fossil record in the early Pennsylvanian. They had already diverged by then into two distinct lineages, captorhinomorphs and pelycosaurs. The two groups are clearly related, but apparently neither is derived from the other. This relationship indicates that early reptiles did not arise from two separate amphibian groups, but had a common reptilian ancestor whose fossil remains are still hidden or permanently lost. The homogeneity of all amniote eggs in the diverse group of reptiles and their bird and mammal offspring attests to the origin of the reptiles from a single common ancestor.

The Amphibian Ancestor

Because of the similarity in skull, vertebral, and limb structure between the early reptiles and the anthracosaurian labyrinthodonts, the latter are the likely reptilian ancestors. Until recently, all known anthracosaurs were too advanced—too divergent—to belong to the reptilian evolutionary line. The discovery of a primitive anthracosaur (*Mauchchunkia*) from the late Mississippian provided a likely ancestor. This moderate-sized (50 cm) amphibian had stout limbs, centra composed of small, crescentic intercentra and large, columnar pleurocentra, and a flat, solid skull that could easily be modified to form the anapsid skull of primitive reptiles.

The Reptile Skull

The anapsid (without openings) skull of early reptiles was solidly roofed over by dermal bones. These bones were arranged in three rows on each side: a median row that included the frontal and parietal bones; a lateral row comprising the postorbital and squamosal bones; and a marginal row in which the jugal and quadrate bones were important elements. The otic notch had disappeared, except perhaps in *Romeriscus,* the earliest known genus. The brain was very small and was encased in a narrow, bony box—

the brain case—that lay beneath the center of the skull roof. The muscles of the jaw attached to the underside of the roofing bones. During the evolution of the reptilian skull, openings developed between the bones of the once-solid skull roof to permit outward expansion of the jaw muscles as they contracted (see Figure 5–2). If the openings developed along the sutures

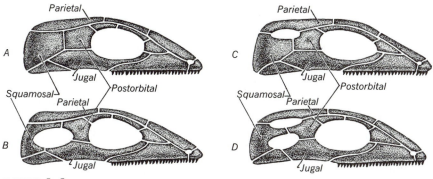

FIGURE 5–2
Diagrammatic view of reptilian skulls showing variation in temporal openings: (A) anapsid, the primitive type without openings; (B) synapsid, with a lateral opening, the type found in the ancestors of mammals; (C) euryapsid, with an upper temporal opening, the type found in plesiosaurs; (D) diapsid, the type found in *Sphenodon*, the crocodilians and other archosaurs, and in modified form in the lizards and snakes.

between bones of the median and lateral rows, a pair of dorsal temporal openings formed; if they developed between the bones of the lateral and marginal rows, a pair of lateral temporal openings appeared. In some reptiles both dorsal and lateral openings developed.

The subclasses into which the reptiles are divided are defined, in part, by the nature of the temporal openings in the skull. Early workers paid most attention to the arches, the bars of bone that lie below the openings. This is why the names of the types of skulls and of some of the suborders are derived from the Greek root *apse,* which means "arch." Thus the diapsids are reptiles with two arches, and hence two pairs of temporal openings in the skull.

RADIATION OF ANCIENT REPTILES

Present-day reptiles are a remnant of the diverse hordes that roamed the world during the Mesozoic. Once they had solved the problems of life on land, reptiles began to radiate rapidly into all environments, land, air, and water. They were better adapted to terrestrial conditions than the amphibians and had, as yet, no competition from birds and mammals. Some took to

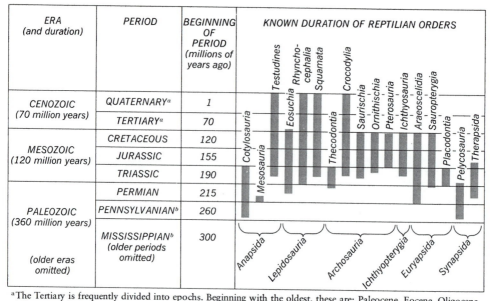

ERA (and duration)	PERIOD	BEGINNING OF PERIOD (millions of years ago)	KNOWN DURATION OF REPTILIAN ORDERS
CENOZOIC (70 million years)	QUATERNARY[a]	1	
	TERTIARY[a]	70	
MESOZOIC (120 million years)	CRETACEOUS	120	
	JURASSIC	155	
	TRIASSIC	190	
PALEOZOIC (360 million years)	PERMIAN	215	
	PENNSYLVANIAN[b]	260	
(older eras omitted)	MISSISSIPPIAN[b] (older periods omitted)	300	

[a]The Tertiary is frequently divided into epochs. Beginning with the oldest, these are: Paleocene, Eocene, Oligocene, Miocene, and Pliocene. The Quaternary is also divided into Pleistocene and Recent.
[b]The Mississippian and Pennsylvanian are often grouped together as the Carboniferous period.

FIGURE 5-3
The geologic time scale, showing the distribution of the reptiles in time.

the air; others returned to the water, though, in a curious reversal of the amphibian pattern, they had either to bear their young alive or to return to land to lay their eggs. The rapid radiation of reptilian species began soon after their first appearance in the Pennsylvanian; by the middle Permian, four of the six major reptilian lineages had appeared and begun to speciate (Figure 5-3).

Subclass Anapsida

The anapsid reptiles include some of the most primitive and some of the most specialized forms (Figure 5-4). They have an anapsid skull with no temporal opening, and the following suite of characters, which clearly identify them as reptiles and not amphibians: two proximal tarsal bones, astragalus and calcaneum; no palantine fangs or teeth with labyrinthodont enamel ridges; tympanum behind quadrate; columella inclined ventrolaterally; temporal series containing small tabular and supratemporal bones but no intertemporal.

Order Cotylosauria. The cotylosaurs first appear in the early Carboniferous, and include the captorhinomorphs, the probable ancestors of all the

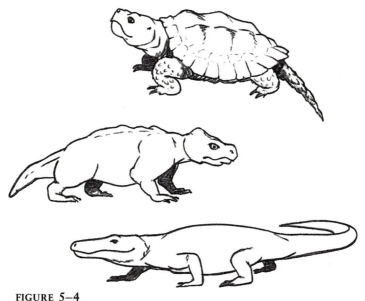

FIGURE 5–4
Anapsid reptiles: (*lower*) a primitive, fish-eating cotylosaur, *Limnoscelis*;
(*middle*) a slow-moving, herbivorous cotylosaur, *Bradysaurus*; (*upper*)
modern Alligator Snapping Turtle, *Macroclemys*.

following terrestrial reptile groups except the synapsids. Most captorhinomorphs were moderate-sized (4–50 cm), stout-bodied, lizardlike
animals with sturdy, sprawling legs.

The more advanced cotylosaurs (procolophonids, pareiasaurs, and
millerettids) replaced the captorhinomorphids in the mid-Permian and survived through most of the Triassic. The procolophonids and millerettids
were small to medium-sized (20–40 cm) reptiles; in contrast, the
pareiasaurs became massive, lumbering beasts (2–3 m) and were the first
to evolve a vertical limb posture and skeleton to support a massive body
on land.

Order Mesosauria. *Mesosaurus,* the only member of the mesosaurian order, was a slender, aquatic animal (1 m) with a long tail, enlarged,
paddlelike hindlimbs, and a slender, toothy skull. It was a short-lived (early
Permian) side branch of cotylosaur evolution, and was apparently confined
to fresh-water lakes and streams. Although it has been suggested that
Mesosaurus may be ancestral to the ichthyosaurs, the strong, paddlelike
hindlimbs indicate otherwise. The back part of the skull in all known specimens is too badly crushed for examination of the temporal region, but in
other characters *Mesosaurus* is close to the cotylosaurs.

Subclass Synapsida

The synapsid reptiles appear as early in the fossil record as the cotylosaurs (early Pennsylvanian) and were the commonest reptiles of the Permian; they began to decline in the Triassic, and had nearly disappeared by its end (Figure 5–5). Before dying out, they gave rise to the mammals.

FIGURE 5–5
Synapsid reptiles: (*upper*) a therapsid, *Cynognathus*; (*middle*) a pelycosaur, *Dimetrodon*; (*lower*) a dicynodont, *Kannemeyeria*.

The synapsids are the only known reptiles with lateral but not temporal openings in the skull. The subclass Synapsida contains two orders, which appeared more or less sequentially.

Order Pelycosauria. The stout-bodied pelycosaurs were common in the Carboniferous and early Permian; a few stragglers survived into the late Permian and, perhaps, the early Triassic. Most were small, though some of the later forms were more than 3 meters long. They all had sprawling legs like the amphibians. Some were long-snouted fish eaters, others were more terrestrial carnivores, and a few became herbivores.

The most spectacular pelycosaurs were the ship lizards (e.g., *Edaphosaurus* and *Dimetrodon*), which had enormously elongated neural spines supporting a saillike fin along the midline of the back. It was once speculated that these reptiles used their large fins to sail across the shallow seas; but it is more likely that the fin acted as a heating and cooling device. When the animal was perpendicular to the sun's rays, the blood circulating through the fin and back to the body would be heated and would raise the body temperature. Once the desired body temperature had been reached, the fin would be held parallel to the sun and, thus, not be heated, but rather act as a cooling surface. Although this was probably an effective thermoregulatory device, it was not the one that was adopted by the pelycosaur's successors. It has also been suggested that the fin was a signaling device used in dominance or courtship displays.

Order Therapsida.   In the mid-Permian, the pelycosaurs were largely replaced by their therapsid descendants, which remained abundant and diversified in the late Permian and early Triassic. Some were carnivores, others herbivores. Some were bulky animals—up to 4 meters long—others small and active. The therapsids are characterized not so much by common structural features as by the trends by which the various lines approached the mammalian condition. These trends include the development of a heterodont dentition, a double occipital condyle, and a vertical limb posture. Therapsids also developed a secondary palate, a bony plate separating the nasal cavity above from the oral cavity below. Ectothermic reptiles can tolerate a low oxygen supply or none at all for a period, because they do not maintain a constant body temperature and thus can survive with a low metabolism. In contrast, mammals must breathe regularly to maintain a high and steady metabolism. A secondary palate, which permits an animal to breathe while eating, was an important evolutionary step toward the mammals.

Since mammalian characters include physiological functions and such soft parts of the body as mammary glands and hair, it is impossible to say when therapsids ceased to be reptiles and became mammals.

Subclass Lepidosauria

The lepidosaurs are diapsid reptiles, with two temporal openings on each side of the skull (Figure 5–6). They appear to have diverged from the cotylosaur stock during the Permian. Of the three orders in this subclass, two—Rhynchocephalia and Squamata—survive.

Order Eosuchia.   The small, lizardlike eosuchians were common in the late Permian and early Mesozoic. They are an important group because of the success of their descendants. The archosaurs appear to have been derived from the earliest eosuchian, probably soon after the latter's origin from the captorhinomorphs. The squamates arose from specialized forms (*Prolacerta*

FIGURE 5–6
Lepidosaurian reptiles: (*upper left*) Tuatara, *Sphenodon*; (*upper right*) Chameleon, *Chamaeleo*; (*lower left*) a mosasaur, *Clidastes*; (*lower right*) Rock Python, *Python*.

and *Macrocnemus* of the Triassic). Eosuchians were largely extinct by the late Triassic, although gaviallike eosuchians *(Champsosaurus)* survived into the earliest Tertiary.

Subclass Archosauria

The archosaurs were the dominant animals of the Mesozoic, as their name implies (*archon* = ruling, *sauria* = reptiles) (Figure 5–7). Aside from a diapsid skull and a tendency toward a larger hind- than forelimb, there are few characters shared by the diverse representatives of this group. With the exception of crocodiles—their only surviving reptilian line—and birds, they are exclusively a Mesozoic group.

Order Thecodontia. The thecodonts are the basal archosaurs that apparently gave rise to the crocodiles, pterosaurs, dinosaurs, and birds. Early thecodonts were small, lizardlike carnivores with a tendency toward bipedalism. Later forms were fully bipedal, but the terminal thecodonts (aetosaurs, phytosaurs) reverted to a quadrupedal posture. The phytosaurs were gaviallike creatures up to 4 meters long. Although superficially like crocodilians, their skull differs in having the nostrils on top of the head and

FIGURE 5–7
Archosaurian reptiles: (*upper left*) a Triassic thecodont, *Saltoposuchus*; (*upper right*) a primitive pterosaur; (*center*) Alligator, *Alligator*; (*lower left*) a saurischian dinosaur, *Tyrannosaurus*; (*lower right*) an ornithischian dinosaur, *Stegosaurus*.

leading directly into the mouth, rather than on the tip of the snout with the air passage separated from the mouth by a secondary palate.

Order Ornithischia.   The ornithischian dinosaurs diverged early from the thecodonts. They are in no way, other than their large size, related to the saurischian dinosaurs, and can be readily recognized by their tetraradiate (four-branched) pelvis. In spite of their early divergence, they remained rare until the late Jurassic and early Cretaceous, when they radiated into four major lines: stegosaurs, ankylosaurs, ceratopsians, and ornithopods.

All four lines were herbivorous. The crested dinosaurs, or stegosaurs, walked on all four legs. The head was usually dwarfed by the massive body and the double row of protective plates or spines on the back and tail. The armored dinosaurs, or ankylosaurs, were giant armadillolike dinosaurs with rows of flat, bony plates covering the body from the neck to the tail. The horned dinosaurs, or ceratopsians, were heavy-bodied quadrupeds somewhat like reptilian rhinoceroses: the large head bore a posteriorly enlarged collar and one or more horns from the snout and orbital region. Similar crests and horns in modern chameleons and ungulates are used in intra-specific display and combat to maintain territories or establish dominance; the ceratopsian horns may have had a similar function. The ornithopods were the only bipedal ornithischians. Most were terrestrial and moderate in size (less than 5 meters), although the aquatic duckbill dinosaurs commonly reached lengths of 10 meters.

Order Saurischia.   The saurischian dinosaurs include the largest of all ter-restrial animals (*Brachiosaurus,* over 78 metric tons), but few other sauris-chians approached this size. Most were less than 10 meters long, and some no more than 1 or 2 meters. The saurischians arose from the thecodonts in the mid-Triassic and remained a dominant reptile group throughout the remainder of the Mesozoic. The saurischians split into two lines, theropods and sauropods, in the late Triassic.

The theropods were carnivorous, bipedal dinosaurs. Two main evolu-tionary trends appear in this line. One group, coelurosaurs, developed small (2–3 m), lightly built bodies with small heads on long necks, and, except for a long tail, were probably ostrichlike in appearance and habits. The other group, carnosaurs, developed large (4–15 m), stout bodies with large heads on short necks. *Tyrannosaurus* was the largest of these spectacular carni-vores and probably fed on larger herbivorous dinosaurs. The sauropods were herbivorous, quadrupedal dinosaurs. They were long-necked, long-tailed, small-headed creatures, probably both browsers and grazers and not restricted to the marshes and lakes as they are often portrayed.

It has been suggested that the dinosaurs, with their large bodies supported by fully vertical limbs, probably had activity patterns like birds and mam-mals and may have been endothermic, able to maintain a high constant body temperature through endogenous heat production.

Order Pterosauria.   The archosaurs gave rise to two aerial groups, the pterosaurs and the birds, which are not closely related. The birds probably originated from an early saurischian ancestor, and the pterosaurs from an unknown thecodont. The long, narrow pterosaurian wings were thin, mem-branous airfoils stretching backward from the arms and the greatly elon-gated fourth fingers. All pterosaurs had long, pointed skulls; the early species had a long, rudderlike tail, which was lost in the later forms. Most probably lived on cliffs near the seacoasts and fed on fish. Although the pterosaurs are commonly thought to have depended on rising air currents for soaring and gliding, recent biomechanical analysis suggests that they were capable of slow, flapping flight and were not completely dependent on gliding. Recently a pterosaur was discovered with the enormous wingspan of 15.5 meters.

Subclass Ichthyopterygia

The fishlike ichthyopterygians were common animals in the Mesozoic seas (Figure 5–8). They possessed a euryapsid skull (one with a single, dorsal temporal opening), but show no other indication of relationship to the Euryapsida or to any other reptilian group. They are assumed to have originated from an early cotylosaur.

FIGURE 5–8
Marine reptiles: (*upper*) a plesiosaur; (*middle*) an ichthyosaur; (*lower*) *Mesosaurus*.

Order Ichthyosauria. Ichthyosaurs were built like streamlined fish and were about the size of small porpoises. They had large heads with long, narrow, tooth-filled jaws. Their eyes were very large, and the nostrils opened just anterior to the eyes. These animals were evidently ecological equivalents to present-day porpoises. Like the latter, they would have been stranded if they had come ashore; thus they must have given birth to young instead of laying eggs. This hypothesis is confirmed by fossils showing traces of well-developed embryos within the mother's body.

Subclass Euryapsida

The euryapsids, like the ichthyosaurs, had a dorsal temporal opening, and the later forms were also marine; but their aquatic specializations were entirely different. There is little doubt that the two subclasses represent independent evolutionary lines.

Order Araeoscelidia. The araeoscelidians, small, lightly built, lizardlike reptiles with long, slender limbs, appeared in the early Permian and seem to have been derived from the captorhinomorphs. They lasted throughout the Triassic, but were never common. All known species seem to have been terrestrial, yet they are the most likely ancestors for the marine sauropterygians and placodonts.

Order Sauropterygia. The dominant sauropterygians were the marine plesiosaurs, aquatic blimps with oarlike limbs and a long, narrow neck and tail. They ranged from 5 to 15 meters in length, and the longer the head, the shorter the neck. They were highly aquatic animals and accomplished swimmers. Plesiosaurs probably "flew" underwater, using the figure-eight limb stroke seen in bird flight and in the swimming of penguins and sea turtles.

Order Placodontia. The placodonts were a group of bizarre, mollusk-eating reptiles. The early, primitive ones resembled long-tailed manatees, and the later, more specialized ones resembled heavily armored sea turtles. They were a short-lived group, with their greatest diversity in the middle and late Triassic.

ANCESTORS OF MODERN REPTILES

Compared with the ancestors of the modern amphibians, the ancestors of present-day reptiles are relatively well known. However, there remain numerous gaps in the fossil record that must be bridged with inferences—many of them rather wobbly.

Subclass Anapsida

Turtles (order Testudines) have existed as such since the late Triassic. The earliest turtle, *Proganochelys* (suborder Proganochelydia), has all the overt structural features of modern turtles, and is only strikingly different in possessing palatine teeth. The gap between *Proganochelys* and its procolophonid or pareiasaur ancestor remains unfilled. The expanded-rib, lizardlike *Eunotosaurus* has now been clearly shown to have no relationship to turtles.

Proganochelys appears to represent an evolutionary side branch. The surviving branch (suborder Casichelydia) has its earliest known members (baenoids) appear in the late Jurassic. The baenoids, primitive in many ways, are cryptodiran (S-necked) turtles that have yet to evolve a retractile neck, indicating a late Triassic divergence between the cryptodires and the pleurodires (side-necked turtles). Aside from baenoids (Jurassic–Pleistocene), cryptodiran turtles do not become common until the Cretaceous, at which time most of the modern families appear. The late Mesozoic seas seem to have allowed an extensive radiation of cryptodiran sea turtles: four families occurred then, though only the cheloniids survived an early Tertiary extinction wave.

The pleurodires never underwent a large-scale adaptive radiation. The two families (pelomedusids and chelids) that live today appeared in the Cretaceous and Eocene, respectively. The one notable feature of pleurodire evolution is that many of the fossil forms were marine or brackish-water forms; yet today, all side-necked turtles are fresh-water inhabitants. Modern turtle families are described in Chapter 14.

Subclass Lepidosauria

Both the lizards and rhynchocephalians arose from early eosuchian ancestors. Rhynchocephalians retained the diapsid skull, whereas lizards lost the lower arch.

Order Rhynchocephalia. A single species of rhynchocephalian, *Sphenodon punctatus,* survives today on a few islands off the coast of New Zealand. The rhynchocephalians were always a small group of no more than two or three families. They first appeared in the Triassic, underwent a minor radiation, and dropped from the fossil record in the early Cretaceous. *Homoeosaurus,* a very close relative of *Sphenodon,* appeared in the Jurassic, illustrating the antiquity of the lone survivor. The modern sphenodontid family is described in Chapter 17.

Order Squamata. The Triassic eosuchian *Prolacerta,* shows a progressive reduction of the lower temporal arch, and thus acts as a firm link between eosuchians and lizards. The first known lizard is *Tanystropheus* of the middle Triassic. This bizarre creature, about 75 centimeters long, had the cervical vertebrae enormously elongated, so that the neck was about as long as the rest of the body and tail combined. The young apparently ate insects; the aquatic adults, though able to come out on land, were largely fish eaters. These grotesque animals (there were several species) have been placed in a separate infraorder, Tanysitrachelia. *Kuehneosaurus* and *Icarasaurus* (infraorder Eolacertilia) appeared in the Upper Triassic. Both had elongated posterior ribs that supported a gliding membrane similar to that of the modern Flying Dragons *(Draco)*. Indeed, these very early lizards seem to have been even better adapted for gliding than *Draco*. In *Icarasaurus*, the "wings" were larger, and the first of the elongated ribs was a strong bar that formed a leading edge. The unexpected specialization of these very early lizards indicates that the saurians must have undergone a remarkably rapid radiation.

The other extinct families from the Jurassic to the early Tertiary are easily placed into the extant infraorders. None of the approximately ten extinct families adds signficantly to our knowledge of the evolution of the modern families, and only the anguinomorph family Mosasauridae is strikingly different from present lizards. The mosasaurs were huge, seagoing fisheaters with long heads and tails and paddle-shaped limbs. They ranged from 5 to 12 meters in length. They were common in all seas during the late Cretaceous, but did not survive into the Tertiary.

Some modern families of lizards (Agamidae, Chamaeleonidae, Scincidae, Anguinidae, Varanidae, Xenosauridae) were present in the Cretaceous. Most of the others appeared in the early Tertiary (Paleocene or Eocene). By the middle of the Tertiary, most of the fossil lizards belong to present-day genera. The worm lizards (suborder Amphisbaenia) do not appear until the Paleocene. However, their structural differences from other lizards suggest a much longer evolutionary history: it is likely that they diverged from the main lizard lineage soon after it had made the transition from the eosuchians. Modern lizard families are described in Chapter 15.

The snakes are the youngest group of reptiles. They appear to be unquestionably derived from lizards, and the varanoid lizards have been proposed as their ancestors. However, the confirming fossil links are missing. The earliest known snakes *(Dinilysia* and a few others) come from the late Cretaceous and show clear affinities to living henophidian snakes. This suggests that snakes had diverged from their lizard ancestors at least by the early Cretaceous, if not before. Snakes are often said to have passed through a burrowing evolutionary phase to obtain their limblessness and special type of retina. This may be true, but other habitats are possible.

The modern Boidae and Aniliidae were present in the late Cretaceous. The Typhlopidae appeared in the Eocene, the Colubridae, Viperidae, and Elapidae by the Miocene. As in the lizards, these Miocene snakes belong to modern genera or their immediate ancestors. Modern snake families are described in Chapter 16.

Subclass Archosauria

The crocodiles (order Crocodylia) have passed through several evolutionary grades during their history, and each grade has undergone a wide adaptive radiation. These grades are represented by suborders (Archaeosuchia, Protosuchia, Mesosuchia, Eusuchia) and show a progressive development of the secondary palate and a general reduction in body armor. The late Triassic archaeosuchians and protosuchians show close ties to the thecodonts, although the typical crocodilian body and limb structure is well developed. The mesosuchians were the crocodilians of the Jurassic and the early Cretaceous. They included dwarf and giant species with the typical crocodilian form and marine species with fishlike tail fins and paddle-shaped limbs. The eusuchians include the two modern families (Crocodylidae and Gavialidae), which had their origin in the late Cretaceous. Modern crocodilian families are described in Chapter 17.

READINGS AND REFERENCES

Carroll, R. L. "Origin of reptiles." *In* C. Gans (ed.), *Biology of the Reptilia.* vol. 1 London: Academic Press, 1969.
_____. "Problems of the origin of reptiles." *Biological Reviews,* vol. 44, 1969.
Colbert, E. H. *Dinosaurs, Their Discovery and Their World.* New York: Dutton, 1961.
_____. *The Age of Reptiles.* New York: Norton, 1965.
Estes, R., T. H. Frazzetta, and E. E. Williams. "Studies on the fossil snake *Dinilysia patagonica* Woodward: Part 1, Cranial morphology." *Bulletin of the Museum of Comparative Zoology,* vol. 140, no. 2, 1970.
Gaffney, E. S. "A phylogeny and classification of the higher categories of turtles." *Bulletin of the American Museum of Natural History,* vol. 155, no. 5, 1975.
Hoffstetter, R. "Revue des recentes acquisitions concernant l'histoire et la systematique des squamates." *Colloques Internationaux du Centre National de la Recherche Scientifique,* no. 104, 1962.
Hotton, N. *Dinosaurs.* New York: Pyramid Publications, 1963.
_____. "*Mauchchunkia bassa,* gen. et. sp. nov., an anthracosaur (Amphibia, Labyrinthodontia) from the Upper Mississippian." *Kirtlandia,* no. 12, 1970.
Joysey, K. A., and T. S. Kemp (eds.). *Studies in Vertebrate Evolution.* New York: Winchester Press, 1972.
Parrington, F. R. "The problem of the classification of reptiles." *Journal of the Linnean Society of London, Zoology,* vol. 44, no. 295, 1958.
Romer, A. S. "Unorthodoxies in reptilian phylogeny." *Evolution,* vol. 25, no. 1, 1971.
_____. "Aquatic adaptation in reptiles—primary or secondary?" *Annals of the South African Museum,* vol. 64, 1974.
Szarski, H. "The origin of vertebrate foetal membranes." *Evolution,* vol. 22, no. 1, 1968.
Tihen, J. A. "Comments on the origin of the amniote egg." *Evolution,* vol. 14, no. 4, 1960.

CHAPTER SIX

REPRODUCTION
AND LIFE HISTORY
OF AMPHIBIANS

MOST OF THE ACTIVITIES of an animal, whether it is feeding, seeking a safe
resting place, or escaping from an enemy, revolve around the preservation of
the individual. Periodically, however, these survival activities are superseded
by another set of activities directed toward the perpetuation of the race. The
urge to reproduce is so strong that the animal may change its customary way
of life completely: it may stop feeding, leave its home, and travel long
distances, exposing itself to enemies on the way, to breed. In this chapter we
are concerned with these activities and their outcome—the production of a
new generation to replace the old.

The most striking thing about amphibian reproduction is its diversity. We
can hardly make a general statement about any phase of it that does not
have exceptions. Fertilization may be internal or external, the animals may
lay eggs or bear their young alive, eggs may be laid in water or on land, there
may be a larval stage or the development may be direct, parents may abandon their eggs or guard them. Speaking in anthropomorphic terms, we
might say that the amphibians have been experimenting, trying to find the
method or methods of reproduction best suited to life on land.

BREEDING ACTIVITIES

Species retain their genetic integrity through the evolution of reproductive isolating mechanisms (see the discussion under Systematics and Taxonomy in Chapter 1). Premating isolating mechanisms operate during breeding activities to prevent the mixing of genes from different species. Postmating isolating mechanisms operate during development and the life of the hybrid animal to prevent it from reproducing. The operation of these mechanisms is nowhere more apparent than in the amphibians, with their diverse reproductive patterns.

Breeding Season

The breeding seasons of amphibians are closely correlated with rainfall and temperature. Since most amphibians, whether tropical or temperate, live in seasonal environments, they have discrete breeding seasons, for their eggs must be laid when there is sufficient moisture to prevent desiccation, and the young must hatch when there is an abundant food supply to allow development and growth. In much of the tropics, the temperature is constant but the climate is divided into a wet and a dry season. Here, most amphibians begin to breed at the beginning of the wet season. In temperate areas, both temperature and rainfall fluctuate seasonally, so temperate-zone amphibians must match their breeding seasons to periods of adequate rainfall and warm temperature.

Obviously, an animal cannot breed if it is not physiologically ready: that is, it must have mature ova or sperm. Therefore, gametogenesis—the production of gametes—must also be timed to match the appearance of the correct weather conditions for breeding. Amphibians appear to ensure this in two ways: breeding is either cyclic or noncyclic. The cyclic pattern is a seasonal rhythm of gametogenesis, with all adult members of a breeding population possessing mature gametes at the same time, and is thus a genetic trait. This intrinsic rhythm is controlled by the effect of changing day length, and possibly temperature, on the anterior pituitary, which, in turn, regulates gametogenesis. The cyclic pattern is common in temperate amphibians and assumed to be largely confined to them. The noncyclic pattern is an aseasonal rhythm of gametogenesis, with different members of the population possessing mature gametes at different times; thus throughout the year a part of the population is always able to breed. This pattern is assumed to be common in tropical amphibians and the temperate "opportunist" breeders, such as Spadefoot Toads *(Scaphiopus)*.

Temperate-zone salamanders are largely cyclic breeders, whereas what little we know of tropical ones suggests they are noncyclic. The former are mostly cool-temperature breeders, mating in either the late fall or the early spring. In those that mate in the fall, such as *Salamandra salamandra* and *Plethodon cinereus,* the female receives the sperm before she possesses mature ova; the sperm are stored in a urogenital pouch, the spermatheca. Oogenesis (production of eggs) continues throughout the winter, and the eggs are fertilized and laid in the spring. If mating occurs in the spring, fertilization and egg deposition follow shortly. *Ambystoma opacum* and *A. maculatum* use the same breeding ponds, but avoid mispairing because *A. opacum* completes its breeding in the fall and *A. maculatum* does not breed until early spring.

Most temperate-zone frogs breed in the spring and early summer. In any given area, the breeding season is divided into a series of sequential, slightly overlapping breeding periods for the different species of a genus, usually with the smallest first and the largest last. For example, in Virginia, *Rana palustris* begins breeding with the first warm rains in early March, and is followed in late March or April by *R. utricularia; R. clamitans* does not begin until the temperature is considerably warmer, in May, and is followed shortly by warmth-loving *R. catesbeiana.* Farther south or north, the sequence remains the same, but the breeding season begins earlier or later; the temperature at which a species begins to breed appears to be fairly constant throughout its range.

In the seasonal tropics and arid areas, breeding begins with the first heavy rainfall. Breeding tends to be explosive, with enormous aggregations of several species of frogs congregating along and in the newly formed streams and ponds. Whereas these "explosive" breeders concentrate their reproductive activities into several days or a few weeks, other species will spread theirs over the entire wet season, so that no large breeding aggregation ever exists, but there are always a few individuals breeding. In those arid areas where rainfall is acyclic, there is no breeding seasonality: the frogs, such as *Scaphiopus,* breed at any time during the year, but only when sufficient rain has fallen.

There are few reports on how many times an individual amphibian will breed during a year. What little evidence we have indicates that the females of cyclic breeders probably mate and lay eggs only once, while the males are able to shed sperm several times. Both sexes of noncyclic breeders may breed several times: the activity of males may be continuous, but that of females is generally sporadic, to allow for periods of oogenesis.

Breeding Sites

Amphibians use a wide variety of breeding sites. This is particularly true of the frogs (Figure 6–1). They may lay eggs in open, standing water (either permanent or temporary), in running water (even mountain torrents), on the ground under stones or logs, in burrows, in mud basins constructed by the

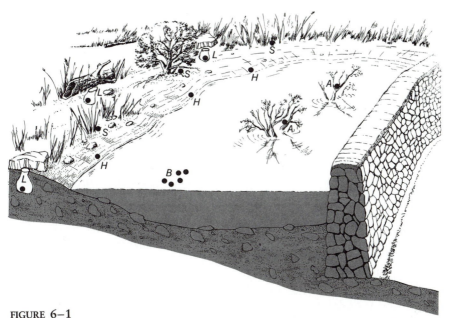

FIGURE 6–1
Segregation of breeding sites in a single pond in northern Mexico: (A) *Agalychnis dacnicolor*; (B) *Bufo marmoreus*; (H) *Hypopachus variolosus*; (L) *Leptodactylus fragilis*; (S) *Smilisca baudini*. [After Dixon and Heyer, *Bull. S. Calif. Acad. Sci.*, 1968.]

males, on the undersides of leaves suspended over water, or in water collected at the bases of the leaves of tropical air plants. Sometimes the eggs or tadpoles, or both, are carried about by one of the parents. The young of *Rhinoderma darwini* go through their posthatching development in the vocal pouch of the male! Even more strange is the Australian *Rheobatrachus*, with its gastric brooding. The eggs are apparently swallowed after fertilization and complete most of their development and larval life in the female's stomach.

Salamanders show less diversity in their choice of breeding sites, but even so, their eggs may be laid in still water, in swift-flowing streams, or on land. No matter how small the difference in breeding sites may seem to us, closely related species of amphibians breeding in the same area at the same time will not use the same sites. For one species to breed in the shallow water at a lake's edge and another in deep water is enough to keep the two reproductively isolated.

Breeding Congregations

Breeding congregations occur largely in those amphibians that are terrestrial or arboreal, yet remain tied to water for a free-living larval stage; and of these, it is primarily the ones with cyclic or seasonally regulated noncyclic breeding patterns that form such congregations. Terrestrial forms with terrestrial eggs, like *Eleutherodactylus* and *Plethodon,* and aquatic forms, like *Xenopus* and *Necturus,* usually mate and lay their eggs in the same place where all their other life activities occur. In contrast, "congregating breeders" are often widely dispersed and far distant from their breeding sites, and must move to their special sites for reproduction. The males of both frogs and salamanders characteristically arrive before the females.

What clues enable widely dispersed animals to locate their breeding sites? Auditory, olfactory, visual, geotaxic (gravity gradient), and hygrotaxic (humidity gradient) clues all play a role in guiding the migration to the breeding site. Their importance varies from species to species, however. Female and late-arriving male frogs can rely on the calling of the early-arriving males to guide them, but these first males have to use other clues, often odors from the breeding site. *Rana temporaria* in England migrates in response to the odor of ripening algae in the ponds. A newt, *Taricha rivularis,* can return to its home using only olfactory clues after being displaced several kilometers. Celestial clues, landmarks, and other visual information are known to guide the movements of amphibians. Frogs and salamanders unquestionably use a combination of senses and a variety of clues to guide their breeding migrations.

Voice

The amphibian voice is almost always associated with reproductive activity and is largely confined to male frogs. A few salamanders and female frogs make escape calls, apparently to startle a predator and permit escape. Male frogs possess well-developed vocal cords and large, inflatable vocal sacs. Air passing outward over the vocal cords produces the sound, and the inflated vocal sacs act as resonators to amplify it. The mating call is unique for each species and serves for species recognition. To enhance this recognition and to avoid confusion with the mating calls of other species, the columella–basilar-papilla hearing complex in each species has its greatest sensitivity at the same frequencies as those occurring in the species' mating call. Thus, for all practical purposes, females hear only the mating calls of males of their own species and mate only with their own kind, so that the call serves as an effective premating isolating mechanism.

Frog calls differ in frequency, pulse rate, duration, and number of notes. The two species of Gray Tree Frogs (*Hyla versicolor* and *H. chrysoscelis*) are indistinguishable in appearance, and may breed in the same pond. The call is a loud, resonant trill, but with a slower pulse rate in *H. versicolor* than in *H. chrysoscelis*. Tests have shown that a female can distinguish between the two and will swim toward the sound of her own species. Aside from maintaining reproductive isolation, the call serves other important roles. It attracts and guides both females and males to the breeding sites, spaces the males evenly around the site, and enables the females to find the males.

In both congregating and noncongregating species, the mating call is used, as in birds, to define a territory. In congregating frogs such as *Hyla crucifer,* the call defines a courtship territory in which no other male is permitted to call. Another male may sit within this territory, but he remains silent and does not interact with any entering females. The silent male becomes subordinate through a series of aggressive bouts, primarily vocal but occasionally with physical contact. When the dominant male clasps a female and they move away to deposit and fertilize the eggs, the subordinate male may begin to call, but will stop upon the return of the dominant male.

In noncongregating species such as *Rana catesbeiana,* the call defines a more inclusive territory, which is used for feeding and egg deposition as well as for courtship and is maintained beyond the breeding season. The territory is defined by vocalization and actively defended with wrestling bouts. Subordinate males and even females must hold low profiles to move through the territory or to approach the male for mating.

Aside from the mating call, the release call is the most important breeding sound. This low, chucklelike call produces large vibrations of the body wall, and tells a male trying to mate to let go—because he has hold of another male, because the female is finished laying her eggs, or because he is on the wrong species. There are numerous other calls, such as the rain call of *Hyla squirella* and the chuckles and growls of *Rana pipiens,* whose purpose is unknown.

Sex Recognition

When we speak of sex recognition, there is no implication that amphibians consciously distinguish between the sexes. It is simply that the presence of a female evokes from the male a given series of reactions that are not evoked by another male. Similarly, a female responds to a male in ways she does not to another female. We know very little of the precise mechanisms of sex recognition. There are, of course, many differences between the sexes besides the primary differences in the genital systems. Some, such as the enlarged tympanum of the male Bullfrog, are permanent; others appear only with the onset of the breeding season. The latter are exemplified by the spiny nuptial pads on the forefeet of many male frogs and toads. These pads help the male grasp and hold the female, but, like the majority of secondary sex characters in amphibians, they seem to have little value in sex discrimination.

Both behavior and the odor (or some other recognizable stimulus) from the secretions of hedonic glands probably function in sex recognition in salamanders. During courtship, the male prods, noses, or otherwise rubs against another animal, showing by his actions that he is a male. A ripe female accepts these advances, another male resists or runs away. Males of some European newts are brightly colored, and display before the females. Sexual color differences may aid in sex discrimination in these forms.

A female frog responds to the voice of a calling male by moving towards him. Males seem to have little ability to discriminate between the sexes, and tend to grasp any object of approximately the right size that approaches. A female whose abdomen is distended with ripe eggs will be held until the eggs are deposited. A male, or a female who has already laid, will be quickly released. On the other hand, a male frog will continue to clasp another male whose abdomen has been artificially distended. The peculiar posture and gait assumed by a female frog during egg deposition apparently help stimulate the male to retain his grasp.

Courtship and Amplexus

Before mating, most animals go through a specific behavioral sequence to stimulate one or both partners to perform the sexual act. If the behavioral sequence is incorrect, the body size mismatched, or the sexual structures of the wrong shape or size, the sexual act will not occur. In amphibians, courtship behavior ranges from simple rapelike activities in many frogs and a few salamanders to elaborate dances, the *Liebespiel* (love play) of most salamanders.

Frogs show little Liebespiel: the male's call alone serves to attract and stimulate receptive females. In some anurans, such as *Bufo,* the male will jump on the back and clasp the body of any passing toad—male or female—in a sexual embrace (amplexus), whereas in *Rana,* the female must come under the male and nudge him in order to be clasped. He may clasp her behind the forelegs (axillary amplexus) or in front of the hindlegs (inguinal amplexus). Since he faces the same way she does, he can shed sperm over the eggs as they are extruded. The simultaneous release of sperm and eggs seems to be synchronized by small, perhaps species-specific body movements of the female.

The simplest salamander courtship occurs in *Hynobius retardatus.* Males arrive at the breeding ponds first. The females follow a day or so later, but are ignored by the males. When a female begins to lay, the male rushes forward and grabs the emerging egg sacs with his forelimbs while pushing the female away with his hindlimbs. There is thus no contact and no apparent mutual stimulation between the pair until egg deposition actually begins. However, there are indications that the male *Hynobius nebulosus* selects a breeding site and the female is attracted to it, perhaps by secretions of the male cloacal glands.

The male of *Eurycea bislineata,* a plethodontid, noses the female and frequently bends his head across her cheek. She finally responds by straddling his tail and pressing her snout against the glands at its base. The tail is bent sharply aside, and the pair walk along in this fashion until mating is completed. In some species of salamander, the male creeps under the female and carries her along on his back; in others, he lashes his tail to waft the secretions of his hedonic glands toward her. Because courtship behavior differs from species to species, it probably functions as an isolating mechanism as well as for sex recognition and stimulation.

FERTILIZATION AND DEVELOPMENT

Fertilization marks the beginning of the hazardous period of development and growth, which will last one to five years or more before the relative safety of adulthood is reached. From fertilization through the embryonic, larval, metamorphic, and juvenile stages, most amphibians are exposed to the vagaries of the environment and to attack by predators. It is the harshness of this development period that has led to the diversity of reproductive adaptations seen in amphibians.

Fertilization

The object of the many complex activities described above is to bring together the male and female gametes so that they can fuse. This fusion of egg and sperm is fertilization, which may take place externally or internally. Some amphibian spermatozoa are illustrated in Figure 6–2.

Fertilization among the primitive salamanders, Hynobiidae and Cryptobranchidae, is external. In all other salamanders it is internal. During courtship, the male deposits several little gelatinous packets containing sperm (spermatophores), and the female picks them up with the lips of her cloaca (Figure 6–3). The male of the salamandrid *Euproctus* deposits his spermatophores on the female's body and uses his feet to move them to the region of her cloaca. He may even stuff them into the cloaca. *Amphiuma* is reported to transfer the spermatophore directly into the female. In the cloaca, the gelatinous cap of the spermatophore dissolves, and the sperm make their way to the spermatheca, a diverticulum of the roof of the cloaca. (Figure 6–4). The eggs are usually fertilized as they pass by the spermatheca, but there is evidence that the sperm of the European *Salamandra atra* mi-

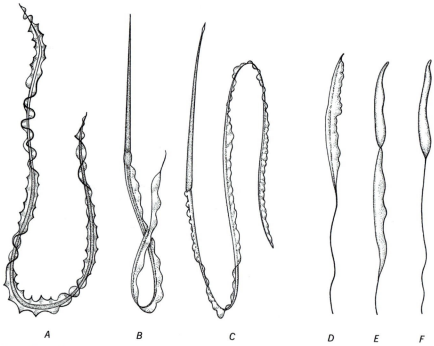

FIGURE 6–2
Amphibian spermatozoa: (A) *Pseudobranchus striatus*; (B) *Triturus marmoratus*; (C) *Amphiuma means*; (D) *Bombina variegata*; (E) *Bufo vulgaris*; (F) *Hyla arborea*. [Redrawn from Noble, 1954; Angel, 1947; and Austin and Baker, *J. Reprod. Fertility*, 1964.]

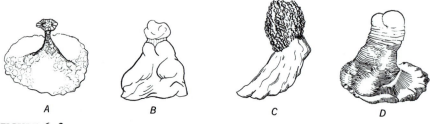

FIGURE 6–3
Salamander spermatophores: (A) *Notophthalmus v. viridescens*; (B) *Ambystoma jeffersonianum*; (C) *Eurycea bislineata*; (D) *Desmognathus f. fuscus*. [Redrawn from Bishop, *N.Y. State Museum Bull.*, 1941.]

grate up the oviduct before the eggs are fertilized. In contrast, the ova of the newt *Notophthalmus* cannot be fertilized until jelly layers have been deposited around them by the glands of the oviduct. Sperm high in the oviduct would be wasted. Most salamanders lay their eggs soon after mating, but *Necturus maculosus,* which mates in the fall, does not deposit its eggs until the following spring.

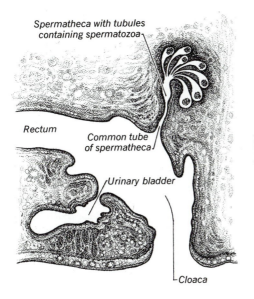

Spermatheca with tubules
containing spermatozoa

Rectum

Common tube
of spermatheca

Urinary bladder

Cloaca

FIGURE 6–4
Relation of the spermatheca to
the cloaca in a salamander.

Frogs usually have external fertilization, the male shedding his sperm over the eggs as they are deposited by the female in water or in some moist environment. However, the live-bearing toad of Africa, *Nectophrynoides,* practices internal fertilization, although it has no intromittent organ for the transmission of sperm. The "tail" of the Tailed Frog, *Ascaphus,* is actually an extension of the cloaca that serves as an intromittent organ (see Figure 13–7). Fertilization is internal in these mountain-torrent forms; perhaps this is to prevent the sperm from being swept downstream before the eggs are fertilized.

Male caecilians have an intromittent organ that injects sperm into the cloaca of the female. Some caecilians are live-bearers. Internal fertilization must be the rule in this group, but we know almost nothing of how the male finds the female or where and how mating takes place. Nothing is known about mating among the sirens.

Eggs

Amphibian eggs are mesolecithal (have a moderate amount of yolk), but the amount of yolk varies widely. Species that lay their eggs on land and undergo direct development have larger yolks than do species that lay in water. Thus the terrestrial egg of the Cliff Frog *(Syrrhophus marnocki)* has a yolk 4 millimeters in diameter, whereas the yolk of the aquatic egg of the Cricket Frog *(Acris gryllus)* is only 1 millimeter in diameter, although both adults are about the same size.

There is enormous diversity in the way eggs are deposited by different species of amphibians. Generally, however, glands in the walls of the oviduct secrete a gelatinous substance around the eggs as they pass down to the cloaca. This substance swells in water and forms a protective jelly. Sometimes many eggs are enclosed in a single jelly mass, sometimes the eggs are separate. The jelly may be in the shape of long strings, in one large, irregular packet, or in several smaller packets deposited separately by the female. Eggs may float on the surface of the water, sink to the bottom, or be attached to the stems of aquatic vegetation or to the undersides of leaves, sticks, or rocks in the water; they may also be laid on land (Figure 6–5).

The number of eggs laid varies greatly, from the single egg of the tiny Cuban *Sminthillus* to the twenty thousand eggs of large species of *Bufo* and *Rana*. As clutch size increases, egg size and amount of yolk tend to decrease, and the eggs are laid in more open and less protected areas. Although salamanders generally lay fewer eggs, the same trend is evident. *Hyla* and *Ambystoma* lay large numbers of eggs in open water and abandon them, and the larvae hatch small and must gather their own food to complete their development. Other frogs and salamanders lay moderate numbers of eggs in protected spots (leaf nests of *Centrolenella* or aquatic nests of *Amphiuma*), and may stay in attendance to protect the eggs and larvae for a short time. A few frogs and salamanders, such as *Eleutherodactylus* and *Plethodon,* lay few eggs in protected spots on land and guard them; the young hatch as miniature adults (direct development) having skipped the free-swimming larval stage. Within an allied group of species, the number of eggs is related to body size, with the bigger species laying the most eggs.

Developmental History

Most amphibians differ from the other tetrapods in having two definitive developmental periods: embryonic development before hatching, and post-hatching development until metamorphosis to the adult form. Young in the posthatching stage are known as larvae among the salamanders and as tadpoles among the anurans. Occasionally there is no larval period, and the young hatch (or are born) as miniature replicas of the adults.

Oviparity, Ovovivipariy, and Viviparity.   Usually the young of amphibians hatch after the eggs have been laid, and the animals are said to be oviparous. In rare instances, however, the eggs are retained in the body of the female while they pass through their embryonic development, and the young are "born alive." If the developing embryo in the mother's body is nourished entirely by food stored in the yolk of the egg, the animal is ovoviviparous. If the embryo obtains part of its food from maternal tissues (the walls of the Müllerian duct), the animal is viviparous. Most members of the genus *Salamandra* are ovoviviparous, but in one, *S. atra,* an extra set of eggs in the oviduct fail to develop and instead become a yolk source for the growing

FIGURE 6–5
Eggs in an opened nest of the terrestrial breeding leptodactylid *Kyarranus loveridgei* of Australia. [Photograph by John A. Moore.]

larvae to feed on when they have used their own supply. The African toad *Nectophrynoides* and the Puerto Rican *Eleutherodactylus jasperi* are also ovoviviparous and give birth to fully metamorphosed young. A few caecilians are viviparous, with the young eating the lining of the oviducal wall. Other caecilians are oviparous.

Embryonic Development. The pattern of early development of the amphibians is typical for vertebrates that have mesolecithal eggs. The stages are: early cleavage, blastula formation, gastrulation, neurulation, tail-bud stage, period of organogeny (formation of organs), period of early heartbeat and development of gill buds, and finally development of gill circulation and hatching. The details of these stages can be found in any text on embryology and need not be discussed here. Figure 6–6 shows the external morphology

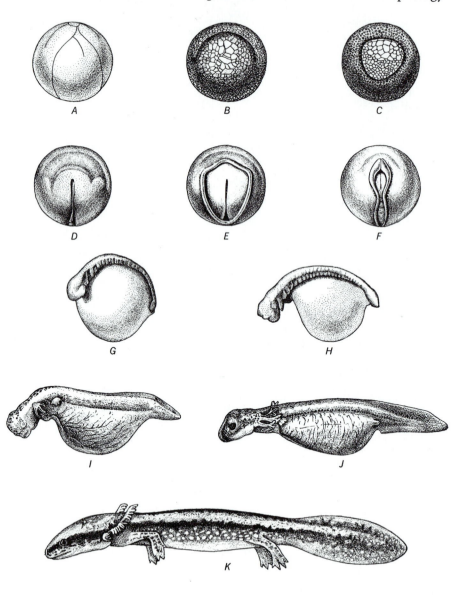

of a developing salamander, *Necturus maculosus;* hatching takes place about at stage J. Table 6–1 lists comparable stages in the development of the Leopard Frog, *Rana pipiens.* As the table shows, there is some variation in developmental rate, which is determined, at least in part, by temperature.

TABLE 6–1
Developmental Stages of *Rana pipiens*

Stage	At 18°C	At 25°C
Fertilization	0 hours	0 hours
Gray crescent	1 hour	½–1 hour
Rotation	1½ hours	1 hour
Two cells	3½ hours	2½ hours
Four cells	4½ hours	3½ hours
Eight cells	5½ hours	4½ hours
Blastula	18 hours	12 hours
Gastrula	34 hours	20 hours
Yolk plug	42 hours	32 hours
Neural plate	50 hours	40 hours
Neural folds	62 hours	48 hours
Ciliary movement	67 hours	52 hours
Neural tube	72 hours	56 hours
Tail bud	84 hours	66 hours
Muscular movement	96 hours	76 hours
Heartbeat	5 days	4 days
Gill circulation	6 days	5 days
Tail-fin circulation	8 days	6½ days
Development of operculum	9 days	7½ days
Operculum complete	12 days	10 days
Metamorphosis	3 months	2½ months

SOURCE: Rugh, *The Frog,* 1951, Blakiston Div., McGraw-Hill Book Co., by permission.

FIGURE 6–6
Development of *Necturus maculosus.* (A) Side view of egg 1 day and 8 hours after deposition, showing second and third cleavage grooves. (B) Bottom view of egg 6 days and 16 hours old. The crescentic blastopore lip sharply separates the large yolk cells from the small cells of the blastodisc. (C) Bottom view of egg 10 days and 10 hours old, showing large circular blastopore. (D) Top view of egg 14 days and 4 hours old; blastopore smaller. Beginning of neural fold formation, especially anteriorly. (E) Top view of egg 15 days and 15 hours old. Yolk plug still visible; neural fold prominent; its free ends reach nearly to the blastopore. (F) Top view of egg 18 days and 15 hours old with three or four pairs of body segments visible. (G) Dorsolateral view of embryo 22 days and 17 hours old; length 8 mm; 16 to 18 body segments. (H) Side view of embryo 26 days old; length 11 mm; 26 to 27 body segments; eye, ear, nasal pits, and mouth well defined. (I) Side view of embryo 36 days and 16 hours old; length 16 mm; 36 to 38 body segments. (J) Side view of larva 49 days old; length 21 mm. (K) Side view of larva 97 days old; length 34 mm. [From Noble, *The Biology of the Amphibia,* Dover, 1954. Reprinted through permission by Dover Publications, Inc., New York.]

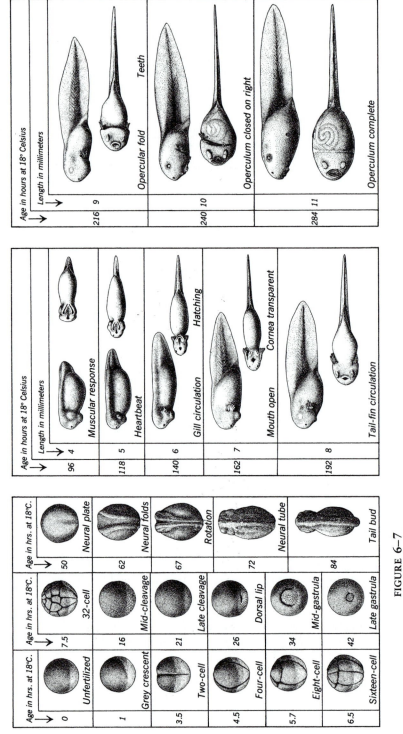

FIGURE 6-7
External morphological features of the developing embryo of the frog *Rana pipiens*. [Redrawn from Shumway, 1940.]

Hatching takes place at about the time the gill circulation develops, or after about six days at 18°C (Figure 6–7).

Posthatching Development.   The degree of development reached at hatching varies greatly in the amphibians. Some frogs and salamanders pass through metamorphosis in the egg, or hatch in an advanced condition so that they are nearly ready to metamorphose. Most frogs hatch as tadpoles, which are quite different in appearance from their parents, and must go through a continuing period of development before they are ready to metamorphose.

The early tadpole of *Rana pipiens* is a small, poorly developed animal, seemingly ill-equipped to meet the vicissitudes of life outside the protecting egg. Its mouth has not yet broken through, its eyes are not formed, its gills are not fully developed. For a few days it remains attached to some object by a pair of V-shaped suckers below the indentation where its mouth will be, and draws nourishment from its remaining supply of yolk. A small, round mouth armed with horny teeth and jaws appears, the eyes take shape, the gills grow rapidly, and the tail elongates. The tadpole is soon ready to take up a free-swimming, food-finding existence, but it still does not look like a frog. An operculum grows posteriorly to cover the external gills, which are replaced by a new set. Gradually the legs develop. The hind legs emerge as tiny buds, and later develop toes and joints. The front legs grow at the same time but are hidden by the operculum. When the front legs do emerge, the left goes through the spiracle, the right through the opercular wall. Shortly before metamorphosis, the tadpole comes frequently to the surface to gulp air. Its lungs are then about to replace the gills as the organs that supply oxygen to the blood. The long, coiled intestine necessitated by the vegetarian diet of the tadpole shortens, because the adult frog is a flesh eater. The gills degenerate, the horny jaws and teeth are shed along with the tadpole skin, the round mouth widens, the tail is resorbed (although a vestige still remains), and the little animal, now a frog, hops onto land.

Salamander larvae are quite different from tadpoles (Figure 6–8). The frog tadpole has one or a pair of suckers on the ventral side of the head that persist from before to shortly after hatching; these are lacking in salamander larvae. On the other hand, the larvae of many salamander species have balancers, long, rodlike structures, one on each side, that protrude from the sides of the head below the eyes. The salamander larva has three pairs of gills, the frog tadpole only two well-developed pairs, the upper or third pair being rudimentary. After hatching, the tadpole develops an operculum that covers the gills (gills are usually evident in salamander larvae until the time of metamorphosis). The legs of salamander larvae are often well developed at hatching. The mouth is like that of the adult. Thus the mature salamander larva resembles the adult much more closely than the tadpole resembles the frog.

SALAMANDER LARVA   FROG TADPOLE

1. SHORTLY BEFORE HATCHING

Long body   Shorter body
No adhesive organs   Prominent adhesive organs
Three pairs of gill rudiments   Only two pairs of gill rudiments distinct

2. SHORTLY AFTER HATCHING

Long body   Short body
No adhesive organs   Prominent adhesive organs
Balancers present in some species   No balancers
Mouth becomes large   Mouth remains small
Three pairs of well-developed external gills,   Only two pairs of well-developed external gills,
    uppermost typically the longest   third pair (uppermost) rudimentary
Opercular folds simple   Opercular folds develop rapidly, soon cover gills

3. LARVAL CHARACTERS FULLY DEVELOPED

Long body   Short, egg-shaped head and body unit
Mouth large, with simple labial folds   Mouth small, with complex accessory parts
External gills present   No external gills
Opercular folds open and loosely overhang   Operculum closed except for a small spiracle
    sides of neck; no spiracle   draining gill chamber
Foreleg buds external, clearly visible   Foreleg buds internal, concealed under closed
    throughout their development   operculum until metamorphosis

FIGURE 6–8
Comparison of salamander larvae and frog tadpoles. [After Orton, *Turtox News*, 1950.]

Larval Types. Differences in posthatching habitat are correlated with morphological differences in the developing larvae. Salamanders have three basic types of life history: some are specialized toward breeding in open bodies of quiet water, some are modified for a mountain stream existence, and some have become specialized for terrestrial life (Figure 6–9).

Warm, still pond water has relatively little oxygen dissolved in it; therefore, a pond-type larva needs large respiratory surfaces to absorb enough oxygen for its needs. For example, the pond-type larva of *Ambystoma* has

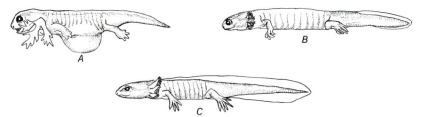

FIGURE 6—9
Principal types of salamander larvae: (A) terrestrial type, *Plethodon vandykei*; (B) mountain brook type. *Dicamptodon ensatus*; (C) pond type, *Ambystoma gracile*. [From Noble, *The Biology of the Amphibia*, Dover, 1954. Reprinted through permission of Dover Publications, Inc., New York.]

long, very filamentous gills and a well-developed tail fin that extends over its back to form a dorsal body fin.

In the cool, rapidly flowing water of mountain torrents, plenty of oxygen is available. The problem faced by larvae in this environment is rather to keep from being swept downstream by the current. The brook-type larva of *Dicamptodon ensatus* is streamlined, with a muscular tail, no dorsal body fin, a reduced tail fin, and small gills.

Finally, the terrestrial salamanders, such as *Plethodon*, have a tiny, short-bodied larva with gills that lack filaments; these gills reduce before hatching. The larva has a rather well-developed yolk mass in the gut and a short tail with no fin. These tiny creatures hatch from eggs laid on land and soon become entirely terrestrial salamanders, without ever going through an aquatic larval stage.

Tadpoles occur in a wider variety of habitats than salamanders, and also show more morphological and ecological diversity. There are pond types (the typical deep-bodied, high-finned polliwogs), mountain brook types with streamlined bodies, strong tails, and reduced fins, and terrestrial or direct-development types. However, unlike the salamanders, the tadpoles have specialized within each type. Within the pond type, there will be tadpoles adapted to feed on the bottom, at midwater, along the surface, at shore edges, on submergent vegetation, and so forth; each of these shows structural modifications of the basic pond-type body form.

In addition to the diversity within these three general habitat types, frogs lay their eggs in a variety of other spots, and the tadpoles have become modified accordingly. Some *Hyla* tadpoles developing in the tiny pools formed in the leaf cylinders of bromeliad plants are slim, with long, whiplike tails and reduced gills. Apparently they get most of their oxygen directly from the atmosphere. They feed on frog eggs, either of their own or of other species, and probably also on other tadpoles. Their jaws are strong, but the larval tooth rows have been reduced. Tadpoles of the African *Hoplophryne* develop in similar habitats and show similar modifications.

Embryos of frogs that lack a free-living larval stage may yet show most of the typical tadpole structures, such as the operculum and the larval teeth and jaws. In others, these structures are reduced and tend to disappear. The developing young of *Rhinoderma darwini* are distinctly tadpolelike, with closed operculum, spiracle, coiled intestine, and larval mouth parts, although the teeth and jaws never harden.

The habitat-related diversity of tadpoles is superimposed on the frogs' taxonomic diversity. Four tadpole types (xenoanuran, scoptanuran, lemnanuran, and acosmanuran) are recognized, and each of these appears to represent a single evolutionary lineage. They can easily be recognized by the structure of the spiracle (respectively paired; single as a long, median tube; single as a short, median tube; and single with a sinistral opening), although there are many other differences.

Paedomorphosis

Many species of salamander retain larval structures, such as gills, into adulthood. This paedomorphosis is of two types: genetically fixed paedogenesis and environmentally induced neoteny. Sirens, proteid salamanders, and many cave-dwelling plethodontids are paedogenetic, for their tissues fail to respond to the secretions of the thyroid gland that initiate metamorphosis in other salamanders.

Some populations of the newt *Notophthalmus viridescens,* the Tiger Salamander *(Ambystoma tigrinum)*, and other members of the genus *Ambystoma* are neotenic. Under normal conditions, the population perpetuates itself without metamorphosis; if the pond begins to dry up, however, the metamorphic mechanism is triggered, and the adults metamorphose and migrate from the drying pond.

True paedomorphosis does not occur in frogs. All paedomorphic anurans reported have been giant, thyroidless tadpoles lacking sexual maturity.

GROWTH AND LONGEVITY

Even after metamorphosis, amphibians must go through a period of growth and development before they become sexually mature. Some species reach breeding age in about a year. Others, particularly the larger forms like the Bullfrog *(Rana catesbeiana)*, are not ready to breed until two or three years after metamorphosis, and the Mud Puppy *(Necturus maculosus)*, not for several years after hatching.

Growth does not stop when an amphibian reaches sexual maturity. Many, perhaps all, continue to grow, although very slowly, for the rest of their lives.

Maximum age is very difficult to determine for animals living under natural conditions; it is much easier with animals reared in captivity. But such sheltered forms probably survive much longer than wild animals. The figures in Table 6–2 are for captive specimens, and do not indicate longevity under natural conditions. Although the table is incomplete, it does indicate that some of these small animals are capable of living for surprisingly long times under favorable conditions. It also shows that the sluggish salamanders apparently live longer than the more lively frogs.

TABLE 6–2
Longevity of Some Species of Amphibians

Order, family, and species	Maximum age (years)
Meantes	
Sirenidae *(Siren lacertina)*	25
Caudata	
Cryptobranchidae *(Andrias japonicus)*	55
Ambystomatidae *(Ambystoma maculatum)*	25
Salamandridae *(Cynops pyrrhogaster)*	25
Amphiumidae *(Amphiuma means)*	27
Proteidae *(Proteus anguinus)*	15
Salientia	
Pipidae *(Xenopus laevis)*	15
Discoglossidae *(Bombina bombina)*	20
Pelobatidae *(Pelobates fuscus)*	11
Leptodactylidae *(Leptodactylus pentadactylus)*	12
Bufonidae *(Bufo bufo)*	36
Pelodryadidae *(Litoria caerulea)*	16
Ranidae *(Rana catesbeiana)*	16
Hyperoliidae *(Kassina weali)*	9
Microhylidae *(Kaloula pulchra)*	6
Microhylidae *(Gastrophryne carolinensis)*	6

We know nothing of the reproductive habits of the early amphibians, but it is doubtful that all the diversity shown by the present-day forms is of recent origin. The ancestral groups must at least have had considerable genetic plasticity to have allowed such diversity to develop. Most of the early forms failed to find a really satisfactory solution to the problem of reproduction on land, and their descendants have remained amphibians. But members of one group, at least, did evolve a practical method of terrestrial reproduction by means of the amniote egg, and they became the reptiles.

READINGS AND REFERENCES

Angel, F. *Vie et Moeurs des Amphibiens*. Paris: Payot, 1947.

Etkin, W., and L. I. Gilbert (eds.). *Metamorphosis*. New York: Appleton-Century-Crofts, 1968.

Florkin, M., and B. T. Scheer (eds.). *Chemical Zoology*, vol. IX, *Amphibia and Reptilia*. New York: Academic Press, 1974.

Gosner, K. L. "A simplified table for staging anuran embryos and larvae with notes on identification." *Herpetologica*, vol. 16, 1960.

Lofts, B. (ed.). *Physiology of the Amphibia*, vol. II. New York: Academic Press, 1974.

Moore, J. A. (ed.). *Physiology of the Amphibia*. New York: Academic Press, 1964.

Rugh, R. *The Frog: Its Reproduction and Development*. Philadelphia: Blakiston, 1951.

Salthe, S. N., and J. S. Mecham. "Reproductive and courtship patterns." *In* B. Lofts (ed.), *op. cit.*

Shumway, W. "Stages in the normal development of *Rana pipiens*." *Anatomical Record*, vol. 78, 1940.

Vial, J. A. (ed.) *Evolutionary Biology of the Anurans*. Columbia: University of Missouri Press, 1973.

CHAPTER SEVEN

REPRODUCTION
AND LIFE HISTORY
OF REPTILES

WHEN A GROUP OF ANIMALS makes a broad shift from one mode of existence to another, there must be a close correlation in evolutionary changes of structure, function, and activity. Nowhere is this more marvelously shown than in the change in reproductive pattern that allowed the reptiles to free themselves completely from the need to return to the water to breed. Terrestrial breeding habits, internal fertilization, shell, and extraembryonic membranes go hand in hand to produce the truly terrestrial egg of the reptile.

Sperm can travel only in a fluid medium. When fertilization takes place away from water, the fluid carrying the sperm must be protected from desiccation. The best method for ensuring this is emission by the male of seminal fluid into the reproductive tract of the female: the act of copulation. Since the ovum is fertilized before it leaves the body of the female, it can be provided with a protective shell by the glands of the oviduct.

Caecilians, most salamanders, and a few frogs evolved internal fertilization, but they never developed a shelled, amniote egg to go with it, and so have been unable to exploit fully the possibilities of this more efficient method of fertilization.

BREEDING HABITS

Breeding is a more solitary event in reptiles than in amphibians, seldom more than a male and a female meeting and mating. Still, as in the congregating amphibians, reproductive isolating mechanisms have evolved to maintain the genetic integrity of each species. However, the premating isolating mechanisms in reptiles are largely visual, unlike the auditory and tactile ones in amphibians.

Sexual Encounters

One advantage of terrestrial reproduction over the aquatic reproduction of most amphibians is that it permits a wide dispersal of egg clutches in space and time. When all the females of a population must place their eggs in the same ponds at the same time, a single catastrophe, such as the too-rapid drying of temporary ponds, will eliminate the entire reproductive effort for the season. Also, concentrations of young attract predators and reduce the local abundance of food. Reptiles thus do not need to lay as many eggs as frogs to insure the survival of enough young to maintain the population.

Most reptiles have discrete breeding seasons, whether in temperate or tropical areas. In temperate areas, the breeding season appears to be regulated by temperature and photoperiod. Mating usually occurs in the spring or early summer, and egg laying follows shortly thereafter. Many tropical reptiles are potentially able to reproduce all year round; however, only in those areas where the climate is relatively stable or homogenous do they do so. In the seasonal tropics, rainfall seems to be the principal regulator of reproduction, at least for females.

Even though reptiles are usually widely dispersed during their breeding season, the meeting of a male and a receptive female is not haphazard. Most lizards occupy well-defined territories. The male's territory tends to be larger and to overlap or include those of several females with whom he mates. A similar arrangement seems to exist for most crocodilians. Although fresh-water and terrestrial turtles do not have well-defined territories, they live in circumscribed habitats and in large numbers, so that males and females are in continual association. The sea turtles are widely dispersed for most of their lives, but each cohort congregates at the nesting beach of their birthplace every several years (2–4) to lay their eggs. This congregation brings the females and males together: the males literally patrol the waters off the nesting beaches, and frequently copulate with the females when they return to the water from nesting forays.

The mating of some snakes also occurs in congregation. Many temperate-zone species aggregate each winter in large hibernacula (wintering dens). When the snakes emerge from hibernation in the spring, mating occurs around the den before they disperse. Thousands of snakes may be involved, as in *Thamnophis sirtalis* of Manitoba, but usually it is no more than a few hundred, and often fewer. In the South Pacific, some sea snakes congregate just for breeding. The majority of snakes, however, do not congregate. Individuals are widely scattered and seldom numerous: males probably locate receptive females by following a scent trail.

Sexual Dimorphism

Sexually mature female and male reptiles often differ strikingly in size, shape, coloration, and other features. Most of these differences are permanent, although differences in coloration are often intensified in sexually active individuals. These sexual dimorphisms are associated mainly with reproductive behavior and with structural modifications to accommodate the reproductive organs.

Male turtles, as a rule, are smaller than females. The larger bodies of females are apparently an adaptation to hold a large clutch of eggs. The tails of males are longer and thicker, and the vent—the opening of the cloaca—is displaced toward the tip of the tail: all modifications to accommodate the penis. Male snakes also have a longer tail with a thicker base than the females, but the males' body is proportionally shorter, as though the vent had moved forward in males to accommodate the hemipenis and backward in females to enlarge the body cavity so as to hold a larger egg clutch without changing the total length. Male and female lizards may be the same size, or males may be larger.

The enlarged and brightly colored heads of male lizards, their enlarged crests and horns, and their brightly colored throats, throat fans (dewlaps), and bodies are all associated with reproductive-territorial behavior. These features accentuate specific body movements and postures, and enable males to hold and defend territories and to attract females. Few other reptiles show such structural or coloration differences. Female and male snakes, turtles, and crocodilians characteristically share the same color and pattern. The cloacal spurs—vestigial hindlimbs—of boid snakes tend to be larger in males, and may perhaps serve a titillative function. The elongated forefoot claws of male Pond Turtles, *Chrysemys*, certainly serve to titillate the female when vibrated alongside her head. Male Box Turtles, *Terrapene carolina*, have ruby red eyes, the females have brown ones; we do not know why.

Sex Recognition

Female reptiles generally play a relatively passive role in sexual encounters. Unlike a male frog, the male reptile must find the female, pursue her, and court her. The males of many reptiles appear not to recognize females as such, but differentiate between "other male" and "not male." That is, another animal—or even an inanimate object—not recognized as a male is regarded as a female. The Banded Gecko (Coleonyx variegatus) depends primarily on behavior for sex recognition; anesthetized males are treated as females by other males. When the blue patches on the sides of the male Fence Lizards (Sceloporus undulatus) are painted out, they are similarly treated as females by other males. Males of marine turtles are sometimes lured into the nets of fishermen by wooden decoys shaped roughly like females.

Such reliance on appearance and behavior for sex recognition cannot be overemphasized, for the initial sex discrimination made by many male reptiles is based on a combination of these two factors. The general behavioral sequence for discrimination appears to be: a male sees an object of roughly the correct size, outline, and color; he responds to this object with a stereotyped sequence of body movements and postures; if the object mimics his behavioral sequence, he recognizes it as a male of his own species; other movements and postures may tell him that it is a female of his own species, or that it is an unidentifiable, nondangerous object and therefore to be ignored. The final male-female interaction preceding copulation probably depends as much on odor and tactile stimuli as on behavior for final recognition. These body movements are unique for each species, and act as premating isolating mechanisms like the calls of frogs.

Head bobbing is the most commonly observed recognition signal, occurring in many lizards, snakes, and turtles. This is a more complex phenomenon than it seems. Each species moves the head up and down a set number of times, and the amplitude of each head movement is fixed. Thus the pattern of head movement is unique for each species and enables both males and females to identify their own kind. The dewlap movements of lizards, such as Anolis and Draco, also show such species-specific stereotypes. These stereotypic displays, at least in lizards, are largely associated with the establishment and defense of territories during the breeding season.

Most male snakes locate and identify females of their own species by scent. Some apparently compete for females through a combat dance, most frequently observed in viperids. Two male vipers will join in an arm-wrestling-like match with the posterior parts of their bodies entwined around one another and the anterior parts raised above the ground, swaying and pushing against each other until one is forced down (Figure 7-1). The bouts may continue for a considerable time before one individual is forced into submission, frees himself, and flees, followed by the victor. Whether the dance contains species-specific movements and whether it is an attempt to establish a territory or social dominance have not been determined, but it does seem to have some sexual significance.

FIGURE 7-1
Combat dance of the Eastern Cottonmouth, *Agkistrodon p. piscivorus*, in a Florida marsh.

Courtship

Most reptiles have a Liebespiel preceding courtship, but it is seldom as elaborate as in many salamanders. Nonetheless, it is sufficient to act as a behavioral premating isolating mechanism. Male turtles of the genus *Chrysemys* court females in the water. Swimming either above the female or backward in front of her, the male vibrates the long claws of his front feet against her cheeks. The speed of vibration is different for each species and enables the female to recognize a male of her own species. Other turtles may bob their heads and nudge and bite the females.

Among diurnal lizards, particularly those in which the males are brightly marked, the male may posture and display in front of the female. Lizards also indulge in nipping and nudging. Sometimes a pair will walk along together, the male straddling the tail of the female and resting his head on her pelvic region when she pauses.

Male and female snakes may glide along side by side, the male caressing the female with his chin and flickering his tongue over her body (Figure 7-2). He may twine his tail around hers and lie on top of her while his body undulates in a series of waves that pass from his tail toward his head (caudocephalic waves).

FIGURE 7–2
Courtship in the King Cobra, *Ophiophagus hannah*, showing how the male and female glide along side by side. [Courtesy of the New York Zoological Society.]

Unlike frogs, most reptiles do not have voices, and in only a few of them does voice play a part in courtship. Many snakes hiss, and one genus, *Pituophis,* has a special membrane on the epiglottis that vibrates to produce a particularly audible hissing sound when air is expelled from the lungs. Geckos have well-developed voices; some can be heard for distances of nearly a hundred meters. But all these sounds are apparently warning or threatening, not amatory. The voice of the crocodilian, by contrast is definitely associated with breeding, though it is not known whether it is a love call, a challenge to other males, or both. A fundamental tone of 57 vibrations per second typically evokes a roar from a half-grown American Alligator *(Alligator mississippiensis).* Indeed, alligators in a lake near the University of Florida are often stimulated to call by fireworks exploded on the campus. The primitive New Zealand Tuatara *(Sphenodon punctatus)* has been heard croaking on cold, misty nights, but again the reason for the call is unknown. It is noteworthy that the best-developed reptilian voices are found among nocturnal forms, in which hearing is presumably better developed than sight.

Some turtles also have voices, and among the testudinids, at least, voice seems to be associated with reproductive activity. The Galapagos Tortoise bellows while mating. The male of the South American *Geochelone denticulata* makes a noise while pursuing the female and while copulating that, to anthropomorphic ears, sounds suspiciously like a chuckle.

Copulation

Courtship is usually, but not always, followed by copulation. Sometimes copulation takes place with little or no preliminary Liebespiel, the male simply grasping the female as soon as he sees her.

Aquatic turtles mate in the water, the others on land. The male approaches the female from behind and mounts her carapace, gripping the front part of her shell with his forefeet and sometimes biting her head and neck. The hind part of his body is pushed below her carapace. The position is very awkward, particularly for a species with a high, domed shell, and sometimes the male loses his balance and falls.

The positions taken during copulation by crocodilians, lizards, and snakes are much alike. The male lies beside or above the female and thrusts his tail under hers so that the regions of the vents are brought together. Among crocodilians, lizards, and some snakes, the male grasps the head, neck, or shoulder of the female with his jaws. Since a receptive female does not struggle, the grasping with the jaws is probably done simply to help the male keep his balance. Snakes and lizards evert the hemipenis on the side next to the female and insert it into her cloaca. It becomes turgid and, held in place by its spines and calyces, usually cannot be removed forcibly without damage to the animals. Copulation may be completed in a few minutes, or it may last for several hours or even a day. Couples in captivity have been known to copulate at repeated intervals for several days.

FERTILIZATION AND DEVELOPMENT

Compared with that of the amphibians, the period of development and growth in reptiles is somewhat safer. This relative safety is reflected in the reptiles' smaller clutch sizes.

Fertilization

All reptiles have internal fertilization. The sperm are deposited directly into the cloaca, eliminating the necessity of mating in a moist environment so that the sperm have an aqueous medium to swim to the ova. Another advantage of internal fertilization is that the sperm are less likely to be washed away or lost through dehydration. From the cloaca, the sperm move into the oviducts and penetrate the ova. Once fertilized, the ovum is encased in a shell secreted by the wall of the oviduct, and is then laid. Since there is a delay between fertilization and egg deposition, embryonic development has often begun before the eggs are released to the outside.

Although fertilization in most reptiles, as in birds and mammals, commonly takes place shortly after emission by the male, many reptiles and a few salamanders have evolved a fail-safe mechanism that allows them to store sperm in an inactive state in the female's reproductive tract. Ova can thus be fertilized months or even years after mating. Such delayed fertilization permits mating in one season and egg laying in another, or allows a female to lay several fertile clutches of eggs even if she has mated only once.

Reptilian sperm, in contrast to those of most vertebrates, which have a life span of a few hours to a few days, can survive for months or years, as is necessary for delayed fertilization. Sperm viability is not indefinite, however. Table 7–1 shows the length of time ten female Diamondback Terrapins *(Malaclemys)* had been separated from males, the number of eggs laid

TABLE 7–1

Viability of Sperm in *Malaclemys* Females
after Separation from Males

Length of separation (years)	Number of eggs laid	Number fertile
1	124	123
2	116	102
3	130	39
4	108	4

each year, and the number of fertile eggs. A few sperm were viable after four years, but the majority had deteriorated after the second year. Female Box Turtles *(Terrapene carolina)* have also laid fertile eggs up to four years after separation from males.

Although turtles possess delayed fertilization, no special sperm storage structures, or seminal receptacles, have been found associated with the oviduct. Such structures do, however, occur in many lizards and snakes. Female lizards have small tubules in the vaginal portion of the oviduct. During the breeding season of *Anolis carolinensis,* sperm are found in the lumen (passageway) of the oviduct as well as in the receptacles, but during the nonbreeding season they are confined mainly to the receptacles. It is assumed but has not been proven that sperm survival is prolonged in the tubules of the receptacle, for viable sperm persist for at least seven months in *Anolis* females, six months in the chameleon *Microsaura pumila,* nearly four

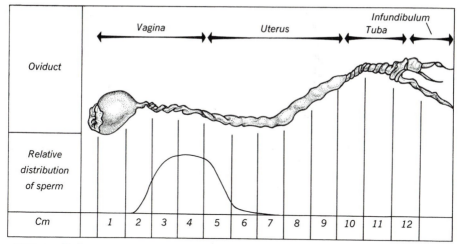

FIGURE 7–3
Diagram illustrating the relative concentration of sperm in various regions of the "ripe" oviduct of a female Prairie Rattlesnake, *Crotalus v. viridis.* [After Ludwig and Rahn, *Copeia,* 1943.]

months in the gecko *Eublepharis macularis,* and nearly three months in *Uta stansburiana.*

In snakes, the seminal receptacles lie at the base of the infundibulum (the funnel-shaped part of the oviduct leading from the ostium), rather than in the vaginal region of the oviduct. Sperm may be found in both the receptacles and the lumen of the oviduct. Figure 7–3 shows the distribution of luminal sperm in the oviduct of a Prairie Rattlesnake *(Crotalus v. viridis)* shortly before ovulation. Table 7–2 shows how long after separation from males sperm remain viable in the oviducts of several species of snake. Delayed fertilization and the presence of seminal receptacles have not been confirmed in the crocodilians or in *Sphenodon.*

TABLE 7–2
Viability of Sperm in Female Snakes
after Separation from Males

Species	Duration of viability (months)
Causas rhombeatus	5
Drymarchon corais	52
Leptodeira annulata polysticta	60
Storeria dekayi	3
Thamnophis s. sirtalis	3

Oviparity, Ovoviviparity, and Viviparity

Most reptiles are oviparous. Some lizards and snakes are ovoviviparous, with eggs hatching either in the oviduct or just after they are laid. There is really no sharp distinction between oviparity and ovoviviparity, since many oviparous forms lay eggs in which the embryos have already begun to develop. Eggs of some colubrid snakes, examined at the time of deposition, contained embryos that ranged from 15 to 55 milimeters in length. Such eggs may not hatch for two or three months. Alligator eggs and those of some lizards also undergo some development before deposition.

Oviparity and ovoviviparity may occur in different species of the same genus, or even in different populations of the same species. The Eurasian species of the Water Snake *Natrix* lay eggs; their North American relatives bear living young. The European lizard *Lacerta vivipara* is a live-bearer throughout most of its range, yet lays eggs in the Pyrenees.

Viviparity implies the exchange of material between the embryonic and maternal bloodstreams. This adaptation has arisen infrequently in reptiles, and then only in the squamates: in the gekkonid, xantusiid, anguinid, scincid, and lacertid lizards and the colubrid, elapid, and viperid snakes. The choriovitelline (yolk-sac) placenta is the most common; the chorioallantoic (urinary-bladder) placenta occurs only in a few skinks—the European *Chalcides ocellatus* and several species of *Sphenomorphus* and *Tiliqua* from Australia. Histological studies of the uterine portion of the oviduct in the elapid *Denisonia* show a partial degeneration of the maternal epithelium, permitting diffusion between the bloodstreams of the mother and the embryo.

The oviducts of lizards and snakes lack well-defined albuminous glands in the upper portions. In a bird's egg the albumen contributes to the amniotic fluid, which protects the embryo from desiccation. Terrestrial eggs that are not provided with a sufficient supply of albumen must be laid in surroundings moist enough to allow water to enter through the permeable shell. Such eggs swell noticeably after deposition. (Even eggs that have albumen are permeable and may lose too much water and desiccate in an arid environment.) It may be that the tendency of lizards and snakes to retain the eggs in the constantly moist environment of the oviduct for varying lengths of time, leading to ovoviviparity and even viviparity, is a compensatory mechanism for the lack of albumen. Such a mechanism may have been a step in the development of the mammals from the reptiles. It has also been suggested that the retention of the eggs in the oviduct by an animal that had developed the ability to regulate its body temperature was a protective device against exposure of the eggs to low temperatures, and that this may have been a significant factor in the evolution of the viviparous mammals from the oviparous reptiles. Snakes and lizards that live at high altitudes or latitudes typically bear living young.

Amplexus, egg deposition, fertilization, and the beginning of embryonic development follow one another so closely in most frogs that a single series of observations may give us a fairly clear outline of the entire reproductive pattern of a species. The pattern of reptilian reproduction is not so apparent. Copulation may take place in the fall, but egg deposition may not occur until the following spring. Even when eggs are laid shortly after mating, it is still possible that the ova were fertilized by sperm received in a previous copulation. Because the beginning of development does not necessarily coincide with the time either of copulation or of egg deposition, it is not possible to calculate precisely the length of the embryonic period. Long, patient hours in the field, the dissection of many specimens in the laboratory, and much luck are needed to study the life history of a reptile.

Egg Structure and Deposition

Reptilian ova are telolecithal (have large amounts of yolk). For this reason, their early developmental stages resemble those of birds rather than those of frogs. The cleavage planes cannot cut through all the yolk, so they are confined to the top of the yolk mass. The embryo forms first as a flat disc, which slowly lifts from the yolk mass and folds into the embryo proper and the extraembryonic membranes.

The fertilized ovum, or zygote, of a reptile is encased in two sets of protective membranes: the shell membranes, derived from maternal tissue, and the extraembryonic membranes, produced by the embryo. The latter are peripheral outgrowths of the embryonic tissue, the chorion, amnion, and allantois (see Figure 3–1). In oviparous species, the extraembryonic membranes usually do not develop until the egg has been laid. The shell is secreted by specialized glands in the oviduct, and is composed of layers of alternating fibers. In oviparous reptiles, the outer layers are impregnated with calcium salts; ovoviviparous forms frequently lack this calcium impregnation, and viviparous forms dispense with the fibrous membranes. The hardness of the eggshell reflects the amount of calcium salts, but even the leathery eggs of most squamates contain some calcium salts. Additional fibrous egg membranes may lie below the calcified eggshell and enclose the albumen (egg white). Albumen is also secreted by the oviducal wall and serves as a water reserve. Only the turtles and crocodilians have an abundant supply of albumen.

All these membranes reduce desiccation and permit the eggs to be laid on land—but not just anywhere, because sufficient moisture and heat are required for normal development. In general, reptiles are more careful in selecting or constructing suitable sites for their eggs than are most amphibians. We see in them a foreshadowing of the elaborate nest-building–incubation–parental-care behavior that characterizes the birds.

Parthenogenesis

Some populations and species of reptiles are unisexual: though an occasional sterile male may be found, all normal individuals are females whose eggs can develop without fertilization. Such parthenogenetic reproduction in reptiles seems to be almost confined to lizards, having been reported in one or more species of gekkonids, xantusiids, agamids, chamaeleonids, teiids, and lacertids. It is suspected in the Braminy Blind Snake, *Typhlops braminus,* in which no males have been found, and may be discovered in other primitive snakes.

Parthenogenesis requires the production of diploid (2*n*) ova instead of the haploid ova that are the normal result of meiosis; this may result from a division of the chromosomes without an accompanying cell division. Normal meiosis follows, pairing the sister chromatids of the preceding chromosome division. Parthenogenesis also requires the activation of cleavage without penetration of the ovum by a sperm.

A parthenogenetic line (clone) can arise within a species, but more frequently it results from hybridization between two species. Sometimes a normally parthenogenetic female will mate with a male of a bisexual species and give rise to a line that is triploid (3*n*). Occasionally, as in the Checkered Whiptail *(Cnemidophorus tessellatus),* a hybrid parthenogenetic form, both diploid and triploid populations are found.

Parental Care

Most reptiles abandon their eggs as soon as they are laid. Indeed, the mother may never even see them. Reports of a female guarding her clutch, based simply on the chance observation of an individual in the vicinity of the eggs, should be viewed with skepticism. Nevertheless, there is evidence that at least a few reptiles have developed care-giving behavior beyond that shown in preparing a proper incubation site.

Female skinks of the genus *Eumeces* stay with their eggs. The mother protects them from small predators, and may turn the eggs periodically and bring them back together if the clutch is accidentally scattered. It has even been suggested that she helps to incubate them by basking in the sun and then coiling her warmed body around them.

Females of several species of snake are known to coil about the developing eggs. Perhaps best known are the pythons. After depositing the eggs, the female python draws them together by moving her tail and body. She heaps them in a pyramidal mass and coils around them, her head capping the spiral formed by her body. She stays with the eggs for six weeks, only rarely moving away to drink, but leaves them about two weeks before they hatch. A brooding female python is able to raise her body temperature as much as

7.3°C above the ambient temperature by spasmodic muscular contractions. In this ability to carry on thermoregulation by endogenous heat production, the brooding python resembles the endothermic birds and mammals. Other snakes that are known to guard their eggs include the cobras and the American Mud Snake *(Farancia abacura)*.

The best-developed care-giving behavior patterns are reported for the crocodilians. The American Alligator *(Alligator mississippiensis)* mother remains in the vicinity of the nest, and is said to moisten it occasionally with the contents of her bladder. When the young hatch, they start a high-pitched grunting that attracts the attention of the mother. If the surface of the nest is packed too hard, she may scrape it away to release the young. She is said to scoop out a wallowing pool for them and to guard them from attack by predators, including other alligators. The young may remain with the mother for a year or more.

One of the most prevalent and persistent of all snake myths is the story of the mother swallowing her young to protect them and later disgorging them. The tale, which is usually told of ovoviviparous species, is of course untrue, but it may have some basis in fact. In captivity, the young of some of these species (for example, the European Viper, *Vipera aspis*) have been seen to remain in the vicinity of the mother for a couple of days after birth and disappear beneath her body when threatened. Whether they do the same in the wild is not certain, but even if they do, such an association must be at best an ephemeral one.

HATCHING, BIRTH, AND YOUNG

Heat is necessary for the development of vertebrate embryos, and the rate of development is in part determined by the amount of heat available. The endothermic birds and mammals provide a relatively high, constant source of heat for their developing young, either by retaining them in the body of the mother or by sitting on the eggs. The length of the embryonic period is fairly constant for each species, though it varies, of course, from one species to another. In contrast, eggs of most reptiles, like their ectothermic parents, are dependent on environmental heat; and since this may vary widely from year to year and from place to place, the length of time it takes reptile eggs to hatch, even within a single species, is extremely variable. The hotter the summer, the sooner the young reptiles appear. In France, young of the Smooth Snake *(Coronella austriaca)* usually hatch in late August or early September, but if the summer is cold they may not appear until October. In the northern United States and Canada, young Painted Turtles *(Chrysemys picta)* may winter in the eggs and hatch the following spring.

Even in live-bearing reptiles, the gestation period is variable. Pregnant females spend much time basking in the sun, and the young are born earlier in a warm summer. In England, young of the Slowworm (a lizard, *Anguis fragilis*) and of one species of viper *(Vipera berus)* may remain in the body of the mother throughout the winter following a cool summer. Eggs of the Tuatara *(Sphenodon punctatus)* are said to take thirteen months to develop. Table 7–3 gives some indication of the range of incubation periods that have been reported for several reptiles.

TABLE 7–3
Length of Incubation Period
for Various Species of Reptile

Species	Incubation period (days)
Caretta c. caretta	31–65
Chelodina longicollis	90–120
Chelydra serpentina	81–90
Eumeces fasciatus	28–49
Lacerta a. agilis	49–84
Sceloporus undulatus	70–84
Alsophis cantherigerus	89–97
Coluber constrictor mormon	51–64
Elaphe guttata emoryi	72–77
Lampropeltis getulus californiae	66–83
Naja naja	69–76
Pituophis m. melanoleucus	59–101
Pituophis m. annectens	64–77

A young reptile ready to hatch needs some means of cutting its way from the egg. Embryonic turtles and crocodilians have a horny projection called the caruncle on the tip of the snout, with which they pierce or slash the shell. Lizards and snakes bear an egg tooth on the front of the premaxillary bone in the upper jaw. The egg tooth is shed within a day or two after birth, but a caruncle may persist for weeks. In the young of ovoviviparous species, which are usually born encased in a thin membrane that can easily be ruptured by the snout and is sometimes broken before the young leave the body of the mother, the egg tooth has degenerated (Figure 7–4).

Young reptiles resemble their parents in both structure and behavior. Baby vipers and pit vipers have well-developed, functional fangs; baby cobras can spread their hoods. The skin is first shed shortly after birth. Similar as they are to the adults in morphology, the young often differ strikingly from their parents in color. They are frequently much more brightly marked, and sometimes have entirely different color patterns. Adult Black

FIGURE 7–4
Hatchling Mountain Patch-nosed Snakes, *Salvadora grahamiae*, as they emerge from the eggs.

Racers *(Coluber c. constrictor)* are a smooth, velvety black above, but the young are vividly marked with brown saddles and spots on a tan background.

GROWTH, MATURITY, AND LONGEVITY

Young reptiles grow rapidly at first, then more slowly. Some species apparently reach a definite size and stop growing; others grow, although very slowly, throughout their lives. Like human children, reptiles seem to grow in spurts, with alternating periods of rapid and slow growth. During hibernation, growth all but stops.

Reptiles are ready to breed before they are full-grown. For some, at least, sexual maturity seems to be a matter of reaching a certain size more than a certain age. Males of the Red-eared Turtle *(Chrysemys scripta elegans)* are sexually mature when the plastron is 9 or 10 centimeters long, but it may take individuals from two to five years to reach this size. Females mature at a plastral length of 15 to 19.5 centimeters which is attained in three to eight years. American Alligators *(Alligator mississippiensis)* do not breed until they are nearly 2 meters long and six or seven years old. Many lizards and snakes take about three years to reach sexual maturity. There is some indication that tropical forms, whose growth is not interrupted by periods of hibernation, mature more quickly. Among the Javanese snakes, the Blunt-head *(Pareas carinatus)* is usually ready to breed in 11 months, the Red-necked Keelback *(Rhabdophis subminiata)* in 13 months, and the Greater Rat Snake *(Ptyas mucosa)* in 20 months.

As with amphibians, our knowledge of the maximum age to which reptiles live is based largely on specimens kept in captivity. Table 7–4 shows the longevity records for a number of species. Apparently, turtles tend to live longer than snakes, and snakes longer than lizards. Again there seems to be a negative correlation between the degree of activity and the duration of life.

Our knowledge of the reproductive habits of many species of reptile either is a complete blank or consists solely of a record of a single clutch of eggs

TABLE 7–4
Longevity of Some Species of Reptile

Order, family, and species	Maximum age (years)
Testudines	
Chelydridae *(Macroclemys temmincki)*	59
Testudinidae *(Geochelone* cf. *gigantea)*	152
Cheloniidae *(Caretta caretta)*	33
Trionychidae *(Trionyx triunguis)*	25
Pelomedusidae *(Pelusios castaneus)*	41+
Chelidae *(Chelodina longicollis)*	37+
Rhynchocephalia	
Sphenodontidae *(Sphenodon punctatus)*	77
Squamata	
Sauria	
Gekkonidae *(Tarentola mauritanica)*	7
Iguanidae *(Conolophus subcristatus)*	15
Agamidae *(Physignathus lesueuri)*	6
Scincidae *(Egernia cunninghami)*	20
Cordylidae *(Cordylus giganteus)*	5
Teiidae *(Tupinambis teguixin)*	13
Anguinidae *(Anguis fragilis)*	54
Helodermatidae *(Heloderma suspectum)*	20
Varanidae *(Varanus varius)*	7
Serpentes	
Boidae *(Eunectes murinus)*	29
(Boa constrictor)	25
Colubridae *(Elaphe situla)*	23
(Drymarchon corais couperi)	25
Elapidae *(Naja melanoleuca)*	29
Viperidae *(Vipera ammodytes)*	22
(Agkistrodon piscivorus)	21
(Crotalus atrox)	22
Crocodylia	
Crocodylidae *(Alligator mississippiensis)*	56

discovered by chance. Yet it is clear that, in spite of the variations in detail, the overall pattern of reptilian reproduction is much less diversified than that of the amphibians: fertilization is always internal; if eggs are laid, they are always laid on land; development is always direct.

The diversity shown by the modern reptiles is just that which must also have been present in the ancestral forms to have allowed the evolution of the two great groups that arose from them. The birds specialized in the development of the hard-shelled egg, well supplied with albumen; the higher mammals specialized in viviparity.

READINGS AND REFERENCES

Andrews, R., and A. S. Rand. "Reproductive effort in anoline lizards." *Ecology,* vol. 55, 1974.

Bauchot, R. "La placentation chez les reptiles." *L'année Biologique,* vol. 4, 1965.

Bellairs, A. *The Life of Reptiles,* vol. II. London: Weidenfeld and Nicolson, 1969.

Cole, C. J. "Evolution of parthenogenetic species of reptiles." *In* Reinboth, R. (ed.), *Intersexuality in the Animal Kingdom.* New York: Springer-Verlag, 1975.

Cuellar, O. "On the origin of parthenogensis in vertebrates: the cytogenetic factors." *American Naturalist,* vol. 108, 1974.

———, and A. G. Kluge. "Natural parthenogenesis in the gekkonid lizard *Lepidodactylus lugubris." Journal of Genetics,* vol. 61, 1972.

Dufaure, J. P., and J. Hubert. "Table de développement du lézard vivipare: *Lacerta (Zootoca) vivipara* Jacquin. *Archives d'Anatomie microscopique et de Morphologie, expérimentale,* vol. 50, 1961.

Fitch, H. S. "Reproductive cycles in lizards and snakes." *University of Kansas Museum of Natural History, Miscellaneous Publication,* no. 52, 1970.

Gibbons, J. W. "Sex ratios in turtles." *Researches on Population Ecology,* vol. 12, 1970.

Licht, P. "Environmental physiology of reptilian breeding cycles: role of temperature." *General and Comparative Endocrinology,* Suppl. 3, 1972.

Saint-Girons, H. "Déplacements et survie des spermatozoïdes chez les reptiles." *Colloques de l'Institut National de la Santé et de la Recherche Médicale,* vol. 26, 1973.

Tinkle, D. W., H. M. Wilbur, and S. G. Tilley. "Evolutionary strategies in lizard reproduction." *Evolution,* vol. 24, 1970.

———. "The role of environment in the evolution of life history differences within and between lizard species." *University of Arkansas Museum, Occasional Paper,* no. 4, 1972.

CHAPTER EIGHT

HOMEOSTASIS

LIKE OTHER ANIMALS, amphibians and reptiles must maintain a certain degree of constancy in their internal environment. Extracellular fluids must be kept in osmotic equilibrium with the cells; pH levels must be maintained within narrow limits; body temperature, at least during periods of activity, must be neither too high nor too low for the proper functioning of the various enzyme systems. The maintenance of a constant internal environment is known as *homeostasis*. Homeostasis is maintained through hormonal regulation of the internal environment and through physiological and behavioral regulation of the exchange of energy and materials with the external environment. The study of internal homeostatic regulation is a subject for the experimental physiologist; exchange with the environment, on the other hand, since it concerns the whole animal in its habitat, properly falls within the realm of herpetology.

In this chapter, we shall be concerned with three of the principal commodities of exchange: heat, respiratory gases, and water. As might be expected, the reptiles, with their dry, scaly integument, have different problems in gaining or losing water and chemicals in solution in water (i.e., gas, nitrogenous wastes, salts, etc.) than have the amphibians.

HEAT

Heat is produced in an animal's body as a by-product of metabolic activity. The respiratory and circulatory systems of amphibians and reptiles are less efficient than those of birds and mammals in bringing an abundant supply of oxygen to the tissues for metabolism. Consequently, herps have low metabolic rates, and cannot generate the heat required to maintain body temperatures high enough for optimal biochemical activity. They must absorb additional heat from an outside source: the sun's rays, the surrounding medium (soil, air, or water), or the surface on which they rest. Nocturnal herps draw upon the solar heat absorbed by the soil substrate during the day. In short, herps are ectothermic: their heat source is external.

Mechanisms of Temperature Control

The fact that herps absorb their heat from the environment does not imply that their body temperature must be the same as that of the environment. Many herps, particularly squamates and some terrestrial anurans, show a remarkable ability to control the level of their body temperature and to maintain it either above or below the ambient temperature during periods of activity. This thermoregulation is achieved through behavioral and physiological mechanisms. An animal can maintain a preferred body temperature behaviorally by shuttling between a hot, sunlit area and a cooler, shaded one, selecting the appropriate time of exposure in each. Physiologically, the animal may lower its temperature by evaporative cooling, or raise its body-core temperature by restricting peripheral circulation. Reptiles and amphibians are also able to adjust physiologically the extremes of temperature at which they can operate normally. As an animal's body temperature approaches tolerated extremes, the animal's primary reaction is to escape to an area of acceptable temperature; this refuge is almost always beneath the ground, where temperature is fairly constant from day to day and usually close to the mean daily temperature.

In spite of the ability to thermoregulate, not all herps do so. If the cost of thermoregulation in energy and time is too great, it is to the animal's selective advantage to allow its body temperature to fluctuate with that of the environment (poikilothermy). Most aquatic herps are poikilothermic, because water is such a strong heat sink that it would rapidly dissipate the animal's heat. Since aquatic herps rarely have effective insulation, they would lose more energy than they could gain by maintaining an elevated body temperature. Burrowing herps are in essentially the same situation as aquatic ones, and also avoid thermoregulation. Animals living in forests or other environments dominated by shade would also expend more effort finding and following patches of sunlight than they could absorb by doing so, so it is to their advantage not to thermoregulate.

Critical and Eccritic Temperatures

Each ectothermic animal has a thermal tolerance range in which its metabolic and other physiological processes can operate normally. At body temperatures above a given point (the critical thermal maximum, CTMax) or below a given point (the critical thermal minimum, CTMin), the animal is incapacitated and unable to perform coordinated movements. These are ecologically lethal points, because although the temperature itself may not kill, it renders the animal incapable of performing its necessary life-preserving activities.

CTMax and CTMin are not fixed temperatures, but are influenced by a number of factors. As environmental temperatures change throughout the year, an animal's critical thermal tolerances change with them (acclimatize). As the weather becomes progressively warmer, CTMax and CTMin are elevated; when the weather cools, they are lowered. Variations of CTMax and CTMin also appear to be associated with the animal's circadian (daily) rhythm. Changes in body form, such as the loss of a tail, will alter the tolerance limits. A rapid rise in body temperature will depress CTMax; a rapid drop will elevate CTMin. Other factors, such as reproductive condition, are also known to alter the tolerance limits in reptiles and amphibians (Figure 8–1).

Since CTMax and CTMin can and do vary for an individual, different tolerance limits for different populations of a single species should not be unexpected. Populations from higher latitudes or altitudes have lower tolerance limits than those from lower latitudes or altitudes. Interspecific comparison of tolerance limits shows a general trend toward narrower tolerance ranges and higher overall limits in tropical species than in temperate ones.

Between CTMax and CTMin, a herp has a preferred (eccritic) body temperature or temperature range in which to perform its activities. Active thermoregulators strive to reach and maintain this eccritic temperature throughout their active period, and consistently do so. The Desert Spiny Lizard *(Sceloporus magister)* feeds well and seems to thrive at a body temperature of 30°C, but defecates frequently only at about 37–38°C. The higher temperature seems to be necessary for peristalsis, and the lizard probably would not survive if it could not operate occasionally at this temperature. At least in reptiles, eccritic temperatures are surprisingly close to CTMax, which in most reptiles is only about 5–6°C higher (Figure 8–2).

Temperature Adaptations in Amphibians

Amphibians tend to function at lower temperatures than reptiles. The critical tolerance limits of salamanders extend from −2° to about 27°C, those of anurans from 3° to 41°C. The highest CTMax among amphibians—nearly

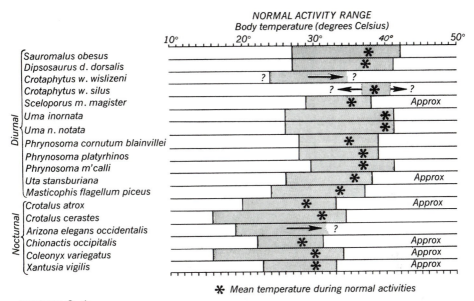

NORMAL ACTIVITY RANGE
Body temperature (degrees Celsius)

FIGURE 8–1

Normal activity ranges of 18 snakes and lizards of the southwestern United States. Diurnal and nocturnal forms both arranged roughly in order of size, with larger species at the top. Bars indicate the normal activity range, that is, the minimum and maximum temperatures voluntarily tolerated by individuals under conditions in the field, in cages set up in the field, or under laboratory conditions. Asterisks (*) indicate mean temperatures recorded when individuals were engaged in normal activities, and hence represent an approximation of the ecological optimum. Data for some species are not extensive enough to be considered conclusive: consequently, portions of this chart are provisional, and revisions may be necessary after controlled experiments in the laboratory have been conducted for the forms in question. [After Cowles and Bogert, *Bull. Am. Mus. Natl. Hist.*, 1944.]

42°C—belongs to *Bufo marinus,* a tropical toad of the open savanna. The CTMax of most other frogs, including tropical ones, seldom exceeds 37°C.

No salamander is known truly to thermoregulate. Most seem to accept the heat available and operate at the ambient environmental temperature, making no attempt to hold their body temperature above that level. This is true, in part, because many salamanders are aquatic or semiaquatic. The terrestrial ones seek shade and, owing to evaporation from their skin, may have body temperatures less than ambient. A few species appear to seek a specific temperature regime, but this is rare. Salamanders simply do not have an eccritic temperature, and apparently neither do many frogs. A few toads and tree frogs have an optimum or preferred temperature and a narrow range of activity temperatures; however, the majority have a wide tolerance range with no clearly preferred temperature. Both salamanders and frogs do use evaporative cooling under stress, and, in that sense, do thermoregulate. A few frogs are known to bask in the sun, and may raise

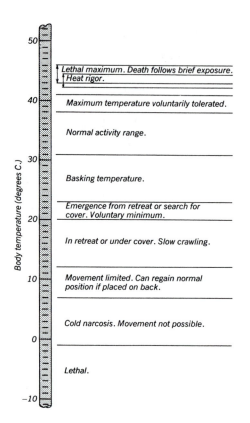

50 —

Lethal maximum. Death follows brief exposure.
Heat rigor.

40 —

Maximum temperature voluntarily tolerated.

Normal activity range.

30 —

Basking temperature.

Emergence from retreat or search for
cover. Voluntary minimum.

20 —

In retreat or under cover. Slow crawling.

10 —

Movement limited. Can regain normal
position if placed on back.

Cold narcosis. Movement not possible.

0 —

Lethal.

−10 —

Body temperature (degrees C.)

FIGURE 8–2
Chart showing the approximate
temperatures for various activities of
Spiny Lizards (*Sceloporus*). The effects
of exposure to heat levels near the
extremes depend on the duration of
exposure. Even temperatures near the
upper limit for the lizards' normal
activity become lethal if exposure is
prolonged. [From "How Reptiles
Regulate Their Body Temperature," by
Charles M. Bogert. Copyright © 1959
by Scientific American, Inc. All rights
reserved.]

their temperatures as much as 10°C above that of the surrounding air or water.

Although most amphibians have no preferred temperature, this does not mean that temperature has no influence on their life. Most temperate-zone frogs have a minimum temperature below which they will not breed. *Scaphiopus bombifrons* is a nonseasonal breeder, and any heavy rainstorm will trigger breeding activity provided the water temperature exceeds 13°C. Even seasonal breeders have a minimum breeding temperature. Embryonic development is, of course, strongly influenced by water temperature; if it is too low or too high, the embryos are killed. Within the tolerance limits, the rate of development is directly proportional to temperature, with the fastest development at the CTMax.

Temperature Adaptations in Reptiles

Most terrestrial and diurnal reptiles actively thermoregulate. In the strictest sense, thermoregulation is the selection of a narrow temperature range in which most life-support activities (feeding, digestion, reproduction, etc.) are performed, and the maintenance of the body temperature within this range in spite of fluctuations in environmental temperature. Under this definition, the squamates are the primary reptilian thermoregulators. They emerge from rest with a body temperature less than that preferred, and bask in a sunlit area or on a heated substrate to elevate their body temperature into the preferred range. Only after the desired temperature has been reached do they begin their normal activity pattern, with constant behavioral and physiological adjustments to maintain the eccritic temperature (see Figure 8-2). If the environment becomes too hot or too cold for them to remain in the preferred temperature range, they will retreat to a relatively stable thermal refuge—usually below optimum temperature—until they can again emerge and thermoregulate.

Aquatic reptiles, such as turtles and crocodilians, are poikilothermic. They do, however, bask by crawling out of the water or floating at the surface, attaining higher body temperatures that may aid digestion. Only the Leatherback *(Dermochelys coriacea),* a sea turtle, appears able to maintain a body temperature above that of the surrounding water. An active swimmer, the Leatherback generates heat through muscle contraction; its oil-impregnated skin acts as an insulator to slow heat loss.

For true thermoregulators, the preferred temperature range tends to run from about 28° to 40°C, with the eccritic temperature near the upper limit of the range. The thermal tolerance range for all reptiles extends from approximately −3° to 50°C, but the actual temperature for active individuals is from about 4° to 46°C (Figure 8–3). The Tuatara is the most cold-adapted species, with much of its activity centered around 10–12°C.

Reproductive activity probably occurs principally in the optimum temperature range. Although this hypothesis remains unconfirmed, male sex

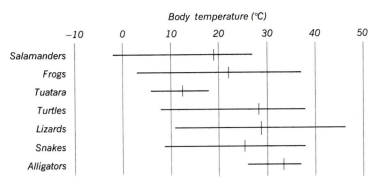

FIGURE 8–3
Ranges *(horizontal lines)* and means *(vertical lines)* of body temperature in active individuals of the major groups of amphibians and reptiles. [After Brattstrom, *Ecology,* 1963; *Amer. Midl. Nat.,* 1965.]

cells seems to be more sensitive to the effects of heat than are other body tissues. Geographic variation in the spermatogenic activity of the Garter Snake *(Thamnophis)* can be correlated with seasonal fluctuations in the number of hours a day the snake can maintain its body temperature within the normal activity range. It has been suggested that increase in temperature causes sterility in male Desert Night Lizards *(Xantusia vigilis)*. Strangely enough, no similar correlation has been found in females.

The rate of embryonic development is also affected by temperature: low temperatures slow development down, high temperatures speed it up. However, the critical thermal limits are largely unknown, as well as whether the rate of development in reptiles is directly proportional to temperature, as it is in amphibians.

GAS EXCHANGE

Probably no group of vertebrates exemplifies the evolution of gas transport systems better than the herp.

Oxygen is more readily available to air-breathing animals than to water-breathing ones; for example, the oxygen provided by water at 20°C is twelve times more rarefied than the atmosphere at the top of Mount Everest. Water-breathing amphibians such as the sirens and salamanders like *Necturus* have to pump a great deal of water over the gills to obtain the necessary oxygen. Terrestrial animals need to pump much less air into their lungs to obtain an equivalent amount. On the other hand, because carbon dioxide is much more soluble in water than is oxygen, water at 16°C is as effective as air in the dispersal of carbon dioxide from the respiratory surfaces. Water flowing over the gills at a sufficient rate to provide the oxygen needed by an aquatic animal also provides for the rapid removal of oxygen from the system. In air-breathing animals, the air flow over the lung surfaces that provides the requisite oxygen removes much less of the carbon dioxide. As a result, the problems of gaining oxygen and dispersing carbon dioxide are quite different in aquatic, semiaquatic, and terrestrial animals.

Respiratory movements may be stimulated either by oxygen depletion or by carbon dioxide stress. The respiratory center in the brain is sensitive to carbon dioxide and initiates respiratory movement if its concentration in the body fluids exceeds a certain level. In frogs, the carotid "gland," a spongy mass located at the base of the internal carotid artery, is sensitive to oxygen concentration in the blood and, if the concentration is lowered, will stimulate respiratory movements. It has been shown in *Rana* that carbon dioxide stimulates respiratory movements even after the carotid gland has been obliterated, but that oxygen deficiency acts as a stimulant only if the carotid gland is intact.

Gas Exchange in Amphibians

Amphibians possess a variety of respiratory surfaces: gills, lungs, buc-copharyngeal mucosa (mouth and throat membrane) and skin. Gills are the principal respiratory organs in larval amphibians, in paedomorphic salamanders such as *Necturus* and *Ambystoma,* and in *Siren.* Lungs assume the primary respiratory role in all adult amphibians except the lungless plethodontid salamanders. The latter rely on the skin, which also serves a small but important role in other amphibians. The buccopharyngeal mucosa is also used by most amphibians, but in a very limited way.

In air-breathing amphibians, oxygen absorption by the lungs increases linearly with temperature, whereas absorption through the skin remains essentially constant (Figure 8–4); the same pattern is followed for carbon

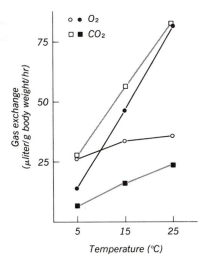

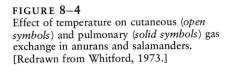

FIGURE 8–4
Effect of temperature on cutaneous (*open symbols*) and pulmonary (*solid symbols*) gas exchange in anurans and salamanders. [Redrawn from Whitford, 1973.]

dioxide removal. The skin accounts for 70 percent of the oxygen intake at 5°C, 50 percent at 10°C, but only 30 percent at 25°C. Lungs are the main respiratory mechanism for active animals; during periods of torpor, such as hibernation, the skin assumes this role.

In the lungless salamanders, the skin is the most important respiratory surface. At no temperature does the buccopharyngeal mucosa account for more than 20 percent of the oxygen intake, and usually much less (Figure 8–5). The skin of the lungless salamanders has become nearly as efficient a respiratory surface as the lungs in other amphibians. This is somewhat surprising, for the increased pulmonary oxygen intake of lunged forms is directly related to increased breathing rate and tidal volume, both of which increase the amount of air flowing over the respiratory surface of the lungs.

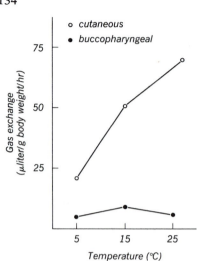

FIGURE 8–5
Comparison of oxygen exchange from buccopharyngeal and cutaneous surfaces in plethodontid salamanders. [Redrawn from Whitford, 1973.]

The lungless forms certainly cannot greatly increase the flow of air over the skin, so they must increase the circulatory flow through the cutaneous capillaries (Figure 8–6).

Respiratory movements in the frog form a cycle of buccal oscillation and lung ventilation. During buccal oscillation the nostrils are open, the glottis is closed, and the floor of the mouth is alternately raised and lowered; this brings fresh air into the buccal cavity. At the start of lung ventilation, the floor of the mouth is lowered and the glottis is opened. Since the air in the lungs is above atmospheric pressure, a stream of air flows from the lungs through the mouth and out the nostrils. The greater part of the buccal cavity

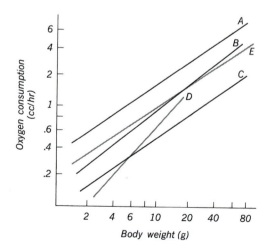

FIGURE 8–6
Effect of temperature and body weight on oxygen consumption in amphibians: (A, B, and C) temperate-zone anurans and salamanders at 5°, 15°, and 25°C, respectively; (D) tropical anurans at 25°C; (E) plethodontid salamanders at 15°C. [Redrawn from Whitford, 1973.]

lies below the opening of the glottis, and there is little mixing of the air in this part of the chamber with the air being expired. After expiration the nostrils are closed, the floor of the mouth is raised, and the buccal air is pumped into the lungs.

Some, but not all, air-breathing amphibians continue buccal movements while submerged. Apparently some oxygen is absorbed from the water. The time that such an amphibian can survive under water probably depends on how well the buccopharyngeal membranes are adapted to picking up oxygen. In water at 15°C, *Rana pipiens* can carry on buccal respiratory movements for as long as seven hours, whereas the more terrestrial *Bufo terrestris* expires after three hours.

Since the relative proportion of gases in the air remains constant at different heights, one of the most important effects of the decreased air pressure at high altitudes is a decrease in the total amount of oxygen available. Oxygen is transported from the lungs to the tissues by hemoglobin in red blood cells; hemoglobin is measured in parts per hundred by weight of the blood. To say, for example, that a frog has a hemoglobin value of 10 means that the weight of the hemoglobin makes up 10 percent of the total weight of the blood. Frogs from the lowlands have lower hemoglobin values than those from mountains. Hemoglobin values ranging from 7.5 to 13.5 have been reported for frogs and toads from the European lowlands, but specimens from the Alps of the toad *Bombina variegata* have values ranging from 14.25 to 15.25. In Guatemala, *Bufo marinus,* which occurs from sea level to elevations of about 1500 meters, has a mean hemoglobin value of 8.66, whereas *Bufo bocourti,* which ranges from about 1700 to 3400 meters, has a mean value of 10.77. If the hemoglobins of these species have similar oxygen-carrying capacities (which is not yet known), the increase in hemoglobin could balance the decrease in oxygen at the higher altitudes.

Although the primitive amphibian lung is fairly adequate for the purpose of maintaining oxygen tension, it is inadequate for keeping the carbon dioxide tension low. Instead, the skin is used to cope with rising levels of carbon dioxide. *Taricha granulosa* disperses 80 percent of the carbon dioxide through the skin at 15°C, and the percentage is only slightly lower at higher temperatures. In the lungless *Desmognathus quadramaculatus,* the skin accounts for 75 percent of carbon dioxide disposal at 15°C and as much as 95 percent at 25°C. *Rana clamitans* loses about 80 percent of the carbon dioxide through the skin at all temperatures.

Gas Exchange in Reptiles

The reptiles were enabled to colonize the land not only by the development of the amniote egg, but also by the development of a relatively water-impervious integument. Respiratory gas exchange through the skin was no longer possible, and even though the improved reptilian lung allowed sufficient oxygen intake, it did not allow for sufficient elimination of carbon dioxide. Physiological adaptations were thus required to make it possible for the animal to tolerate higher carbon dioxide concentrations in its body fluids. Of the carbon dioxide that enters the bloodstream, most combines with water to form carbonic acid (H_2CO_3), which then dissociates to form bicarbonate ions (HCO_3^-). Adaptation of the reptiles to a fully terrestrial existence required adjustment to higher levels of bicarbonate ions in the blood.

In the truly terrestrial reptiles, three types of respiratory movement occur: lung ventilation, buccopharyngeal pumping, and panting. Air is pumped into the amphibian lung by buccal force, but is drawn into the reptile lung by suction. Most reptiles take in air by muscular expansion of the rib cage and body wall. Since this lowers the pressure in the lungs and abdominal cavity below atmospheric pressure, air moves into the lungs. There is then a respiratory pause, with the glottis closed and the lungs inflated. Air is forced from the lungs by the elastic recoil of the lung tissue and by contractions of the body-wall musculature. The turtle, with its rigid shell, to which the ribs are fused, changes the size of the body cavity, and hence the intracavity pressure, by alternate contractions of antagonistic muscles in the flank and shoulder region.

It was thought for a long time that buccopharyngeal pumping in most reptiles functioned to ventilate the lungs, but it is now known that, at least in the lizards and turtles, pharyngeal pumping takes place only during respiratory pauses and functions primarily as an olfactory mechanism.

Gas exchange in reptiles, as in amphibians, is affected by temperature. Oxygen absorption increases linearly with temperature. This is to be expected in ectothermal animals, since the metabolic rate also increases with temperature. The rate of oxygen intake in most reptiles would be shifted even farther to the right in Figure 8–6 than that of tropical anurans, since they operate at higher body temperatures.

Many of the aquatic turtles are known to be able to stay submerged for long periods without breathing. In turtles of the genera *Trionyx* and *Sternotherus*, enough oxygen is gained from pumping water in and out of the pharynx to maintain metabolic processes for long periods if the turtle is not physically active. Thus the animal may be able to stay submerged indefinitely. Not all diving turtles have this ability, though: *Chrysemys scripta* and related species seem to be incapable of deriving appreciable oxygen from the water.

Among reptiles, turtles seem especially adapted to tolerate anoxia. Table 8–1 shows the mean time in minutes that members of different families of reptiles are known to survive in an atmosphere of pure nitrogen.

TABLE 8–1
Tolerance of Anoxia
in Various Families
of Reptiles

Order and family	Number of species tested	Mean time (minutes)
Testudines		
Chelydridae	1	1050
Pelomedusidae	2	980
Testudinidae	14	945
Kinosternidae	5	876
Trionychidae	1	546
Chelidae	2	465
Cheloniidae	2	120
Squamata		
Sauria		
Iguanidae	6	57
Gekkonidae	1	31
Anguinidae	1	29
Scincidae	4	25
Teiidae	1	22
Serpentes		
Viperidae	3	95
Boidae	3	59
Elapidae	1	33
Colubridae	22	42
Crocodylia		
Crocodylidae	1	33

SOURCE: Data from Belkin, *Science*, 1963.

The ability of aquatic turtles to stay submerged relies on physiological reactions in addition to the ability to obtain oxygen from the water. These reactions include inhibition of heartbeat, anaerobic glycolysis (breakdown of sugars without free oxygen), increase in plasma bicarbonate, and possibly the use of hydrogen acceptors other than pyruvic acid in glycolysis. Turtles other than the sea turtles have approximately two and a half times as much plasma bicarbonate as other reptiles; sea turtles have half again as much.

During typical anaerobic respiration in animals, pyruvic acid resulting from the breakdown of glucose accepts hydrogen from the hydrogen acceptor nicotinamide-adenine-dinucleotide (NAD), and is thereby converted into lactic acid. The possibility that turtles are able to use hydrogen acceptors other than pyruvic acid in glycolysis should not be discounted, although it remains to be tested. Carp, when exposed to long-term anoxia, form fat rather than lactic acid, and it seems probable that turtles are also able to form substances less disturbing to the tissues than lactic acid.

WATER BALANCE

Because animals that live in fresh water have body fluids that are hyperosmotic to the surrounding medium, water tends to move into their bodies by osmosis. In contrast, animals that live in the sea are usually hypo-osmotic to seawater, and thus tend to lose water to the environment. Terrestrial forms lose water to the surrounding air. In addition to the water needed to keep the body in osmotic balance, there must also be water available to excrete nitrogenous wastes. These wastes are predominantly ammonia, urea, and uric acid. Ammonia is highly water-soluble and very toxic, urea is also highly water-soluble but less toxic, uric acid is relatively insoluble and non-toxic.

Osmoregulation in Amphibians

The amphibians, with their diverse habitats and life histories, have varying problems of water balance, and have met these problems in diverse ways. In general, osmoregulation is achieved by a balance between water intake through the skin and water loss through the kidneys.

Most amphibian larvae develop in fresh water. They produce copious amounts of dilute urine to eliminate the excess water that moves into their bodies by osmosis. Their nitrogenous wastes are mostly in the form of ammonia, which is rapidly flushed from the body in the urine. During metamorphosis from the aquatic larva to the terrestrial adult, there is a sudden, dramatic shift in the form in which nitrogen is excreted. Urea becomes the chief excretory product. The Bullfrog, *Rana catesbeiana,* excretes 84 percent of its nitrogenous wastes in the form of urea as an adult, but only 10 percent as a tadpole. Amphibians that remain aquatic as adults usually retain the larval excretory pattern. The aquatic African Clawed Frog, *Xenopus laevis,* excretes about 75 percent ammonia and 25 percent urea as a tadpole. During metamorphosis the excretion is 46 percent ammonia and 54 percent urea, but the adult reverts to the tadpole pattern. Since urea is less toxic than ammonia, it does not need to be removed so rapidly from the body, and the terrestrial amphibians are able to reduce their rate of urine production during periods of water shortage.

Amphibians are caught in an adaptive dilemma. They rely on cutaneous respiration for a significant portion of their gas exchange, and thus must maintain a moist and permeable skin. Yet to be successful terrestrial animals, they must reduce and regulate water loss. They have struck a balance between the two needs. Their skin is not as water-permeable as is often assumed. Many frogs, particularly the more terrestrial species, have a definite layer of ground substance (layer G) between the stratum compactum and stratum spongiosum of the dermis (Figure 8–7). This noncellular layer, which consists of calcium and acid mucopolysaccharides (polysaccharides with attached amino groups), presumably plays an important part in reducing water loss.

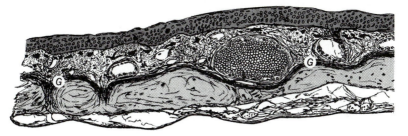

FIGURE 8–7
Layer G as it might appear in *Rana arvalis*. [Redrawn from Elkan, *J. Zool.*, 1968.]

Although exactly how the G layer controls water transport is unknown, its differential development in different species of frogs indicates that it does serve that function. Layer G is essentially absent in ascaphids, pipids, and aquatic leptodactylids, but is well developed in most hylids, in terrestrial leptodactylids, and in all pelobatids, bufonids, and ranids that have been examined—a total of 42 species in the last three families. All salamanders examined so far seem to lack layer G.

Under extremely dry conditions, several anuran species have evolved special antidesiccation mechanisms. Two Argentinian toads, *Ceratophrys ornatus* and *Lepidobatrachus ilanensis,* and the Australian Burrowing Frogs, *Cyclorana,* become dormant, and their shed skin forms a cocoon. The cocoon reduces water loss to nearly one tenth that of uncovered animals. An African *Chiromantis* and an Argentinian *Phyllomedusa* have recently been discovered to secrete a lipid compound, which they spread over their entire body with an elaborate foot-wiping behavior; the lipid barrier is as effective as reptilian skin in reducing evaporative water loss.

Amphibians vary considerably in their ability to tolerate dehydration, but all tend to have a higher tolerance than most other vertebrates. The tolerance of an amphibian shows a clear relationship to its preferred habitat. Table 8–2 shows that frogs occupying arid habitats are more tolerant to dehydration than species from aquatic or moist habitats. Furthermore, many arid species seem to be able to slow their rate of evaporative water loss, and thus increase the time that they can tolerate dehydrating condi-

TABLE 8–2

Vital Limits of Water Loss for Ten Species
of Frogs from Various Types of Habitat

Species	Habitat	Tolerable loss in body weight (%)
Scaphiopus hammondi	Terrestro-fossorial	48.8
Scaphiopus holbrooki	Terrestro-fossorial	48.1
Bufo boreas	Terrestrial	43.6
Bufo terrestris	Terrestrial	43.0
Hyla regilla	Terrestrial	39.0
Hyla cinerea	Terrestro-arboreal	37.3
Rana pipiens	Terrestro-semiaquatic	36.6
Rana aurora	Semiaquatic	34.0
Rana utricularia	Semiaquatic	32.4
Rana grylio	Aquatic	31.2

SOURCE: Thorson and Svihla, *Ecology,* 1943.

tions. One means of doing this is to store water in the urinary bladder and reabsorb it when stressed.

Amphibians gain water by absorbing it through their skin, by extracting free and metabolic water from their food, and by drinking. Most amphibians can rehydrate fairly rapidly when sitting or lying in water. Arid-land anurans are able to absorb water faster than their relatives in more mesic (moderate moisture) habitats; this adaptation allows them to gain water rapidly at the end of a drought and begin reproduction. Apparently in many anurans there are regional differences in skin permeability. The Spotted Toad, *Bufo punctatus,* absorbs nearly 70 percent of its water through a "rump patch," the skin surface most easily exposed to a small and transient water source. Some frogs, such as *Scaphiopus,* are able to draw water directly from moist soil. The larger the grain size of the soil, the easier it is for the frog to absorb the water. When the soil dries to a point where water cannot be removed from it, the frog adjusts its osmotic balance as if it were in salt water.

For the most part, amphibians do not live in brackish or salt water, but one species, *Rana cancrivora,* regularly inhabits brackish water, and others may be exposed to it on occasion. *Rana cancrivora* is able to maintain osmotic balance with the environment in the same manner as the elasmobranchs: that is, the animal retains enough urea in the body fluids and tissues to bring it into osmotic balance with the saline water in which it lives.

It is well known that several species of *Bufo* (e.g., *B. bufo, B. viridis,* and *B. calamita*) can live up to several months in water of higher salinity than typical fresh water. *B. viridis* has the ability to adjust to as much as 75 percent seawater. In this species this ability seems to be due to a temporary increase in urea, inorganic ions, and, to a lesser extent, amino acids in the plasma.

Osmoregulation in Reptiles

Reptiles convert their nitrogenous wastes into insoluble, nontoxic uric acid, which can be temporarily stored without the addition of water. This makes it possible for the embryo to excrete nitrogen into the allantoic sac within the confines of the eggshell. It also means that little water needs to be expended by the adult for the removal of nitrogenous wastes. Indeed, reptile urine is often a semisolid paste. Until a few decades ago, it was generally assumed that reptiles lost most of their water through kidneys and respiratory surfaces, and that the skin was relatively impermeable to water. Actually, two thirds of the water loss is through the skin. More recently, in experiments with scaleless snakes (mutant individuals of *Natrix sipedon* and *Pituophis melanoleucas*), scales proved to be no more effective in retarding water loss than the skin between the scales. Nonetheless, the rate of water loss through the skin in reptiles is much less than in amphibians. This impermeability, together with the ability to excrete a hyperosmotic urine, has enabled the reptiles to survive and diversify in extremely arid areas and in the oceans. In fact, lizards and snakes form a significant part of deserts' vertebrate faunas.

Reptiles obtain most of their water by drinking or from free and metabolic water in their food. Desert and marine species obviously do not drink, and must obtain all their water from their food. Although reptiles lose water readily through the skin, few are able to absorb it by the same pathway. Only *Caiman sclerops* and a few species of lizard have been reported to be able to absorb water through the skin, and then only if immersed in water or in completely saturated air. The skin of a few desert lizards *(Moloch, Uromastyx,* and *Cordylus)* has a network of very narrow open channels. When the animal is placed in a moist or wet area, water moves through these channels by capillary action to the mouth and is swallowed. Snakes will absorb no water from supersaturated soil unless their heads are in contact with the surface, permitting them to drink.

In reptiles, the rate of water loss is closely associated with the species' habitat. Terrestrial species lose water more slowly than aquatic species (see Table 8–3 for an example in turtles), desert species more slowly than forest species. This relationship holds for all reptiles—turtles, snakes, lizards, and amphisbaenians, The differences are significant: in legless reptiles, the rate of water loss in rain-forest species is as much as a hundred times that in desert species.

Most marine reptiles (turtles, crocodiles, snakes, and lizards) and a few desert lizards have specialized head glands—salt glands—which actively

TABLE 8–3
Water Loss in Several Species of Turtle

Species	Number of individuals	Habitat	Mean loss in body weight (%)
Terrapene c. carolina	28	Terrestrial	3.6
Clemmys guttata	30	Semiaquatic	8.9
Clemmys insculpta	7	Semiaquatic	11.1
Chelydra serpentina	4	Aquatic	15.6
Sternotherus odoratus	24	Aquatic	20.0

SOURCE: Ernst, *J. Herpetol.*, 1968.

secrete salt. The salt glands are an obvious adaptation to aid the kidney in maintaining a positive water balance where the environment (seawater) or the foods cause an excessive salt load. The glands secrete sodium and potassium salts in a nearly dry state.

READINGS AND REFERENCES

Bentley, P. J., and J. W. Shield. "Respiration of some urodele and anuran Amphibia. II: In air, role of the skin and lungs." *Comparative Biochemistry and Physiology*, vol. 46A, 1973.

Brattstrom, B. H. "Thermal acclimation in Australian amphibians." *Comparative Biochemistry and Physiology*, vol. 35, 1970.

Cloudsley-Thompson, J. L. *The Temperature and Water Relations of Reptiles*. Watford: Merrow, 1971.

Gans, C. "Strategy and sequence in the evolution of the external gas exchangers of ectothermal vertebrates." *Forma et Functio*, vol. 3, 1970.

———, and W. R. Dawson (eds.). *The Biology of the Reptilia*, vol. 5, *Physiology*. London: Academic Press, 1976.

Huey, R. B. "Behavioral thermoregulation in lizards: Importance of associated costs." *Science*, vol. 184, 1974.

Seymour, R. S., and A. K. Lee. "Physiological adaptations of anuran amphibians to aridity: Australian prospects." *Australian Zoologist*, vol. 18, 1974.

Shield, J. W., and P. J. Bentley. "Respiration of some urodele and anuran Amphibia. I: In water, role of the skin and gills." *Comparative Biochemistry and Physiology*, vol. 46A, 1973.

Weiser, W. (ed.). *Effects of Temperature on Ectothermic Organisms*. Berlin: Springer-Verlag, 1974.

Whitford, W. G. "The effects of temperature on respiration in the Amphibia." *American Zoologist*, vol. 13, 1973.

Whittow, G. C. (ed.). *Comparative Physiology of Thermoregulation,* vol. 1, *Invertebrates and Nonmammalian Vertebrates.* New York: Academic Press, 1970.

Wilson, K. J. "The relationships of maximum and resting oxygen consumption and heart rates to weight in reptiles of the order Squamata." *Copeia,* no. 3, 1974.

Zweifel, R. G. "Reproductive biology of anurans of the arid Southwest, with emphasis on adaptation of embryos to temperature." *Bulletin of the American Museum of Natural History,* vol. 140, 1968.

CHAPTER NINE

RELATION
TO BIOTIC
ENVIRONMENT

NOT ONLY DO THE amphibians and reptiles maintain homeostasis by a more or less constant exchange of such things as heat, water, and respiratory gases with the physical environment, but they constantly interact with biotic factors as well. These factors include food, competitors, predators, and parasites. It is the combination of physical factors and biotic factors that determines the nature of a habitat.

The environment not only provides the means of existence for the animal—oxygen, water, food, and shelter—but it also imposes restraints upon the animal. It sets the bounds within which the animal must live, and largely determines how many of each kind will survive. Every species has an inherent power of reproducing itself, a biotic potential, in numbers far in excess of the number that actually can or do survive. The existing population of a species at any one time reflects a balance between its biotic potential and the resistance of the environment. When they are in equilibrium, just enough young reach maturity each year to replace the adults that die, and the population remains constant. The restraining factors of the environment themselves fluctuate: one year may be too wet, another too dry; one year the food may be abundant, the next may bring famine; another year predators may be numerous, but later there may be few. As a result, populations of animals also fluctuate. If environmental resistance is reduced, the population grows; if environmental resistance increases, the population becomes smaller. Generally these fluctuations cancel each other, so that over the years the numbers of a given species remain fairly constant. Drastic or long-term changes in the environment, however, produce permanent changes in populations. European settlers, with their guns and their custom of draining swamps to make farmland, materially changed the environment of the American *Alligator mississippiensis* and permanently reduced its numbers.

The study of ecology is in part a study of the factors contributing to environmental resistance. These factors are many, complex, and interrelated, but they are susceptible to analysis. From such studies certain basic principles have emerged. In 1840 Justus von Liebig first clearly expressed what is now known as Liebig's Law of the Minimum. This law simply says that of all the factors necessary for life, the one that is present in minimum quantity is the one that offers the greatest restraint to a population. For example, if the food of a given species is very scarce in a region, that scarcity determines how many individuals of the species can survive in that region. If food is plentiful but homesites are few, it is the scarcity of available homesites that limits the size of the population.

Liebig's law has been modified and expanded through the years. Victor E. Shelford has pointed out that it is not always a minimal factor that limits the distribution of an animal: sometimes it is a maximal factor. Too high temperatures, too much water, too much light, too many predators or parasites restrain a population just as effectively as does too little of something. Thus Shelford expanded the Law of the Minimum into the Law of Tolerance, which says that a species has a range of tolerance for a given environmental factor, bounded on one side by a minimum and on the other by a maximum. An animal may have a wide range of tolerance for one factor and a narrow range for another. Those animals with the widest ranges for all factors are the ones most likely to be widely distributed. The period of reproduction and development is likely to be the critical one, for many environmental factors are more limiting on eggs, embryos, and larvae than they are on adults.

Since it is relatively simple to set up a controlled experiment in which only a single physical factor is varied, the precise effects of many physical factors can be determined more readily than can the effects of the less easily manipulated biotic factors. But no animal lives in a biotic vacuum. Herps are subject to restraints imposed by the living things around them—the predators that prey upon them, the parasites that live at their expense, the competitors that vie with them for the limited necessities of their existence. Many of these restraints, although superficially obvious, are difficult to study quantitatively.

POPULATION COMPONENTS AND INTERACTIONS

A *population* is a group of like animals living in the same area. Each population is an adaptable unit of life adjusting to the pressure of its environment; however, this adaptability is merely the by-product of the individual members' struggle for survival and for the perpetuation of their genetic lines. The population adjusts through alteration of the content and time span of its components.

Age Distribution

Three classes of individuals occur in a healthy population: juveniles, subadults, and adults. The adults are the sexually mature individuals capable of reproducing; although it is not common in wild populations, a few senile individuals may be present who have lost their ability to reproduce. Juveniles comprise all younger members of the population—from hatchlings to, but not including, the adult-sized yet reproductively immature, subadults. The actual age and number of individuals in each age class vary according to the longevity of the species and how the population structure is adapted to its specific habitat. Long-lived species, such as the Galápagos Tortoise *(Geochelone elephantopus)*, have more adults than juveniles. Because the adults are so long-lived, few juveniles and subadults are needed to replace the few dying adults. In contrast, some Side-blotched Lizard *(Uta stansburiana)* populations are predominantly juveniles and subadults in late summer and early fall and predominantly adults in late winter and early spring. The population is an annual one, with a complete turnover of members every year. Most herps have an age structure somewhere between these extremes. The age structure of a stable population will always be determined by the life span and age of sexual maturity of its members. Expanding populations will tend to have a preponderance of juveniles, declining populations a preponderance of adults.

Sex Ratio and Reproductive Potential

In most herp populations the numbers of males and females are equal or the females are more numerous. Although superficial appearances often show a preponderance of males, this tends to occur only because male behavior makes them more visible. Male frogs sit and call night after night at a pond; females arrive, lay their eggs, and depart in one evening. Since females are the producers of the next generation, there is an obvious advantage in having an abundance of females and just enough males to assure fertilization.

The reproductive potential of a female is the product of the size of the clutch (number of young produced at each reproductive period) and the number of reproductive periods during her life span. In amphibians and reptiles, reproductive potentials range from less than a dozen possible offspring in a small-clutch, short-lived species, such as *Uta stansburiana,* to more than a hundred thousand in large-clutch, long-lived species, such as *Bufo marinus*. Differences in reproductive potential do not indicate that one species is any more successful than any other, but simply reflect different adaptive pathways to the maintenance of a viable population structure.

Clutch size and body size are frequently associated, since the larger a female is, the more eggs her body cavity can hold. These characteristics are in turn associated with the hatchlings' probability of survival to adulthood. High egg and juvenile mortality will select individuals capable of producing more offspring, and thus larger individuals will be favored. Sea turtles and large frogs and lizards characteristically have large clutches. *Sphaerodactylus,* the Dwarf Geckos of the West Indies, and *Eleutherodactylus,* leptodactylid frogs with direct development, can be small and produce small clutches of one or a few eggs, because juvenile mortality is not so severe for them. The Dusky Salamanders, *Desmognathus,* of the Appalachian Mountains show this trend. The species range from fully aquatic to fully terrestrial, the latter being the smallest and having the least severe environment for juveniles. As would be expected, as body size increases so does clutch size.

Abundance and Density

Abundance is the number of individuals in a population; density is the number per unit area. Abundance conveys little biological information, for it presents no data on the animals' interaction with their environment, unless to say of a rare or endangered animal that there are few left and extinction is imminent. Density, on the other hand, reflects the carrying capacity or the available resources of the habitat; it is usually expressed as the number of individuals per acre or hectare. As would be expected, density is generally inversely related to adult size and tends to be highest in tropical species. Density also tends to be inversely related to species diversity: the greater the number of species sharing or partitioning the available resources—no matter how abundant the resources are—the fewer individuals of each species there can be. In most cases, densities of herps fall below

250 individuals per hectare; surprising densities have been reported, however, such as 500 adults of the gecko *Hemidactylus garnoti* on a 4,200 square-foot plot (more than 13,000 per hectare) on Tinian Island in the Pacific.

The size of a population is regulated by the population's birth rate (natality) and death rate (mortality). In a stable population natality and mortality are equal; in an expanding population natality exceeds mortality; and in a declining population mortality is greater. Survivorship curves show when death occurs relative to birth (Figure 9–1). In *Eumeces fasciatus* and *Uta*

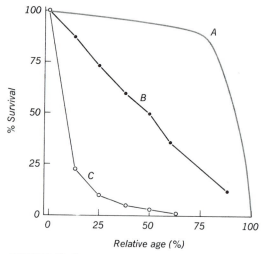

FIGURE 9–1
Survivorship curves for selected reptiles. (A)
Hypothetical curve for high survivorship of early- and
middle-life stages. (B) *Xantusia vigilis.* (C) *Eumeces
fasciatus.* *Uta stansburiana* has a similar curve. [After
Pianka, *Evolutionary Ecology,* Harper & Row, 1974.]

stansburiana, mortality is high early in the life of each generation; once adulthood is reached, most individuals survive and reproduce until eliminated by old age. This type of survivorship curve is the most common for amphibians and reptiles. The Desert Night Lizard, *Xantusia vigilis,* has a survivorship curve in which the mortality rate is the same throughout the entire life cycle of each generation. As yet, no herp population is known that has a low mortality in its growth and early reproductive stages with mortality increasing in older age classes.

COMMUNITY INTERACTION

Populations are regulated largely by the availability of food and by the intensity of predation and competition.

Food

Like all animals, amphibians and reptiles are dependent on other living things as sources of food. Most are carnivores, obtaining their energy by eating other animals. Herbivory is largely confined to lizards, turtles, and larval anurans. Among adult anurans, only the Marine Toad, *Bufo marinus,* is known to have populations dependent on plant material as a food source, and this appears to be a diet forced on them by insufficient animal food resources. Larval anurans, on the other hand, are generally herbivores. They feed largely on algae and bacteria suspended or adhering to objects in the water. Xenoanuran and scoptanuran tadpoles are predominantly filter feeders, whereas lemnanuran and acosmanuran tadpoles are mainly grazers or browsers, rasping off vegetable material with their keratinous mouth parts. In the latter group, a few forms, such as *Anotheca, Leptodactylus pentadactylus,* and *Scaphiopus bombifrons,* have evolved active carnivorous tadpoles that feed on other tadpoles. Herbivorous tadpoles occasionally act as scavengers, feeding on the carcasses of dead or dying animals.

Among reptiles, herbivory is not uncommon in the turtles. The terrestrial testudinids (tortoises—*Testudo, Geochelone*) and emydines (Box Turtles—*Terrapene*) eat mainly plant material, although animal material is not excluded from their diet. Many of the larger aquatic turtles—for example, *Chelonia, Chrysemys, Batagur, Podocnemis, Dermatemys,* and *Carettochelys*—feed nearly exclusively on plants; however, this herbivory is restricted to the adults, the young eating insects and other invertebrates. Herbivory in lizards is confined to larger species of the agamids, cordylids, iguanids, and scincids. Apparently herbivory is a necessity for these larger lizards: their food-catching ability is not sufficient to support their total metabolic needs, so they have shifted to a more readily available food resource.

Snakes are exclusively carnivores, and many have evolved restrictive diets and specialized feeding mechanisms. The egg-eating snakes of Africa (*Dasypeltis*), slug eaters (*Pareas, Dipsas),* and toad eaters (*Xenodon, Heterodon)* are largely dependent on one food resource. Some of the sea snakes have perhaps the most restrictive diets, limiting their diet to one or two species of fish. Most snakes are generalist in their diet, although they may confine themselves largely to favorite food items. Many pit vipers feed largely on warm-blooded prey, other snakes eat mainly fish and amphibians, and a few (King Cobra, *Ophiophagus;* King Snake, *Lampropeltis getulus*) prefer other snakes. Crocodilians are another exclusively carnivorous group, usually feeding on any morsel of animal flesh that comes their way. A few, such as *Gavialis,* restrict their diet to fish.

The carnivorous turtles eat invertebrates (crustaceans and mollusks) and vertebrates (fish and amphibians) as adults, mostly insects as juveniles. Lizards and salamanders are mainly insectivorous, although a few have specialized diets, e.g., the mollusk diet of the Caiman Lizard *(Dracaena guianensis)* (Figure 9–2).

Herps generally eat when and what they can. The advantage of such a generalist approach is that the abundance of any given food item varies seasonally (Figure 9–3). By eating a wide variety, a species can always select the most abundant item, which requires the least energy to catch. Some herps, such as lizards and frogs, eat daily if food is available; others, like snakes, eat infrequently, since their intake is larger at a single feeding.

Prey–Predator Interactions

Although most amphibians and reptiles are predators, they are also prey in turn for other animals. Tree frogs eat insects, snakes eat frogs, and hawks eat snakes. Few herps can escape from such an intermediate position in the food chain. Although adults of the giant snakes, giant tortoises, and other large herps may no longer be prey items, all their growing stages are preyed upon, often quite heavily. The high mortality of young sea turtles is a case in point.

Predation is not usually deleterious to a population or a species, though it is, of course, destructive to the individual. A natural balance is maintained between the prey and predator populations. Predators act as one of the factors limiting population density. As a particular prey population increases in number, more individuals are visible to the predators, which begin to crop the prey population, usually the weaker and less fit individuals. Eventually, the prey population is reduced to a level where the predators must spend considerable effort and time in search and capture, so that they switch to a new prey population. Although this is an oversimplified explanation, it illustrates the constant interaction between prey and predator.

The prey and its predator are in a constant adaptive race. The prey attempts to become inedible or attack-resistant by physiological, behavioral, and morphological modification, and the predator changes its feeding habits and structures to maintain its ability to capture and eat the prey. Some of these adaptations, such as protective coloration and toxicity, will be discussed in the last section of this chapter.

The effect of predation on prey populations can be seen by reexamining the survivorship curves of reptiles (Figure 9–1). In the type C curve, where

FIGURE 9–2
The Caiman Lizard, *Dracaena guianensis*, feeding on a snail.

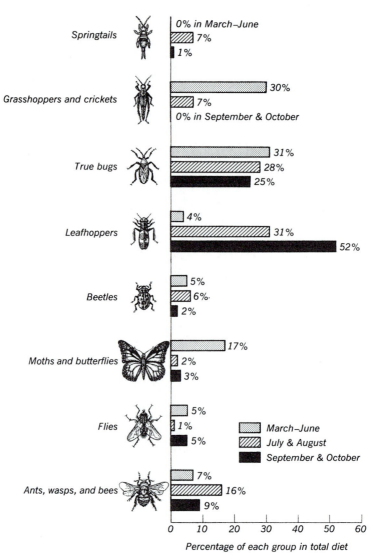

FIGURE 9–3
Seasonal change in food composition of the Side-blotched Lizard (*Uta stansburiana*) in Utah. Percentages are those of total number of items represented. Note particularly the increase in the consumption of leafhoppers (mainly the Beet Leafhopper, *Eutettix tenellus*) as the season progresses and these insects become more abundant. [Data from George F. Knowlton, after Oliver, *North American Amphibians and Reptiles*, Van Nostrand, 1955.]

mortality is high in the early stages of life, predation is the primary cause of death. Generally, large numbers of unprotected young are produced, but only a few survive to adulthood, where because of size (sea turtles) or escape behavior (lizards) they are less susceptible to predation until they reach old age. In type B populations, all age classes are equally susceptible to predation, so that predation is present but low in all life stages. In type A populations (unknown among herps), most young and adults are able to avoid predators, which catch mainly the older individuals.

Competition

Since the supply of any resource is finite, each individual must compete with others of its own species and with other species for its share of the resource. It is easy to invoke the idea of competition to explain intra- and interspecific interactions; but since competition is seldom seen in action in wild populations, only the evolutionary end products, which supposedly have resulted from competitive interaction, are apparent.

The sexual dimorphism of head and body size in many lizards, particularly the anoles, is thought to be the result of intraspecific competition for food. Most anoles show three distinct size classes: adult males, adult females and subadults, and juveniles. Adult males have the largest bodies and heads, females and subadults are intermediate in size, and juveniles are, of course, the smallest. Members of each class capture insects in a wide range of sizes, but tend to concentrate on the largest size they can handle. Thus, although the ranges for the classes may overlap, the tendency to maximize food size enables each class to feed on its own food resource, reducing intraspecific competition for food. Competition among adults is further reduced by the establishment of territories, which provides each individual with its own feeding reserve.

Interspecific competition is reduced by each species' occupying a unique microhabitat in the community. For example, in forest anoles, space is partitioned among species into a lower-trunk and ground habitat, a trunk habitat, a canopy habitat, and so forth. Anuran tadpoles partition an apparently homogenous habitat into different feeding zones. One species will feed on the surface film, another at midwater on the plankton, another on the bottom detritus, and still others on submerged vegetation or the algal coat on submerged objects. In this way, the pond supports more tadpoles, with a minimum of interference among species.

Competition for resources may lead to the exclusion of one species from the resource, rather than the evolution of a mechanism to partition the resource. One species gains a competitive advantage over the other by being a better exploiter of the limiting resource or by interfering with the other's access to it. Such exclusion appears to be occurring between the Red-backed Salamander *(Plethodon cinereus)* and the Shenandoah Salamander *(P. nettingi shenandoah)* in the mountains of Virginia. *P. cinereus* is a widespread species, whereas *P. n. shenandoah* is restricted to a few isolated talus slopes. Investigation shows that *P. cinereus* is a better predator of the arthropod fauna in the soil and leaves little for *P. n. shenandoah*; further, when both occur together, *P. cinereus* tends to occupy the deeper and more humid parts of the burrow systems. The only reason that *P. n. shenandoah* survives on the dry talus slopes is its greater food-catching ability and tolerance of desiccation under dry conditions.

Symbiotic Interactions

Symbiosis is a special and close association between two kinds of organisms literally "living together." Several degrees of symbiotic interaction are possible between organisms. Mutualism is an interaction in which both individuals benefit; in commensalism, one benefits and the other receives neither benefit nor harm; in parasitism, one benefits at the expense of the other. Because these categories grade into one another, it is often difficult to classify a given association.

Mutualism.   Only a few instances of true mutualism have been reported among the herps. A species of green alga grows inside the outer membranes of the eggs of the Spotted Salamander *(Ambystoma maculatum)*. This is apparently beneficial to both organisms: eggs that are inhabited by algae produce larger embryos, hatch earlier, and have a lower mortality rate than eggs that do not have algae, and the algae seem to grow more vigorously in eggs containing embryos than in those from which the embryos have been removed. The mechanisms by which plant and embryo affect each other are unknown.

Some aquatic turtles are true mossbacks, their shells being covered with a dense growth of algae. The shell provides a place of attachment for the plants, and the plants help conceal the turtle.

One type of mutualism is "cleaning," in which an organism removes—and eats—deleterious organisms from the body of another. On the Galápagos Islands, large red crabs have been seen crawling over the bodies of Marine Iguanas *(Amblyrhynchus)* and pulling ticks from their skins. Nearly 2,500 years ago, the Greek historian Herodotus reported that crocodile birds enter the mouths of basking crocodiles to remove leeches from their gums. The story has never been verified, but if true it is undoubtedly the oldest record of this type of symbiosis.

Commensalism.   Commensalism is a one-sided symbiotic interaction in which one member derives some advantage—protection, transportation, or shelter—and the other is not affected. One kind of commensalism is inquilinism, in which one species lives in a domicile constructed and occupied by another. Herps may be guests of, or hosts to, many other animals. Perhaps the best known inquilines are the ones that live with the various species of Gopher Tortoise *(Gopherus)*. Sidewinders *(Crotalus cerastes)*, Great Basin Rattlesnakes *(Crotalus viridis)*, Spotted Night Snakes *(Hyp-*

siglena torquata), and Banded Geckos *(Coleonyx variegatus)* have all been found hibernating in dens of the Desert Tortoise *(Gopherus agassizi).* The burrows of the Gopher Tortoise *(Gopherus polyphemus)* are occupied, occasionally, by Eastern Diamondback Rattlesnakes *(Crotalus adamanteus),* and, typically, by Gopher Frogs *(Rana areolata capito).* The Tuatara *(Sphenodon)* often makes its home in the burrow of a bird, the Sooty Shearwater.

Many different kinds of protozoans are present in the intestines of herps. They are usually considered parasites, but since many seem to do no appreciable injury to the host, they should perhaps be classed as commensals. It is even possible that some of them are beneficial.

Parasitism. Parasitism is a symbiotic association in which one organism spends most or all of its life on or in the body of another, derives its food from its host, and frequently damages the tissues of the host. Parasitologists have found the herps fruitful sources of parasites, both external and internal. Mites, ticks, leeches, copepods, fly larvae, flukes, tapeworms, threadworms, thorny-headed worms, and protozoans have all been reported to be parasites of the herps.

Outside those banes of the zookeeper, the mites, perhaps the best known parasites of reptiles are the flukes (trematodes) that are found in the respiratory and digestive systems of snakes (Figure 9–4). Most snake autopsies

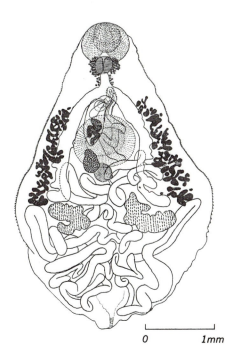

FIGURE 9–4
The lung fluke *Pneumatophilus leidyi,* from a Banded Water Snake, *Natrix fasciata.* [After Byrd and Denton, *Proc. Helminthol. Soc., Wash. D.C.,* 1937.]

0   1mm

show heavy infections by these trematodes, but such infestations are seldom fatal. Indeed, the presence of trematodes does not seem to have much effect on the well-being of the snakes, although studies of the poisonous *Denisonia* and *Notechis* of Australia indicate that parasites may diminish their venom yields.

There are some interesting correlations between trematode infections and the habits of snakes. Two species of Garter Snake *(Thamnophis sirtalis* and *T. butleri)* differ quite pronouncedly: the average specimen of *T. sirtalis* is heavily infected with trematodes, *T. butleri* seldom has them. These parasites pass through the early stages of their development in the body of a snail. If, after they leave the snail host, the larvae are ingested by the amphibian, they undergo further development in its body. If the amphibian is then eaten by a snake, the snake becomes infected and the trematodes reach adult form within it. Since *T. sirtalis* feeds largely on amphibians, it is usually heavily parasitized; since *T. butleri* eats earthworms and leeches rather than amphibians, it is not. The presence of the parasite provides a clue to the diet of the host.

For both an animal species and its parasites to survive, there must be an adjustment between them. If the members of a parasitic species are so numerous and so deadly as to kill all available hosts, then the parasite itself will become extinct. Parasitism therefore tends to evolve toward commensalism. Adjustment between the trematodes and snakes seems to be an adjustment between individuals—that is, the body of an infected snake appears able to withstand the damage done by the parasites. Adjustment may also take place between two populations. In the vicinity of Gainesville, Florida, Carolina Anoles *(Anolis carolinensis)* are parasitized by the larvae of a fly of the family Sarcophagidae. In some years, during May and June, the bodies of dying anoles are found on the ground, their insides eaten away by maggots. As many as 34 maggots have been taken from a single anole. During other years, however, no evidence of such parasitism is found. It is probable that the periods of heavy infection are periods in which the anoles are particularly abundant, and that the numbers of the two populations fluctuate together, so that there are never enough flies to reduce the number of lizards below the level necessary to maintain the population.

One of the most interesting studies of amphibian parasitology was the attempt by Dr. Maynard M. Metcalf to correlate the evolution of the ciliate protozoans of the family Opalinidae with the evolution of the frogs they parasitize. Since he found that opalinids are not strictly host-specific, but sometimes transfer from one family to another, studying them did not reveal much about the evolution of the frogs. The main interest of Metcalf's work lies in his method of gathering and combining data from the organisms and their parasites. Since host and parasite evolve together, this method should be useful in future evolutionary studies.

Defense Mechanisms

The "evolutionary goal" of each population is to increase its number. This, of course, means only that each individual produced must have a greater probability of survival. To attain this goal, reptiles and amphibians have evolved a variety of defense mechanisms that protect them from predators.

Structural Modifications. The bony shell of a turtle is the most obvious and most elaborate morphological modification for protection against predators. This bony box of endochondral and dermal bones encases the entire trunk and still leaves sufficient space for the retraction of the limbs and head into its protective covering. Only the young, before they have attained adult size and complete ossification of the shell has occurred, are subject to heavy predation. A few families of lizards (Anguinidae, Cordylidae) have developed an armored skin of interlocking, platelike osteoderms. The osteoderm-embedded back of crocodilians is a similar protective device. The fusion of the skin and dorsal skull bones in some anurans (e.g., the Cuban Tree Frog, *Osteopilus septentrionalis*) allows the animal to use its head to plug the entrance of its hole, thus protecting its body.

The epidermal scales of reptiles also provide some degree of defense against predator attack, although their main function is probably as an antiabrasion mechanism. However, the enlargement and increased keeling of the caudal scales in a few lizards *(Uromastyx, Enyaliosaurus)* make the tail a defensive weapon. Further, many lizards and salamanders use the tail as a decoy to permit escape. The vertebrae and muscles of the tail have cleavage planes that allow it to break easily from the body (caudal autotomy). While the broken tail is twitching and attracting the predator's attention, the animal can escape unnoticed.

Toxicity. A discussion on poisons in herps usually centers on the venomous snakes; but many amphibians are just as toxic. The difference is that snakes use their poison to obtain food, whereas amphibians use theirs to avoid becoming food. All frogs and salamanders possess poisonous skin glands. The toxicity of the different species depends on the number and concentration of the poison glands and the potency of the poison, which varies from deadly to no more than an irritant. The salamandrid salamanders, particularly *Salamandra* and *Notophthalmus,* have the most toxic skin secretion. In many anurans (Poison Arrow Frogs, *Dendrobates;* Mink Frog, *Rana septentrionalis;* Mexican Tree Frog, *Pternohyla fodiens*), the poison glands are uniformly distributed over the entire skin surface. In others, some areas of the body bear large concentrations of poison glands, such as those on the dorsal caudal surface of *Ambystoma maculatum* or the paired paratoid glands of *Bufo.*

Color and Pattern.   Colors and their arrangement on the body can serve as defense mechanisms against predators in several ways. Cryptic coloration hides the prey from the predator's view by matching the prey's shape and color to its background (camouflage) or breaking up the prey's outline on its background (disruptive coloration). The uniform green of many arboreal herps makes them indistinguishable from their leafy surroundings; many also adjust their locomotor behavior to slow, methodical movements so that their camouflage is as effective during motion as at rest. Forest-floor species are often somber shades of brown or irregular blotches of brown, orange, yellow, and green, both of which serve to camouflage the animal. Other herps, with their contrasting rings or stripes of color, seem particularly noticeable when placed on a uniformly colored background, but in their native habitat, these contrasting bands of color join with their surroundings and obscure the prey's outline or silhouette. The striped or lined pattern may also cause a predator to misdirect its attack, because in a discontinuous environment only part of the prey will be seen, and during movement the striped pattern gives the impression of a stationary target. Other brightly colored amphibians and reptiles advertise their presence with a coloration that won't be forgotten, because they are either distasteful or toxic. Once a predator has caught and attempted to eat one member of the prey population, it is not likely to attack other members. This is called warning or aposematic coloration. Bright colors are also used to frighten or startle a predator in order to provide an extra moment for escape. For example, some frogs have inguinal flash patches. In these cases the animal is cryptically colored when at rest; the bright color is usually hidden until it is suddenly exposed.

Mimicry.   *Mimicry* simply means the evolution of one species to resemble another, thereby gaining some advantage. Two main kinds of mimicry are Batesian and Müllerian. In Batesian mimicry, a harmless, edible species (the mimic) comes to look like some harmful or distasteful species (the model). Predators learn to avoid not only the model but also any other animal that looks like it, and thus the mimic is protected. In Müllerian mimicry, two or more harmful or distasteful species come to resemble each other: the number of victims lost as each new generation of predators learns to avoid prey of a particular appearance remains the same, but is spread over several species.

Mimicry undoubtedly occurs in herps. The mimicry of both the behavior and the coloration of viperine snakes by some nonpoisonous and poisonous colubrid snakes in Africa and the Near East is likely a mimetic complex. Mimicry also occurs in salamanders of the eastern United States, with the red eft stage of the newt *Notophthalmus* as the model and the Red Salamander, *Pseudotriton ruber,* the mimic. Experiments have shown that vertebrate predators learn to reject red efts as prey after one or two experiences, and will then also reject the nonpoisonous *Pseudotriton*.

The most widely discussed mimetic complex in the amphibians and reptiles is that of the coral snakes (*Micrurus* and relatives). Many of the species resemble each other in being brightly marked with rings of black, white or yellow, and red. Within the ranges of these species are found similarly marked nonpoisonous and mildly poisonous forms (examples are *Pseudoboa* and *Erythrolamprus,* respectively). It looks like classical mimicry—Batesian where only a single species of coral snake is found in a given place, as in the United States, Müllerian in South America where several species are found together. Two objections have been raised, however. Coral snakes are largely secretive, burrowing forms, probably seldom attacked by diurnal predators that rely on visual cues to identify prey. Also, coral snake venom is so deadly, it has been argued, that few predators would survive a bite long enough to profit by the lesson. An alternative explanation for the situation in South America is that it represents a modified Müllerian mimicry, in which both the coral snakes and the nonpoisonous forms are mimics, with some mildly poisonous form the model. But this explanation will not suffice for the United States, where mildly poisonous forms with the coral snake pattern are absent.

READINGS AND REFERENCES

Arnold, S. J. "Species densities of predators and their prey." *American Naturalist,* vol. 106, 1972.

Brandon, R. A., and J. E. Huheey. "Diurnal activity, avian predation, and the question of warning coloration and cryptic coloration in salamanders." *Herpetologica,* vol. 31, 1975.

Greene, H. W., and W. F. Pyburn. "Comments on aposematism and mimicry among coral snakes." *Biologist,* vol. 55, 1973.

Heyer, W. R., R. W. McDiarmid, and D. L. Weigmann. "Tadpoles, predation and pond habitats in the tropics." *Biotropica,* vol. 7, 1975.

Jaeger, R. G. "Competitive exclusion: Comments on survival and extinction of species." *Bioscience,* vol. 24, 1974.

Milstead, W. W. *Lizard Ecology: A Symposium.* Columbia: University of Missouri Press, 1967.

Moll, E. I., and J. M. Legler. "The life history of a neotropical turtle, *Pseudemys scripta* (Schoepff) in Panama." *Bulletin of the Los Angeles County Museum of Natural History,* no. 11, 1971.

Pianka, E. R. "The structure of lizard communities." *Annual Review of Ecology and Systematics,* vol. 4, 1973.

———. *Evolutionary Ecology.* New York: Harper and Row, 1974.

Pough, F. H. "Lizard energetics and diet." *Ecology,* vol. 54, 1973.

Savage, R. M. *The Ecology and Life History of the Common Frog.* London: Pitman, 1961.

Schoener, T. W. "Resource partitioning in ecological communities." *Science,* vol. 185, 1974.

Tinkle, D. W. "The life and demography of the side-blotched lizard, *Uta stansburiana.*" *Miscellaneous Publications, Museum of Zoology, University of Michigan,* no. 132, 1967.

————, H. M. Wilbur, and S. G. Tilley. "Evolutionary strategies in lizard reproduction." *Evolution,* vol. 24, 1970.

Wickler, W. *Mimicry in Plants and Animals.* New York: World University Library, 1968.

Wilbur, H. M. "Competition, predation, and the structure of the *Ambystoma–Rana sylvatica* community." *Ecology,* vol. 53, 1972.

CHAPTER TEN

BEHAVIOR

IF YOU APPROACH a pond on whose banks a number of frogs are resting quietly, you will be greeted by a series of splashes, as the frogs jump into the water and swim to hide under the detritus at the bottom. They have been stimulated by the sight or sound of your approach and have responded with an escape reaction. In so doing they have exemplified one definition of behavior—that it is a response to a stimulus. A stimulus, in turn, is defined as some change in the external or internal environment, to which the organism responds. Thus the definition of behavior is really circular. It is also too broad: a neotenic salamander may respond to an increase in iodine content of the water by metamorphosing, but this is not usually regarded as behavior. Instead of attempting to give a precise definition of behavior, we shall simply list some of the characteristics of those responses that are considered behavioral.

CHARACTERISTICS OF BEHAVIOR

Behavior is overt. The frog's response to a stimulus is visible to an observer. Usually, the response entails movement and follows immediately after the stimulus; but the response, or part of it, may be delayed—the next time the frogs emerge from the pond, they may sit closer to its edge or react more quickly to your approach.

Behavior is selective. Out of a number of possible responses to a stimulus, the animal makes one. The frogs could have remained in place, or turned to face you, or made threatening displays. This is not to say that a frog deliberately chooses which response to make. Its response is determined by a complex of internal and external factors beyond its control.

Behavior is usually reversible. If you sit quietly at the pond awhile, one by one the frogs will appear at the surface, swim to the bank, and resume their former positions.

Behavior may be more or less modified by experience. If you approach the ponds several times a day for several days, at least some of the frogs will no longer respond to your approach by jumping into the water. They have become habituated—that is, they have learned not to respond to the stimulus.

DETERMINANTS OF BEHAVIOR

Behavior is determined by several biological variables: genetic inheritance, learning from previous experience, and physiological factors, including age and sex.

Inheritance

An animal is a product of its genetic makeup. Every aspect of its life is influenced to a degree by its heredity. All behavioral responses are determined, at least in part, by heredity, though many of them can also be modified by experience (learning). Behavior extends across the entire spectrum, from completely inherited to completely learned responses. There is no clear division between the two, and we must be careful in classifying behavior not to overemphasize one to the exclusion of the other.

Some behavioral patterns appear to be inherited in the same way a nerve or muscle is inherited. These patterns cannot be modified by experience and are species-specific—characteristic of the species to which the animal belongs and frequently differing from similar patterns in related species. As is typical of iguanids, a male Fringe-toed Lizard (Uma) will challenge another lizard on approach. The challenge display consists of a series of foreleg push-ups that move the front part of the body up and down. The series in the Mojave Fringe-toed Lizard U. scoparia) consists of two (sometimes three) rapid push-ups, a pause with the body held low, a push-up, a pause with the body low, and an elevation of the body to half-height. In another Fringe-toed Lizard (U. notata), the series begins with two slow push-ups, a third push-up with a return only to half-height, a small rise and decline, and a final complete push-up (Figure 10–1). These differences in movement pattern enable males and females of one species to recognize their own kind and ignore other species.

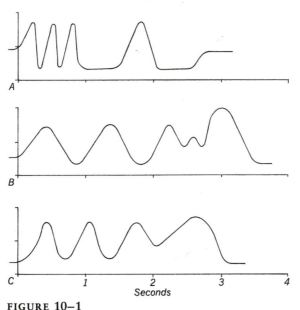

FIGURE 10–1
Push-up sequences in Fringe-toed Lizards. The height of the line indicates the relative elevation of the forepart of the lizard's body during a single sequence: (A) *Uma scoparia*; (B) *U. notata*; (C) some individuals of *U. inornata*. [Redrawn from Carpenter, *Copeia*, 1963.]

Innate behavior patterns are not invariable. Just as inherited structural characteristics show variation, innate behavior also shows intraspecific differences. Some members of one race of *Uma notata*, for example, omit the slight rise and dip. Such differences, however, can be expected to be small within a given population, because, like other inherited characteristics, behavior is molded by natural selection. Thus if a movement pattern is used for species recognition, any individual whose behavior deviates too widely from the norm will be less likely to obtain a mate.

Learning

Learning is defined as a more or less permanent modification of behavior as the result of experience. Almost all animals can become habituated—that is, learn to ignore a neutral stimulus, one that promises neither reward nor punishment. Herps are capable of learning much more than this. An individual learns its surroundings, the limits of its home range, where to find food or shelter. A tree frog, after a night's foraging, can find its way back to the particular leaf on the particular tree under which it sleeps during the day.

There is a strong genetic component in what an animal is capable of learning. It learns most readily those things that are important to its particular way of life. A Bullfrog *(Rana catesbeiana)* can learn to avoid a bad-tasting insect in one or two trials; it may be very slow to learn a maze. A Mole Salamander *(Ambystoma talpoideum),* which lives in an underground burrow system, might be much quicker at learning the maze. Because of these differences in what animals need to be able to learn, it is difficult to design experiments that adequately test differences in intelligence between groups, and there have been very few real comparative studies of learning in herps. We expect learning ability to increase with increasing size and complexity of the brain, and it is possible that reptiles are indeed better able to learn than are amphibians.

Physiological Factors

The physiological state of an animal is important in determining how it behaves. A hungry snake pays more attention to a mouse in its cage than does a recently fed one. Desert Iguanas *(Dipsosaurus dorsalis)* have been shown to learn the way through a maze faster as their body temperature approaches the eccritic temperature. A dehydrating frog will assume a posture that reduces the exposed body surface. Among the physiological factors that may influence an animal's behavior are hormones, endogenous rhythms, age, and sex.

Hormones.   The relationship between hormones and behavior is most obvious in the differences in an animal's activities in the breeding and in the nonbreeding season. Combat between rival males is common in many courting reptiles. Injection of testosterones has been found to elicit aggression in reproductively quiescent males of the Texas Tortoise *(Gopherus berlandieri)* and in many species of iguanid lizards.

Endogenous Rhythms.   Most animals show rhythmic changes in activity known as endogenous rhythms. The changes are correlated with physiological changes in the animal and are timed to correspond to changes in the environment, such as day and night, lunar phases, seasons, and tides. The cycles are usually independent of external stimuli. When an animal that normally is active during the day and rests at night is placed in constant darkness or constant light, it usually retains its cyclic activity pattern, although it may drift slightly out of phase with the 24-hour day. That is, it may show cycles of 23 hours or 27 hours. These are called *circadian* (= about a day) rhythms. Because circadian rhythms are not set precisely to a 24-hour cycle, the animal is able to adjust to seasonal changes in the length of daylight and darkness. Apparently the innate "biological clock" on which such rhythms are based is continually reset by environmental cues. The nature of the clock is not known; it seems to be correlated with changes in the rate of hormone secretion. The innateness of circadian rhythms, yet their tendency to lose their precise timing, are seen in Wall Lizards *(Lacerta)* hatched and reared under conditions of constant temperature, humidity, and darkness (Figure 10–2).

Occasionally a rhythm does seem to depend on external stimuli. The Slimy Salamander *(Plethodon glutinosus)* is normally most active at sunset, when it emerges from its burrow to feed; but if it is kept in constant darkness, its activity periods become arhythmic after several days. If the animal

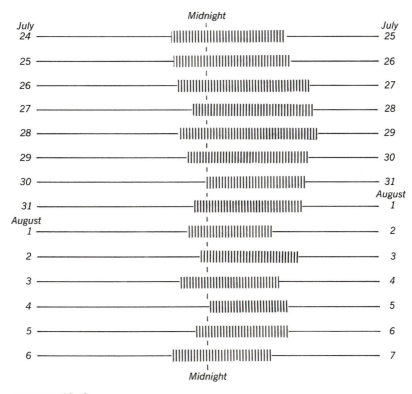

FIGURE 10–2

Periods of activity in a lizard, *Lacerta*, incubated and reared in total darkness. [After Hoffman, Z. *Vergleich. Physiol.*, 1959.]

is then exposed to a period of light, the cycle will assume a new rhythm timed from the beginning of the new period of darkness.

Seasonal changes in activity are apparent in many herps. Hibernation is an inherent, regular, and prolonged period of inactivity during the winter, in contrast with retraherence, which is a temporary retreat from adverse weather conditions. In northern Florida, Broad-headed Skinks (*Eumeces laticeps*) hibernate early in October and do not emerge until March, even though in this region there may be many winter days when the temperature is well within the activity range of these lizards. Carolina Anoles (*Anolis carolinensis*) in the same region are active on warm days throughout the winter, remaining in shelter during cold spells. The anoles have no rhythm, but simply become dormant in response to an external stimulus. The skinks have an endogenous seasonal rhythm, whose biological clock may be adjusted by decreasing day length, as in some mammals and birds.

Estivation is a regular period of inactivity similar to hibernation, but occurring in the summer, as a retreat from hot, dry climatic conditions. The Sardinian Cave Salamander (*Hydromantes genei*) has been reported to estivate in summer even in a laboratory, under presumably constant, humid conditions. Captive Australian Long-necked Turtles (*Chelodina longicollis*) tucked themselves away for weeks in the summer without feeding, even though their ponds were kept filled with water. Here again, endogenous rhythms seem to be at work.

Age and Sex.   The effects of age and sex on the behavior of an animal are frequently obvious. For example, adult male frogs call during the breeding season; females and juveniles do not. Many such differences are mediated by hormones; but size alone, which is often correlated with age and sex, can affect behavior. In the Brown Anole, *Anolis sagrei,* males are larger than females and choose higher perch sites. A small male will be unable to establish and maintain a territory against a larger competitor.

ELEMENTS OF BEHAVIOR

Behavior can be analyzed in terms of the muscular movements made by the animal, the stimuli that elicit them, and the motivational states that produce one type of behavior rather than another.

Fixed Action Patterns

The motor components of behavior consist of fixed action patterns (FAPs). A fixed action pattern is a highly coordinated muscular movement that an animal can perform without previous exercise or learning. The sequence of muscular contractions is innate and constant. The lizard push-up movement is a fixed action pattern. It is neither a simple reflex nor a chain of reflexes. In a reflex, the same stimulus always elicits the same response. Whether an animal performs a given FAP, by contrast, depends on several factors, including the animal's motivational state, its previous experience, and the length of time since the action was last performed.

When a Hognose Snake *(Heterodon)* is threatened, it rears up, flattens and spreads its neck, hisses, and often lunges with closed mouth at the intruder. If this display does not drive away the supposed attacker, the snake "plays dead": it flops on its back, lolls out its tongue, and goes limp. Both the threat display and the feigning of death are sequences of FAPs. Most Hognose Snakes kept in captivity soon stop responding to teasing with this particular behavior. The stimulus that once evoked the response no longer does so.

Sign Stimuli

A given FAP sequence is usually performed in response to a specific, typically very simple stimulus known as a sign stimulus. The animal generally reacts to a single aspect of a situation, rather than to the total situation. During the breeding season, a male lizard of the *Sceloporus torquatus* group will challenge and then attack another male engaged in courtship activities. A courting male of this genus approaching a female gives a characteristic series of short, fast, bobbing motions of the head. When an investigator attached a string to the head of a carved wooden model and bobbed it in this way, it elicited a challenge display from a male. Rubber models, dead females, and live individuals of many other species were also effective, so long as they appeared to give the courting bob. The characteristic motion of the head is a sign stimulus that produces challenge and attack in these lizards.

When such a sign stimulus is presented by one animal and evokes a response in another, it is also called a releaser. Releasers usually serve for intraspecific communication, but they may also be directed toward members of other species. Many toxic amphibians, such as the Fire-bellied Toad *(Bombina)*, have brightly colored undersides. When attacked, such an animal may arch its back and spread its legs wide to display its ventral side (unken "reflex"). The flash of bright color thus shown is a releaser that deters attack by the predator.

Motivational State

Animal behavior is frequently classified on the basis of motivational states, or drives. An animal is constantly bombarded by a great number of stimuli, and it has a repertoire of behavioral responses at its disposal. At any given time, which stimulus it responds to, and how, are determined by its motivational state. This, in turn, is determined by the animal's internal condition—how hungry or thirsty or hot it is, the level of hormones in its blood, and so forth—and by stimuli from the environment. An approaching predator can drive a basking lizard into a cool shelter.

It is convenient to classify behavior as feeding, reproductive, thermoregulatory, etc., but it should be stressed that this categorization may conceal differences in the mechanisms concerned. For example, salamanders and lizards both show thermoregulatory behavior. A salamander in a stream is stimulated to move about by adverse temperature conditions; under favorable conditions, such movement is inhibited. The result is that salamanders tend to aggregate where the stream temperature is closest to their preferred body temperature. A lizard, on the other hand, will move directly to a basking perch and orient its body toward the sun. A movement such as that of the salamander, which results from an adverse stimulus but is not directly oriented toward or away from the source of the stimulus, is called a kinesis. An oriented movement like the lizard's is called a taxis.

Sequences of Behavior.   Patterns of behavior often form functional sequences associated with given motivational states. Frequently such a sequence begins with appetitive (searching) behavior: a hungry snake searches for food, a female turtle laden with eggs searches for a nesting site. This behavior exposes the animal to a stimulus that initiates the next phase, an orientation of the body in relation to the stimulus (a taxis). The sequence is terminated by a consummatory act—eating the prey, depositing and covering the eggs—which alters the animal's motivational state.

Simultaneous Arousal.   Ordinarily, an animal acts at any given time under the impetus of a single motivational state. For example, the escape drive normally takes precedence over feeding or thermoregulation. It frequently appears, however, that two conflicting drives are present at the same time. This is especially true in the breeding season, both during aggressive encounters between males and during courtship. A Common Toad *(Bufo bufo)* at a breeding pond will attempt to mate with a moving object of a size that, in the nonbreeding season, would cause it to flee. Although the tendencies to approach and to escape may both be present, the drive to reproduce suppresses the survival instinct. In mammals and birds, various types of behavior have been found to be associated with the simultaneous arousal of conflicting drives. Thus when two birds meet at the boundary between their territories, the drives to fight and to escape come into conflict. A bird in this situation may peck at twigs instead of its adversary (redirected response) or show inappropriate behavior such as feeding or preening (displacement). During the course of evolution, these inappropriate movements may be incorporated into the threat or courtship display. They then come to be performed in a highly stereotyped way, and are frequently emphasized by the development of bright colors or special structures. They are said to have become ritualized. The effects of simultaneous arousal in herps have been almost completely neglected. Their elaborate threat and courtship displays, especially those of the lizards, offer a fertile field for research.

Directional Travel

In contrast to tactic movements (changes in posture or short-distance movements in direct response to a stimulus), animals also make longer-distance movements guided by external cues. Such directional travel falls largely into two classes, homing and migration. Homing movements are the daily movements of an animal within a limited home range. The term is also used to designate the return of a displaced animal to its home range, but this is really an unnatural situation imposed on the animal by a researcher. Migration is long-distance movement from one area of activity to another and eventual return to the original area; it is usually associated with travel to breeding or hibernation sites. Like most biological categories, homing and migration are not always clearly separable, but form a continuum.

Homing movements seem to be largely guided by visual and olfactory cues and landmarks. The animal learns its home range, and uses landmarks to find its way from resting site to feeding site. If removed from their home range, some animals take a circular path until they strike a familiar landmark, then return directly home. Others immediately orient in the direction of home. Evidence indicates that the latter animals use celestial cues to orient correctly.

A number of herp species (e.g., *Ascaphus truei, Acris gryllus, Rana catesbeiana, Taricha torosa, Chrysemys picta*) have been shown experimentally to use celestial cues for directional guidance. Cricket Frogs *(Acris gryllus)* removed from their home shore and released in the center of a test pen, from which only the sky was visible, were able to get their bearings and move in a direction perpendicular to the home shore, both during the day and on moonlit nights. It was not necessary for the frogs to see the sun; but when the sky was completely overcast, their movement was random (Figure 10–3). The frogs seem to have both a sun compass and an internal clock that allows them to compensate for the changing position of the sun during the day.

Such celestial orientation is common among animals, and has been proposed as the basic mechanism for directional travel, particularly migration, among herps. However, it would be foolish to insist that any animal uses one mechanism to the exclusion of all others. There is no reason to rule out any sensory input as a possible directional cue. Terrestrial migrants, such as

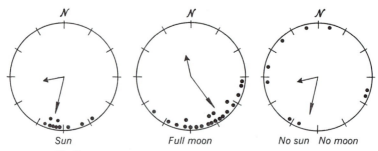

FIGURE 10–3
Orientation in Cricket Frogs, *Acris gryllus*. The long arrow indicates the line parallel to the frogs' home shore; the short arrow indicates the direction of the home pond. [Redrawn from Ferguson, Landreth, and McKeown, *Anim. Behav.*, 1967.]

toads, may find their breeding pond by using the slope of the land, the odor of the pond, the humidity gradient, the vocalization of other toads, the memory of movements made as postmetamorphic toadlets, and other cues instead of, or in addition to, celestial orientation. The champion long-distance migrants are the sea turtles (Cheloniidae and Dermochelyidae). Atlantic Green Turtles, *Chelonia mydas*, regularly travel over 2,000 kilometers from their feeding grounds off the coast of Brazil to lay their eggs on Ascension Island. We still don't know how these animals can return so precisely to the beaches where they were hatched. Celestial cues, currents, temperature gradients, the smell and taste of seawater?

Social Behavior

Members of a population in a given area are frequently found, not scattered at random throughout the area, but gathered together in more or less compact groups or colonies. If the animals happen to be in the same place because of the presence of food or some other physical factor in the environment, they form an aggregation. When two or more animals congregate and remain together as a result of their interactions with one another, they form a social group.

Many congregations of amphibians and reptiles have been recorded. The breeding choruses of frogs and "balls" of hibernating snakes are perhaps the best known. The tadpoles in a given pond are sometimes found to be gathered in a dense mass, rather than scattered at random. This has most often been reported in the Spadefoot Toads *(Scaphiopus)*, but has also been noted in other forms. The Marine Iguana of the Galápagos Islands *(Amblyrhynchus)* forms closely packed colonies. It is frequently assumed that such assemblages are aggregative, rather than social. Almost always, though, when the assemblage is investigated in detail, it is found that each individual derives some benefit from the presence of the others. It has been shown, for example, that salamanders of the genus *Ambystoma* tend to huddle together when exposed to dehydration, and that the rate of water loss for a single individual placed in a container is much greater than when two or more individuals are placed in the container together. Under such conditions, natural selection would favor the development of social groups. It is probable that many of the so-called aggregations of amphibians and reptiles have a social basis.

Ringneck Snakes *(Diadophis punctatus)* are very gregarious. In the field, half a dozen or more may be found under one small log or rock. A group of forty of these snakes were tested in a laboratory arena provided with ten flat discs as sources of cover. Physical and chemical conditions under and around all the discs were the same. The tendency of the snakes to congregate is shown in Table 10–1. The snakes seemed to trail each other across the

TABLE 10–1

Dispersal of Forty Male Ringneck Snakes *(Diadophis punctatus)* in an Arena Containing Dry Sand and Ten Twelve-Inch Discs

| Days after release | Number of individuals | | | | | | | | | | | |
| | Under disc | | | | | | | | | | Exposed | Hidden |
	1	2	3	4	5	6	7	8	9	10		
2	1	1	1	5	6	4	0	5	1	2	14	0
3	0	2	1	2	4	7	0	6	0	2	16	0
4	1	2	0	1	4	14	0	6	0	4	8	0
5	1	2	2	0	4	17	0	4	0	2	8	0
6	1	1	1	2	4	12	0	6	1	2	10	0
7	0	2	0	0	3	24	1	6	0	0	4	0

SOURCE: Dundee and Miller, *Tulane Stud. Zool.,* 1968.

sand, and showed a decided preference for locations that had previously been occupied by other Ringnecks. These snakes are thus truly social, and the organized hibernating masses of other temperate-zone snakes are probably also social.

Sociality is not an all-or-none proposition. An animal may be social for part of its life, asocial at other times. Some snakes are only social during courtship and mating; Spadefoot Toads *(Scaphiopus)* are social as tadpoles and breeding adults, solitary the rest of the time; many lizards live in more or less closely knit colonies all their lives. We have already alluded to the increased chance of finding a mate as one of the advantages of group life.

But there are many others, including environment alteration, protection from predators, and social facilitation.

Environmental Alteration. The presence of a number of individuals in one spot may change conditions in a way that is beneficial to the animals. Elevation of humidity and consequent reduction in the rate of dehydration are but one example.

In a cluster of about 180 tadpoles of the Pacific Tree Frog *(Hyla regilla)*, about three fourths of the animals were oriented with their tails toward the sun. This resulted in the maximum exposure of their dark dorsal surfaces to the sun's rays, and increased the amount of heat absorbed. The temperature of the water in the part of the pond where the tadpoles had congregated was higher than in other parts of the pool. An increase in temperature increases the metabolic rate, and so speeds metamorphosis.

Spadefoot tadpoles develop in temporary, shallow ponds, where there is much danger that the water will evaporate before they can metamorphose. In rapidly drying ponds, tadpoles gather in a group on the bottom and, by lashing their tails, make or deepen a depression to hold water long enough for them to metamorphose (or at least to survive until the pond is filled again by rain).

Protection from Predators. Although a group of animals is more apt to attract the attention of a predator than is a solitary individual, the protection provided by the group usually outweighs this disadvantage: the escape behavior of one individual alarmed by an approaching predator releases the escape reaction in others. That a predator is more likely to attack an isolated individual than a large congregation constitutes selection pressure favoring the animals' tendency to remain close together. Predacious beetle larvae have been reported to avoid entering social groups of Spadefoot tadpoles, though they readily attack isolated individuals.

Social Facilitation. In social facilitation, a stimulus provided by the activity of one member of the group elicits similar behavior in other members. Laboratory tests of male Spring Peepers *(Hyla crucifer)* showed that over 90 percent of the frogs called in response to a tape recording of calling Peepers, but less than 5 percent called when the tape recorder was not played. It seems clear that the buildup of the large breeding choruses characteristic of this species depends on the frogs' tendency to respond to the calling of other individuals.

Social feeding groups of Spadefoot tadpoles have frequently been observed. The combined activity of the tadpoles stirs up the bottom detritus, so that more bits of organic material are made available to all members of the group.

Social Organization

Animals that live in social groups usually develop special behavioral mechanisms that tend to minimize aggression among members of the group. These lead in turn to the development of patterns of social organization. Two types of social organization found among many diverse vertebrates are territoriality and social dominance.

Territoriality.   An animal does not wander completely at random, but remains within a home area, a restricted stretch of familiar territory within which it conducts its normal activities. Animals that undertake extensive migrations often have more than one home area. Thus a male frog may have one home area away from the breeding pond during part of the year and, when he arrives at the pond during the breeding season, establish a new home area along the bank. Turtles and snakes frequently have extensive home areas, but those of lizards are usually small. When a Syrian Fringe-toed Lizard *(Acanthodactylus tristrami)* was repeatedly frightened away from its burrow, it seemed to be thoroughly familiar with the terrain for about 40 square meters around the burrow's entrance. If it was driven to the limits of this area, it became uncertain, stopped, and changed direction (Figure 10–4). The experiment suggests that this animal has a very small home area.

Within the home area, the animal usually has a homesite to which it regularly returns when not foraging for food or carrying on other activities. The home may be a burrow, a particular leaf on a tree (as for some tree

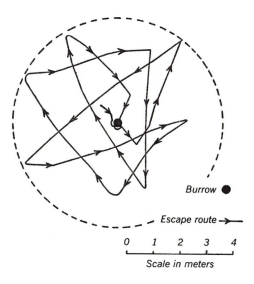

FIGURE 10–4
Route taken by a Syrian Fringe-toed Lizard, *Acanthodactylus tristrami*, when attempts were made to force it away from its burrow. [After Riney, *Copeia*, 1953.]

Burrow ●

Escape route →

0   1   2   3   4

Scale in meters

frogs), or simply a restricted area, such as a strip along the bank of a stream, within the home area. When an animal defends the vicinity of its homesite against intruders of the same species, it is said to show territoriality; the defended area is the territory. This type of behavior occurs most often among males, who frequently drive away other males during the breeding season; but sometimes females, and even young, show territoriality.

Defense of territority at the breeding site has been observed in males of a number of species of frog (e.g., *Rana clamitans, Hyla faber, Leptodactylus insularum, Dendrobates galindoi*). The frog may simply warn the intruder by a threatening display, but sometimes shoving or wrestling bouts develop. Territoriality is well developed in the South American frog, *Phyllobates trinitatus*. One frog will hop on top of another in an effort to drive it away. Adult females are the most aggressive, but even very young frogs may have territories. The territories are feeding, not breeding, areas.

Defense of territory is common in lizards. The Marine Iguanas (*Amblyrhynchus cristatus*) of the Galápagos Islands live along the coast, rarely going more than 12 to 15 meters inland. Along the shore, the young and females mass together, often piled atop one another two or three deep, all within about 9 meters of the water's edge. As many as 76 have been observed within a space of 9 square meters. The older males, which are much larger than the females, take positions between the massed females and young and the water. They generally keep about 1.5 to 3 meters apart, and each keeps to his own sunning territory. Any trespass by one male into the sunning terrain of another is the occasion for a fight. The contestants butt each other, each endeavoring to get his horny, knobby head beneath his opponent's chin. On some of the islands, the iguanas are found assorted into family groups composed of one large male with two to four females. The families are separated from one another by 18 to 45 meters of shoreline.

Usually the possessor of a territory has the advantage in an aggressive encounter, and is able to drive away the intruder. Territory is more often maintained by display than by fighting. The display is usually highly stereotyped and ritualized. A male Bullfrog (*Rana catesbeiana*) in his calling territory floats high in the water, with his head raised to display his yellow throat, and calls frequently. If another frog in the high position approaches, the resident frog challenges by giving a short, staccato, hiccupping call and swimming a short way toward the intruder. If the second frog continues to call or approach, the first repeats the challenge. In 58 of 79 such encounters observed, the incident ended when one of the frogs left the area. The rest culminated in pushing and wrestling bouts. The frog that outpushed his opponent remained high in the water and continued to call; the other adopted a low position by deflating his lungs and swam away. Frogs that had not succeeded in establishing a territory consistently maintained a low position in the water and were not challenged or attacked. High position apparently acts as a release that evokes aggression; low position inhibits attack.

Territoriality tends to space the animals out and thus maximize the use of the area's resources, while at the same time retaining the advantages of the group. For the Bullfrog, warning against predators and social facilitation may both be important advantages of sociality. Females are attracted to large choruses, but apparently not to individuals calling alone. As with many other animals, males that fail to establish territories do not mate. These are presumably the weaker, less well-adapted members of the population. Territoriality thus tends to maintain the genetic fitness of the species.

Social Dominance.   Many animal societies are organized into dominance hierarchies. If one individual is more aggressive, more successful in attacking others, an individual lower in the hierarchy will seek to avoid aggressive encounters by fleeing or by adopting a submissive posture that inhibits attack. The second individual may in turn be dominant to another still lower in the hierarchy. As with territoriality, dominance, once established, is more often maintained by ritualized display than by actual fighting.

Lizards are usually territorial, but when they become overcrowded, as when a number are placed in an artificial enclosure or when the available habitat is restricted in size, they tend to shift to a dominance hierarchy. After a series of aggressive encounters, one male establishes dominance over all the rest, and they flee or show submissive postures at his approach. Since only the dominant is able to reproduce, the effect should be to reduce population size to a level where individual territories could again be established.

Many breeding male frogs show a type of dominance hierarchy in their calling sequence. Choruses of the Southern Spring Peeper *(Hyla crucifer bartramiana)* are composed of many individual trios. A trio is initiated by a single individual, who calls a varying number of times. Soon he is answered by another, and the two call in sequence. Then a third member joins to form a trio. Apparently the same individual initiates a trio each time. He is considered the dominant male. Figure 10–5 is a musical rendition of a trio of Spring Peppers. Similar duets, trios, and quartets have been reported for a number of other species. The Foam-building Frog *(Physalaemus pustulosus)* has such a choral structure. During the night, males drop from the call sequence in reverse order: that is, the low frog in a sequence leaves first. Females that enter the pond in the pauses between sequences of calls move to the dominant male as soon as he initiates a new sequence. Thus he has a greater chance of reproductive success, both because he calls for a longer period of time and because he has a greater chance of attracting females.

Ethological research on vertebrates has been conducted mostly on birds and fishes, and to a lesser extent on mammals. Reptiles and amphibians

FIGURE 10–5
Musical rendition of the formation of a calling trio of the Southern Spring Peeper, *Hyla crucifer bartramiana.* [After Goin, *Quart. J. Florida Acad. Sci.*, 1948 (1949).]

have been almost ignored in this research; but the many examples given in this chapter indicate that the principles of behavior formulated for the other groups apply to the herps as well, and that these animals deserve further ethological study.

READINGS AND REFERENCES

Adler, K. "The role of extraoptic photoreceptors in amphibian rhythms and orientation: a review." *Journal of Herpetology,* vol. 4, 1970.

Brattstrom, B. H. "The evolution of reptilian social behavior." *American Zoologist,* vol. 14, 1974.

Burghardt, G. M. "Chemical perception in reptiles." *In* J. W. Johnston, *et al.* (eds.), *Communication by Chemical Signals.* New York: Appleton-Century-Crofts, 1970.

Duellman, W. E. "Social organization in the mating calls of some neotropical amphibians." *American Midland Naturalist,* vol. 77, 1967.

Dundee, H. A., and M. C. Miller III. "Aggregative behavior and habitat conditioning by the Prairie Ringneck Snake, *Diadophis punctatus arnyi.*" *Tulane Studies in Zoology and Botany,* vol. 15, no. 2, 1968.

Eibl-Eibesfeldt, I. *Ethology: The Biology of Behavior.* New York: Holt, Rinehart and Winston, 1970.

Galler, S. R. *et al.* (eds.). *Animal Orientation and Navigation.* Washington: National Aeronautic and Space Administration, 1972.

Harris, V. A. *The Life of the Rainbow Lizard.* London: Hutchinson, 1964.

Jaeger, R. G., and J. P. Hailman. "Effects of intensity on the phototactic responses of adult anuran amphibians: a comparative survey." *Zeitschrift für Tierpsychologie,* vol. 33, 1973.

Storm, R. M. (ed.). *Animal Orientation and Navigation.* Corvallis: Oregon State University Press, 1967.

Twitty, V. C. *Of Scientists and Salamanders.* San Francisco: W. H. Freeman and Company, 1966.

CHAPTER ELEVEN

SPECIATION
AND GEOGRAPHIC
DISTRIBUTION

WHEN HUMANS FIRST BEGAN to explore the world, they became aware that different kinds of animals lived in different regions: crocodiles were found in the river Nile, but not in the Thames; many snakes lived in Europe, but there were none in Ireland. As long as the doctrine of Special Creation held sway, it was easy to explain these facts by saying that each region had its own fauna, created especially for it and adapted to it. But this explanation was never really satisfactory. For one thing, there is too much overlap: not all animals are restricted to a single region. Many of the animals the early explorers found in the New World were different from anything they had ever seen, but others were strikingly similar to those they had known at home. It was also observed that animals could adapt well to regions in which they did not naturally occur. Attempts to introduce animals into a new region—for example, the European Edible Frog *(Rana esculenta)* into England—were sometimes successful.

When the true nature of fossils was finally recognized, and their orderly succession became apparent, it was realized that some animals had once lived in regions where they no longer occurred, and that many forms had become extinct. Dinosaurs had once roamed every continent, but they no longer existed. It became necessary to postulate a whole series of special creations, the next to last presumably destroyed by the biblical Flood. The attempt to maintain the doctrine of Special Creation eventually became absurd.

It was his observations on the distribution and local variation of animals in South America and the nearby Galápagos Islands that led Charles Darwin to his ideas on the origin of species by the gradual change of populations through time. The pattern of distribution of animals is one of the strongest arguments for evolution, and offers many clues to the course evolution has taken. Conversely, to understand the present distribution of animals we must take account of their evolutionary history as well as their ecological requirements. Thus speciation and distribution patterns supplement and reinforce each other.

ORIGIN OF NEW SPECIES

New species arise from preexisting species through the process of evolution. Evolution simply means a change through time, but when we use it to denote a biological process, it means a change in the gene frequency of a population through time. Although old species may be constantly evolving to adapt to changes in their environment, new species are not constantly appearing. If you recall the definition of a species in Chapter 1, you will remember that a species is a group of like organisms that are reproductively isolated from other groups. Thus, for new species to arise, populations of an existing species must evolve separately, and this discontinuity must remain until the difference in gene frequency and composition of the populations results in their inability to produce fertile offspring—that is, until the populations are reproductively isolated.

Since discontinuities may develop in several ways, there are several ways new species may arise:

     I. Allopatric speciation—divergence of populations
        in different geographic areas.
    II. Sympatric speciation—divergence of populations
       in the same geographic area by
       A. Ecological segregation
       B. Temporal segregation
       C. Polyploidy and parthenogenesis

Allopatric Speciation

No plant or animal species occurs continuously throughout its entire geographic range. A species is broken into a number of small populations, each centered in an area of preferred habitat. Since the environmental conditions differ from area to area, each population is exposed to different selective pressures and evolves to meet them. Although the exchange of genetic material between adjacent populations may be low, immigration and emigration of just a few individuals is enough to maintain the genetic continuity of the species. However, if a population becomes isolated from neighboring populations by a physical barrier, such as a mountain range, the populations begin to diverge genetically, since there is no gene flow. The rate of divergence depends on the nature and intensity of the different selective pressures to which the populations are exposed. If the populations are separated long enough, genetic differences appear that prevent the production of viable or fertile offspring. The isolated population has evolved into a new species, for it is now reproductively isolated from its ancestral species.

Although we have never seen a new species arise, the steps in speciation are visible in extant species. Many of the herp populations of peninsular Florida differ from their mainland relatives. In some instances, the peninsular populations are reproductively isolated from the mainland populations (Florida Black-headed Snake, *Tantilla relicta*); in others, reproductive isolation has not occurred (Florida Mud Turtle, *Kinosternon subrubrum steindachneri*). Whether the differences are specific or subspecific, the change in population characteristics usually occurs in the area of the Suwannee Straits, an ancient seaway that once cut Florida from the mainland.

Sympatric Speciation

In contrast to allopatric speciation, sympatric speciation means the segregation of populations living in the same geographic area. Isolation of populations occurs, not by distance or geographic barriers, but by habitat differences (ecological segregation), different seasons of reproduction (temporal segregation), or instantaneous reproductive isolation (polyploidy and parthenogenesis). Allopatric speciation appears to be the most common mechanism in reptiles and amphibians, as it is in other animals; however, the sympatric speciation mechanisms also appear to have produced new species of herps.

In ecological speciation, an ancestral population occurs in two contiguous habitats that are strikingly different and impose different selective pressures on their inhabitants. Adaptations for survival and successful reproduction in one habitat will be unsuited for the other. Those individuals who breed with members of their own habitat cohort will produce more viable offspring, and if this condition lasts long enough the populations of the two habitats will become reproductively isolated. The Florida Scrub Lizard *(Sceloporus woodi)* and Eastern Fence Lizard *(S. u. undulatus)* live in contiguous but microclimatically distinct habitats, and may be an example of ecological speciation. Such segregation of individuals is also seen in the

Lesser Earless Lizard, *Holbrookia maculata:* the populations living on white sands are lightly colored, those on the adjacent black rock outcrops are dark. These populations are not reproductively isolated, but do show how ecological segregation can evolve.

The potential for temporal speciation exists in herps, but we have no direct evidence that it has in fact occurred. All that is required is for a population to divide its breeding season into two separate periods, with the members of each breeding-period cohort and their offspring confining their reproductive activities to that period. One can imagine a frog species that breeds during the wet season confining its reproduction to the beginning and end of the season and evolving into two species, or sea turtles, with their two- to four-year breeding cycles, segregating into populations breeding in different years, which then evolve into new species through random genetic drift.

New species can arise instantaneously by polyploidy (multiple sets of chromosomes). The Gray Tree Frog, *Hyla versicolor,* of the eastern and central United States is a tetraploid (four chromosome sets), morphologically indistinguishable from the sympatric *H. chrysoscelis,* but reproductively isolated from it. A number of other polyploid frogs are known. Some populations of the South American *Odontophrynus americanus* are tetraploid; *Ceratophrys dorsata* is octoploid (eight chromosome sets). These polyploid frogs are bisexual, and are reproductively isolated from the diploid (two chromosome sets) species from which they evolved.

Speciation by polyploidy can also result from the breeding of two species that do not normally interbreed (hybridization). The salamanders *Ambystoma tremblayi* and *A. platineum* are triploid (three chromosome sets) hybrids of *A. jeffersonianum* and *A. laterale. A. tremblayi* has two sets of *A. laterale* and a single set of *A. jeffersonianum* chromosomes, whereas *A. platineum* has the reverse. The triploids are all females, and reproduce gynogenetically. The eggs are formed by mitosis; development is initiated by penetration of the sperm from either *A. jeffersonianum* or *A. laterale,* but the sperm and egg pronuclei do not fuse, and thus the sperm contributes no genetic material to the offspring. Twenty-six all-female species of lizard are known—some triploid, others diploid. This phenomenon is discussed further under "Parthenogenesis" in Chapter 7.

ISOLATING MECHANISMS

No effort is needed to maintain reproductive isolation if two originally united populations are separated in either time or space. However, such temporal and spatial barriers are often breached, and mating between the two populations becomes a possibility. If they have not diverged greatly, they will merge back into a single population; if there is sufficient genetic divergence, they will be prevented from successful reproduction by one or more isolating mechanisms:

I. Premating mechanisms
    A. Differences in breeding behavior
        1. In breeding sites
        2. In breeding season
        3. In courtship behavior
    B. Physical differences
        1. In recognition characters
        2. In body size and form
        3. In form of genitalia and related structures
II. Postmating mechanisms
    A. Hybrid sterility
    B. Hybrid inviability
    C. Primary sterility

Isolation mechanisms are described briefly in Chapter 1, and many examples are given in Chapters 6 and 7. The point to be made here is that postmating mechanisms are all related to genetic incompatibility. Such incompatibility develops only through genetic divergence arising from the processes of speciation. Postmating isolation mechanisms are usually the ones that operate when two related species first come into contact under breeding conditions. Although such mechanisms maintain the genetic integrity of each species, they are of no benefit to the cross-mating individuals. These individuals essentially waste their gametes, for their offspring cannot survive or are not capable of reproduction. Thus there is a selective pressure for the evolution of premating isolation mechanisms to avoid such loss of gametes. Premating mechanisms dominate in the segregation of sympatric species.

OTHER EVOLUTIONARY PROCESSES

It is frequently assumed that all evolution comprises random mutations and the accumulation of different gene frequencies in different populations. But other processes are also at work: the chromosomes themselves evolve.

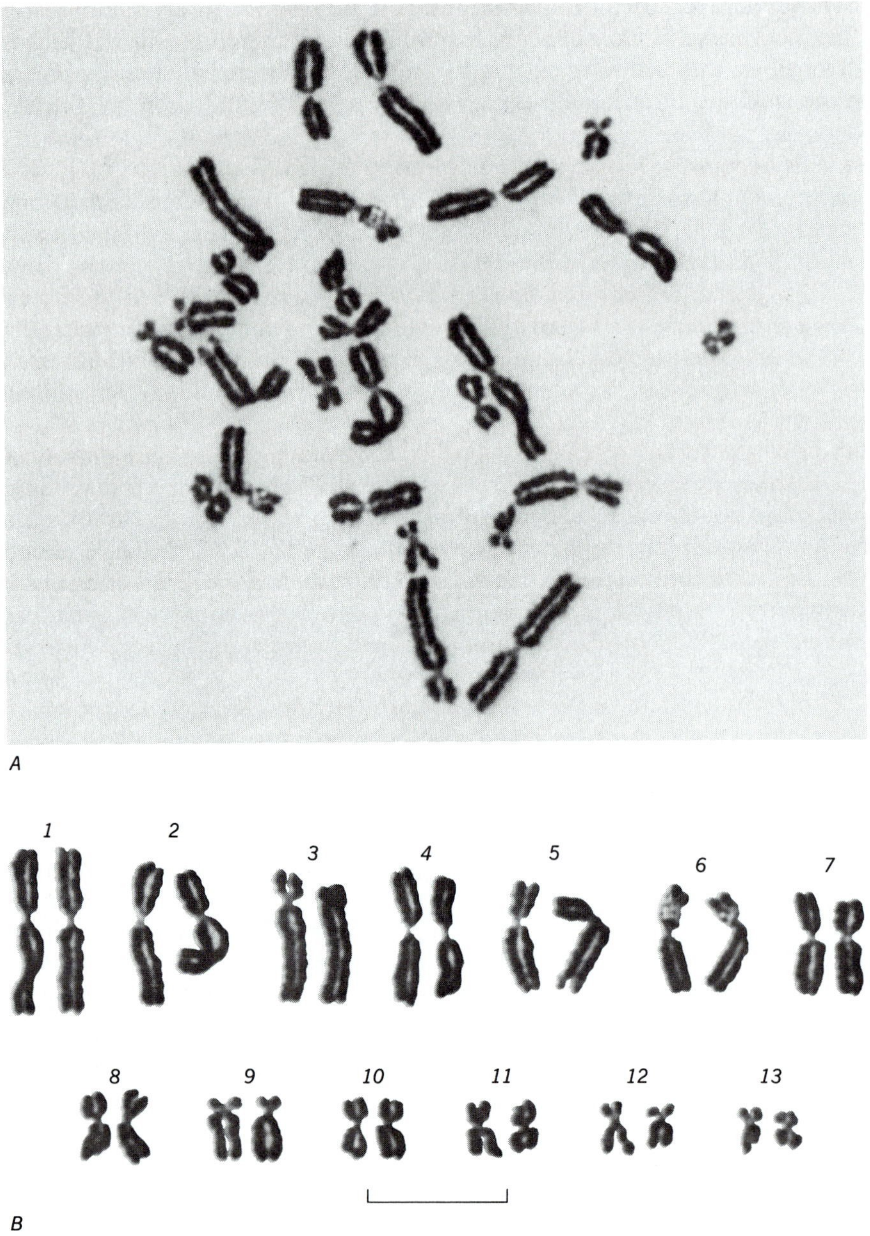

A

B

FIGURE 11–1

(A) Metaphase chromosomes of a male *Pachymedusa dacnicolor* as they appear under the microscope. (B) The same, arranged in order of decreasing length. The reference line represents a length of ten microns (10^{-5} m). [Courtesy of Charles J. Cole and *Systematic Zoology*.]

Evolution of Karyotypes

The karyotype of an organism is a description of its chromosome comple-ment. It is usually represented by a photograph or drawing of the metaphase chromosomes, arranged by pairs in order of decreasing length (Figure 11–1). Sometimes the decrease in length is gradual throughout the series, but sometimes the chromosomes are readily divided into two groups, large (macrochromosomes) and small (microchromosomes). The chromosomes differ not only in length, but also in the position of the centromere (the point to which the spindle fibers attach during mitosis). If the centromere is midway between the two ends, the chromosome is said to be metacentric. If it is closer to the midpoint than to one end, the chromosome is submetacen-tric. If it is closer to the end than to the midpoint, the chromosome is subtelocentric. These chromosome types may be spoken of collectively as bi-armed. If the centromere is at the end, the chromosome is telocentric (or acrocentric or uni-armed) (Figure 11–2). Chromosomes also differ in whether secondary constrictions are present, and in the pattern of banding revealed by special staining techniques.

There are various ways in which a karyotype can be modified. Two telocentric chromosomes can fuse to form a single, larger, bi-armed chromosome, a process known as centric fusion. Centric fission also occurs: a bi-armed chromosome can break in the region of the centromere to form two telocentric ones. The tree frogs Osteopilus septentrionalis and O. brun-neus are closely related. O. septentrionalis has a karyotype of $2n = 24$ bi-armed chromosomes, a typical hylid pattern. O. brunneus has a karyotype of $2n = 34$ chromosomes, of which 7 pairs are bi-armed and 10 pairs uni-armed. The 20 uni-armed chromosomes correspond to the sepa-rate arms of the 5 largest chromosome pairs of O. septentrionalis. Appar-ently the karyotype of O. brunneus arose from one like that of O. septentrionalis by centric fission (Figure 11–3).

Another process of karyotype modification is pericentric inversion. The part of the chromosome containing the centromere breaks loose and is reinserted upside down. In this way the shape of the chromosome can be changed. For example, a metacentric chromosome can be converted into a subtelocentric one.

Among amphibians, most chromosome evolution has apparently been by centric fusion and the loss of microchromosomes. Primitive forms tend to have a large number of chromosomes, more telocentric ones, and many microchromosomes; advanced forms have fewer chromosomes, most of which are bi-armed, and lack microchromosomes. The hynobiid salaman-ders have chromosome numbers of $2n = 40$–62 and may have as many as 50 telocentrics, including up to 12 pairs of microchromosomes. The salamandrids have $2n = 22$–24, with no telocentrics or microchromosomes.

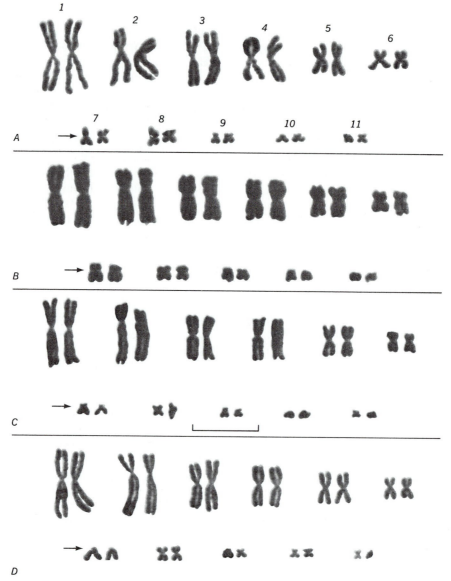

FIGURE 11–2

Karyotypes of four populations of *Sceloporus undulatus*, showing geographic variation in centromere position on chromosome number 7: (A) female *S. u. elongatus*; (B) male *S. u. hyacinthinus*; (C) male *S. u. hyacinthinus*; (D) female *S. u. tristichus*. Reference line = 10 microns. [Courtesy of Charles J. Cole and the American Museum of Natural History.]

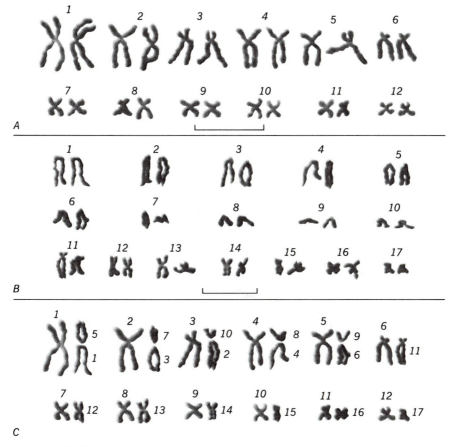

FIGURE 11–3

(A) Karyotype of *Osteopilus septentrionalis*. (B) Karyotype of *Osteopilus brunneus*. (C) The same as B with the chromosomes arranged to show their possible evolution from those in A. Reference lines = 10 microns. [Courtesy of Charles J. Cole and *Herpetological Review*.]

A similar pattern appears in frogs, though chromosome numbers are smaller ($2n$ = 14–38) and centric fission appears to occur more frequently. The sirenids have unusual karyotypes: like the primitive hynobiids, they have large numbers of chromosomes ($2n$ = 46–64), but they have no microchromosomes and few or no telocentrics. It has been suggested that the ancestor of the sirenids arose as a polyploid of a form that already had an "advanced" amphibian karyotype. Too few caecilians have been studied as yet to permit reliable interpretation.

Similar processes may have occurred in reptiles, but microchromosomes seem usually to have been retained. Among turtles, the cryptodires, usually considered more advanced, have more chromosomes ($2n$ = 48–66) than do the pleurodires ($2n = 26–34$). Many lizards and snakes have $2n = 36$ chromosomes.

We are far from understanding the significance of such changes in karyotype. Still, they have proved very useful in evolutionary studies. The chromosomes of *Amphiuma* are much larger than those of *Ambystoma*, but are otherwise nearly identical, suggesting the derivation of the amphiumids from the ambystomatids. When a question arose about whether the Sand Dune Lizard, *Sceloporus graciosus arenicolor*, might really be a race of *S. undulatus*, examination of the karyotypes resolved the question. Differences in karyotype can serve as isolating mechanisms by preventing normal meiosis.

Quantitative Changes in DNA

Throughout the course of organic evolution, there have been periodic, step-like increases in the amount of DNA per genome. These increases are not associated with increases in ploidy, but occur within the existing chromosomes. They seem usually to have occurred in lineages that have made fundamental adaptive shifts. For example, fishes have more DNA per nucleus than the lower chordates, and amphibians more than fishes. The modal value for fishes (except the lungfishes, which are a case apart) is about 1.7 picograms per nucleus. (A picogram = 10^{-12} grams.) For frogs, the modal value is about 8.4 pg. Such increases are believed to have provided the raw material for new gene complexes, which in turn gave the lineage the genetic plasticity to make a fundamental change in its way of life.

Once such a significant adaptive shift has occurred, though, much of the added DNA may be lost. Reptiles have less on the average than amphibians, about 3–5.7 pg, and birds less than reptiles, about 1.5–3.5 pg.

As a class, the amphibians are unusual in the enormous interspecific variation they show in the amount of DNA per nucleus. The quantity of DNA is correlated with several other characters. The correlation is not exact, but in general, the more DNA, the larger the cell, the lower the metabolic rate, and the slower the rate of development. It has been suggested that, in the amphibians, it is selection for precisely these effects of the amount of DNA on metabolic and developmental rates that is responsible for the great intraclass variability. The salamanders, with their inefficient oxygen intake mechanisms, have very large to enormous amounts of DNA. The primitive hynobiids have about 33–43 pg/nucleus. The forms with the highest amounts, 90–165 pg, are all paedomorphic *(Andrias, Necturus, Proteus, Amphiuma)*. The paedomorphic sirenids also have DNA amounts of over 90 pg. The more advanced salamandrids and plethodontids have intermediate amounts. Evolution in the salamanders and sirens seems to have entailed large increases in the amount of DNA per nucleus. Frogs, with their better-developed lungs, have less DNA (3–30 pg/nucleus). The forms with the smallest amounts are the desert-adapted species of *Scaphiopus,* which breed in temporary ponds and have very rapid development rates.

PATTERNS OF DISTRIBUTION

During their evolutionary history, animal groups pass through certain typical stages. Initially the group is small in number and occupies a small area. It then expands its range while simultaneously diversifying. Next comes a contraction stage, with a decrease in diversification and area of occurrence. These stages apply equally well to a species, a genus, or a higher taxonomic group. The distribution of any group of animals is the result of the interplay of two sets of factors:

    I. Extrinsic factors
        A. Distribution of favorable environments
        B. Changes in environments through geologic time
            1. Climatic
            2. Biotic
        C. Formation of highways permitting dispersal or of barriers preventing dispersal
    II. Intrinsic factors
        A. Physiological requirements of group
        B. Time and place of origin
        C. Potential rate of spread
            1. Biotic potential
            2. Vagility
        D. Genetic plasticity of group

Extrinsic Factors

An animal's distribution reflects an area of favorable environment. The environment must be favorable for all life stages, for if one life stage cannot survive, the population will soon become extinct. Both amphibians and reptiles are limited by temperature, because their ectothermic physiology requires heat from an external source. Proceeding north or south from the tropics, the number of herp species declines rapidly with the annual mean temperature. Amphibians are more cold-tolerant than reptiles, and the true frogs *(Rana)* reach the Arctic in both the Old and the New World. In the Old World, a salamander *(Hynobius keyserlingi),* a lizard *(Lacerta vivip-*

ara), and a snake *(Vipera berus)* extend into the Arctic in areas where the temperature is moderated by warm currents. The reptilian osmoregulatory system is more resistant to water loss than that of amphibians, so that reptiles are able to live in drier areas and cross extensive salt-water barriers. A favorable environment implies not only favorable physical conditions, but favorable biotic ones as well. There must be a balance between predators and prey, hosts and parasites, competitors and food.

Most environments are stable only over relatively short periods of time by geologic standards. One of the best-known changes that has taken place is the warming of the northern hemisphere over the last ten thousand years. This trend is demonstrated by the retreat of the glaciers since the close of the Ice Age. The process appears to be continuing to the present day, and many groups of animals are still extending their ranges northward. Further back in geologic time, the coal beds in Pennsylvania show that semitropical swamps once flourished in what is now a hilly, well-drained, temperate region.

Climatic aspects of the environment are not the only ones that change: biotic factors are also constantly shifting, since the forces of evolution are continually at work on all forms of life. New sources of food become available to animals able to take advantage of them; new enemies appear that must be evaded. Perhaps most important of all, new and better-adapted competitors for food and breeding sites either move into the area or evolve within it.

Although the main continental land masses have probably remained relatively constant, at least since the Cenozoic, there have been many changes in the connections between them. South America was cut off from North America by an arm of the sea for the greater part of the Cenozoic era. During this time, distinct faunas evolved in the two regions. When the land connection was reestablished during the Pliocene, it became possible for North American forms to invade South America and vice versa. The mingling has been very incomplete, and the two faunas are still essentially different. For example, only one group of salamanders *(Bolitoglossa* and its allies of the family Plethodontidae) has invaded South America.

Of course, the same topographic feature may serve as a highway of dispersal for some forms and as a barrier to others, depending on the physiological requirements of the groups in question. Broad lowland valleys are barriers to salamanders adapted to dwelling on mountaintops, but may serve as highways to other forms, such as toads.

Intrinsic Factors

Even closely related forms often show differences in their physiological requirements. Furthermore, some animals have broad ecological tolerance and are able to adapt to conditions over a wide area, whereas others are limited in ecological tolerance and hence are restricted to a narrow range. This sort of difference is probably reflected in the distribution of two of the North American Rat Snakes: the ecologically adaptable *Elaphe o. obsoleta* ranges from Ontario and northern New England south to Georgia and west to Minnesota and Texas, whereas *E. subocularis* is limited to the arid region of trans-Pecos Texas, southern New Mexico, and adjacent Coahuila.

Besides physiological requirements, which determine which environments a group can occupy and what highways of dispersal are available to it, the group's time and place of origin play an important part in determining its geographic distribution. If a group arises in a region that is blocked from an adjacent region by some barrier, it will not be able to spread into that region—even though there may be environments there that are well suited to its needs. The frog family Leptodactylidae is very widespread in South America and contains many genera, but only three of these are present in the United States; this is probably because it was only recently, in the geologic sense, that a passageway was opened.

If a group is of very recent origin, it may not have had time to spread very far from its center of origin. How fast a group will spread depends in part on its biotic potential. Animals capable of producing large numbers of off-spring in a relatively short time will, other things being equal, be able to occupy new areas more rapidly than forms with a low rate of reproduction. The vagility (inherent power of movement) of the species may also affect the rate of dispersal. Both of these factors are probably relatively minor, since many forms with low biotic potentials and limited vagility have been able to occupy large areas of the earth's surface.

Finally, the genetic plasticity of the group will determine whether it will be able to occupy new environments, whether it can adapt itself to changes in the environment *in situ,* or whether it must follow receding belts of its old environment or become extinct.

MOVEMENT OF SPECIES

The movement of a species is a populational phenomenon. An entire species does not emigrate *en masse,* abandoning its old range and establishing a new one. Instead, a few individuals move into new regions and establish new populations; populations elsewhere become extinct. The species moves slowly over decades or centuries like a huge amoeba, withdrawing from unfavorable areas (local extinction) and expanding into favorable ones (colonization). The net result may be nothing more than a continual adjustment in the edge of the species' range, or it may be an actual shift in the species' distribution, its adaptation to a different environment, or the generation of a new species from a population that has become isolated.

Expanding Populations

Most animals produce more offspring than can survive within the distributional confines of the parent population. Thus overcrowding is an ever-recurring phenomenon in any but a dying population. Individuals near the periphery of the range are constantly being pushed into new and unfavorable environments. Undoubtedly, most of them perish; however, occasionally such colonists find a favorable area and establish a new population, or adapt to an area unfavorable to their parents.

Relict Populations

Under favorable conditions, a species will thus increase its range by founding new populations along the periphery of the range. Eventually, conditions will change, and the species must retract its range if its populations are unable to adapt to the changing environment. Populations often become isolated in small pockets of favorable environment as the main range of the species retracts from them. The survival of such relict populations depends on the stability of these pockets of the original environment. The existence of relict populations does not necessarily indicate that a species is becoming extinct, but simply that the range of suitable habitat is not as great as it was. The cluster of prairie species *(Kinosternon flavescens, Heterodon nasicus, Pseudacris streckeri)* in Illinois shows that the prairie once extended that far east. The relict populations of *Salamandra salamandra* in northern Africa indicate the past existence of a continuous humid land connection to Europe. Other relict distributions, such as the confinement of *Sphenodon* to a few small islets in New Zealand, do indicate that a species is becoming extinct.

Waif Populations

A waif population is one whose ancestors reached an area (usually an island) as strays or castaways rather than through the normal dispersal of an expanding population. In times of flood, an uprooted tree or a fragment of riverbank may be swept downstream and carried out to sea. It may be caught by ocean currents and eventually stranded on the shores of a distant island. Such a natural raft may harbor a clutch of eggs, a pregnant female, or a pair of individuals of the same species. If they survive the hazards of the journey, they may be able to establish a population on the island.

It is not always easy to tell whether a population is waif or whether it reached the island at a time when the island was still connected to the mainland. In either case, speciation may have occurred, so that the population differs from any now found on the mainland. Islands that have been formed entirely by volcanic action, or that have been completely submerged since their last previous connection with another land mass, must *ipso facto* be populated by waifs.

In the past few thousand years, man has, either accidentally or intentionally, carried many animals from one part of the globe to another. It is customary, though, to speak of populations arising from animals transported by human agency as introduced rather than waif species.

The island of Bermuda is inhabited by a species of lizard, *Eumeces longirostris,* that differs greatly from all other species of its genus and is found in no other place; it is undoubtedly waif. All the other herps of the island seem to have been introduced by man.

DISTRIBUTION OF FAUNAS

Animal and plant species naturally occur where they can survive and reproduce. A species' distribution indicates its ecological and physiological requirements and limits. The edge of its range shows the presence of a barrier. Where the barrier is common to most species of a fauna, it is likely to be a geographic barrier of long standing, and nearly impossible to cross. The areas on either side represent centers of speciation and dispersal for unique faunas. The world was recognized to comprise six of these faunal regions, or zoogeographic realms, by P. L. Sclater in 1858. Although the names of these zoogeographic realms have changed somewhat, their delineation remains basically the same (Figure 11–4).

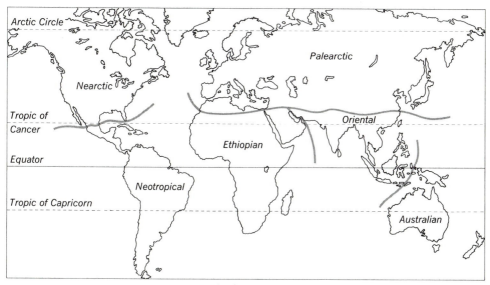

FIGURE 11–4
Zoogeographic regions of the world.

The New World is divided into the Nearctic region (North America, including the Mexican Plateau) and the Neotropic region (Central and South America). All of Europe, Africa north of the Sahara, and Asia north of the Himalayas constitute the Palearctic region. Since there is a high degree of faunal similarity between the Nearctic and Palearctic, Sclater combined them as the Holarctic region. Africa south of the Sahara, the Arabian Peninsula, and Madagascar form the Ethiopian region. The Oriental region includes all of Asia south of the Himalayas and extends southeastward through the Malay Archipelago to Celebes and the Philippines. The Australian region comprises Australia, New Guinea, New Zealand, and adjacent islands. The scattered islands of the Pacific are sometimes recognized as the Oceanic Islands region.

Although it is best not to try to define these regions too sharply, they are useful to describe gross distributional patterns and are, furthermore, deeply entrenched in the zoological literature.

CONTINENTAL DRIFT

The faunal composition of any area is a product of the area's biotic, climatic, and geologic history. The geologic history delimits the area's boundaries: when, for how long, and to what other areas it was connected, its longitudinal and latitudinal movement, and the topography of the land. The climatic history determines the past living conditions, and thus which animal groups could have survived *in situ,* which could have immigrated and become successfully established, and when. The biotic history tells which animal groups were initially present, which ones could invade, and how each adapted to the changing environment.

The continental masses have undergone considerable movement and change since the modern orders of amphibians and reptiles first appeared. Each continent consists of a plate or a series of plates of low-density rock, floating on the denser but "fluid" mantle of the earth's outer core. These continental plates have probably been sliding slowly over the surface since early in the earth's history. Approximately 300 million years ago, in the Carboniferous period, all the plates formed a single supercontinent, Pangaea, divided by a long east-west embayment, the Tethys Sea. Pangaea remained a single giant continent through the Permian.

Then, in the Triassic, the continental plates began to move apart (Figure 11–5). The North American plate drifted away from the South American and African plates, as the North Atlantic Basin began to form. The North American and Eurasian plates remained together as a large northern continent, Laurasia. The southern continent, Gondwana, began to split: the South Atlantic appeared as a narrow rift developing from the south, though

South America and Africa were still broadly connected to each other and to the Indian plate. These three combined plates began to move northward, while the combined Antarctic and Australian plates rotated eastward. Rifting continued throughout the Jurassic and the Cretaceous. The Indian plate, containing the future India and southwest Asia, separated from Africa and moved northward. Madagascar drifted away from Africa, and later, near the end of the Cretaceous, Africa and South America completed their separation.

Early in the Tertiary, the North Atlantic rift split North America from Eurasia, Australia began to drift away from Antarctica, and South America joined with North America. Later the Indian plate collided with the main Asiatic land mass and slid partly under it, thrusting up the Himalayas. The world was beginning to assume its present appearance. The classic zoogeographic realms described so many years ago correspond remarkably well to the continental plates delimited by modern geologists.

Climatic conditions on the early continents differed considerably because of the differences in continent sizes and ocean currents. Even more important were the positions of the continents on the globe. Initially, much of Laurasia was tropical and much of Gondwana was either temperate or subtropical. (During the Permian, a massive ice sheet covered much of Gondwana.) The general movement of all the plates except the Antarctic has been northward, so that the Gondwanan components have become warmer and the Laurasian ones cooler.

GEOGRAPHIC DISTRIBUTION
OF AMPHIBIANS AND REPTILES

Amphibians and reptiles are among the best animals to use in the study of terrestrial geographic patterns. Since they are ectothermic, like the majority of the world's animals, their distributional patterns are likely to be more widely applicable than those of the endothermic birds and mammals. They are more limited in vagility than the winged insects and birds or the wide-ranging larger mammals. The outlines of their primary patterns of distribution are thus less liable to be blurred by rapid, wide, and essentially random dispersal. Because of their unshelled eggs and unprotected skins, amphibians are ecologically bound to regions where fresh water is available. They are quite intolerant of seawater. We can be reasonably sure, then, that they have not fortuitously crossed extensive areas of the sea or arid desert lands *en masse*. Where essentially similar faunas occur, on either side of such a barrier, we can safely assume that they reached those regions before the barrier formed.

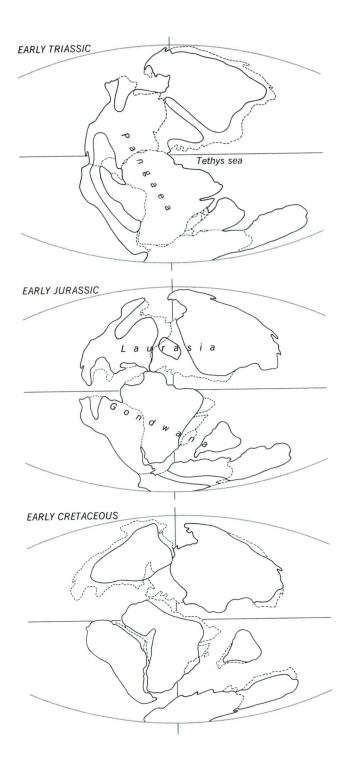

EARLY TRIASSIC

Tethys sea

P a n g a e a

EARLY JURASSIC

L a u r a s i a

G o n d w a n a

EARLY CRETACEOUS

Geographic distributions may be analyzed at different levels, and each level will yield a different historical picture. Familial distributions indicate the center of origin of a family and its probable dispersal routes. Specific patterns of geographic variation yield data on the evolution of a species and the effect of relatively recent climatic and geologic events on its movements. No matter what level is analyzed, it is essential that the true relationships of the group be known; otherwise the geographic interpretation will be incorrect.

Distribution of Amphibian Families

The caecilians are tropical animals, and probably have always been so. Their greatest diversity and most continuous distribution occurs in Africa and South America; they also occur discontinuously in the Oriental region, from the Seychelles Islands, India, and Ceylon to Java, Borneo, and the southern Philippines. An early Tertiary fossil has been found in South America, and a Pleistocene one in Europe. Caecilians probably arose in the tropics of Gondwana, and moved into Central America and the Oriental region in the early Tertiary.

The salamanders are predominantly Holarctic in distribution, and presumably originated and diversified in Laurasia. Only one subgroup of plethodontid salamanders has invaded the tropics, and this appears to have been a recent (late Tertiary and Pleistocene) radiation. The general pattern of salamander distribution today is discontinuous and relict: cryptobranchids in eastern North America and eastern Asia, proteids in eastern North America and Adriatic Europe, plethodontids in the New World and the eastern Mediterranean. These families must have enjoyed widespread distribution throughout the Holarctic at one time, perhaps comparable to the present Holarctic distribution of salamandrids. The Asian hynobiids and the North American ambystomatids may have risen from a cryptobranchid stock. The amphiumids may in turn have been derived from an ambystomatid stock, and today share a relict distribution in the southeastern United States with the sirens.

In contrast to the salamanders, the frogs probably arose in Gondwana. The xenoanurans (pipids and rhinophrynids) are Gondwanan derivatives:

FIGURE 11–5

The continents at three stages during the Mesozoic era. Solid lines indicate the shorelines of the ancient continents, broken lines the present continental shorelines. Vertical and horizontal lines represent 0° longitude and 0° latitude, respectively. Even though the continental plates were abutting, the continents may have been separated by shallow continental seas. These seas appeared and disappeared with great regularity in the past. [Composite after Dietz and Holden, *Sci. Amer.*, 1970, and Tarling and Tarling, *Continental Drift*, Doubleday, 1975.]

their present and fossil occurrence is associated with the South American and African plates. The scoptanurans (microhylids) are also Gondwanan derivatives. In view of their diversity and wide distribution in the Neotropic, Ethiopian, and Oriental regions, they may originally have been residents of the warm temperate and subtropical areas of Gondwana, exclusive of the Antarctic and Australian plates. Although they occur in the Australian region today, their greatest diversity there is in New Guinea, and they have very limited distribution in Australia proper—which suggests that they penetrated into the Australian region by invasion from the East Indies.

The lemnanurans (ascaphids, discoglossids) are Holarctic forms, and probably originated from an early penetration of primitive anurans into temperate Laurasia. Although *Leiopelma* is often considered a member of this group, its New Zealand distribution and derived characters suggest that it represents another lineage of primitive frogs from temperate Gondwana. The acosmanurans (bufonids, ranids, hylids, leptodactylids) are largely southern-hemisphere groups, but their supposed ancestral group, the pelobatids, are Holarctic—suggesting a Laurasian origin for this family, which must have expanded into tropical Gondwana and there given rise to the modern frogs. The ranids and their relatives (hyperoliids, rhacophorids) are primarily Ethiopian and Oriental. They may have arisen in Africa and dispersed several times, the older dispersals giving rise to new families, such as the rhacophorids, the more recent to the genus *Rana*, with sufficient time for high speciation, but with no great specialization of any species group. The bufonids show a predominantly Neotropic and African radiation pattern. The hylids, including pelodryadids, and the leptodactyloids (leptodactylids, heleophrynids, and myobatrachids) appear to be of temperate or subtropic Gondwanan origin. The remaining small families probably all arose *in situ*—the sooglossids on the Seychelles Islands and the others in South America—and have never spread to other areas.

Distribution of Reptile Families

The rhynchocephalians are a declining group. The single surviving member, the Tuatara, shows a relict distribution restricted to New Zealand; fossil forms are found in Africa, Eurasia, and North America. Although the crocodilians are more widespread, their present distribution, compared with that in the late Cretaceous and early Tertiary, is relictual. Fossil evidence suggests that the modern subgroups of crocodilians originated in subtropical and tropical Laurasia and spread into South America and Africa, where they underwent a secondary radiation.

Modern and fossil turtle distributions indicate a Gondwanan origin and radiation for the pleurodires and a Laurasian origin for the cryptodires, with radiations occurring in both the northern and southern continents. Many of

the early pleurodires appear to have been marine, and although a few penetrated into the northern hemisphere, their major radiation occurred in the
southern. Today, South America is the center of their distribution, since
both living families occur there. The history of the cryptodires is somewhat
more complex: their fossil forms are predominantly from the northern
hemisphere, yet some have had successful radiations in the Oriental, Ethiopian, and Neotropic regions. The trionychids and testudinids have their
greatest diversity in Africa, the kinosternids in Central America, and the
batagurines in Southeast Asia. The emydines and chelydrids are largely
Nearctic. The sea turtles (cheloniids and dermochelyids) are tropical, and
may have had their origin in the ancient Tethys Sea or the early Atlantic
Ocean.

The squamates, with their great diversity and relatively poorly known
fossil record, present an even more complex history. Today, the amphisbaenians are tropical or subtropical animals; but they have a strong fossil
record in North America, and may have risen there under tropical conditions. The gekkonids and scincids occur throughout the tropics, and both
have primary centers of radiation in Africa and the Indoaustralian area. The
iguanids are mainly a Nearctic group, yet have isolated members in
Madagascar and the South Pacific; they probably once had worldwide distribution and have been replaced in the Old World tropics by their relatives,
the agamids. The anguinids are probably Laurasian derivatives that have
expanded into the southern-hemisphere tropics and undergone a secondary
adaptive radiation. The teiids are strictly a New World group, the lacertids
Palearctic, and the cordylids Ethiopian. The chamaeleonids are now centered in Africa, but have been reported from the upper Cretaceous of eastern
Asia. The varanids are widespread in southern Asia, Africa, the East Indies,
and Australia, and are known as fossils from North America and Europe.
The xantusiids and helodermatids are North American, but fossils of both
are reported from Europe. If all these fossils are correctly allocated, these
families must once have been more widely distributed. The dibamids and
xenosaurids clearly have relict distributions, the former in eastern Asia, the
East Indies, and Mexico, the latter in Mexico and China. No fossils of
Lanthanotus of the East Indies or the pygopodids of Australia have ever
been found, so there is no evidence they ever were more widespread.

The fossil record of snakes is very poor. All three scolecophidian families
are clearly southern-hemisphere groups, and their primitiveness suggests
that they arose when Gondwana was a single continent. The henophidians
are also presumably Gondwanan. The acrochordids are Australian, the
uropeltids Indian, and the xenopeltids southwestern Asian. The boids are
widespread in the tropics and subtropics, reaching the temperate zone in
North America, and have a fossil record in North America and Europe. The
aniliids have a relict distribution in Mexico, Central America, and Southeast

Asia. They have been reported from the Miocene of North America. Like the anguinids, the viperids are probably Laurasian derivatives that underwent a secondary adaptive radiation in the southern-hemisphere tropics. Since the relationships of the colubrids remain unknown, no sensible distribution pattern can be discerned, except that the Holarctic was one center of adaptive radiation and each of the other zoogeographic regions except Australia was a radiation center. The elapids either replaced the early Australian colubrids or evolved earlier and did not permit the colubrids' establishment and radiation. The Australian and Oriental regions are centers of distribution and radiation for the elapids and their sea-snake derivatives. It is strange that the New World elapids have been so adaptively conservative. Are they really elapids, or are they separately derived from the colubrids? Such questions are one of the reasons for studying distributional patterns, for patterns that do not fit the known dispersal routes may indicate errors in the accepted classification.

READINGS AND REFERENCES

Cole, C. J. "Evolution of parthenogenetic species of reptiles." *In* Reinbroth, R. (ed.), *Intersexuality in the Animal Kingdom*. New York: Springer-Verlag, 1975.

Cracraft, J. "Continental drift and vertebrate distribution." *Annual Review of Ecology and Systematics,* vol. 5, 1974.

Darlington, P. J., Jr. *Zoogeography: The Geographical Distribution of Animals*. New York: Wiley, 1957.

———. *Biogeography of the Southern End of the World*. Cambridge: Harvard University Press, 1965.

Dobzhansky, T. *Genetics of the Evolutionary Process*. New York: Columbia University Press, 1970.

Gorman, G. C. "The chromosomes of the Reptilia." *In* A. B. Chiarelli and E. Capanna (eds.), *Cytotaxonomy and Vertebrate Evolution*. New York: Academic Press, 1972.

Laurent, R. F. "La distribution des amphibiens et les translations continental." *Memoires du Muséum National d'Histoire Naturelle,* ser. A, vol. 88, 1975.

Lewontin, R. C. *The Genetic Basis of Evolutionary Change*. New York: Columbia University Press, 1974.

MacArthur, R. H. *Geographical Ecology: Patterns in the Distribution of Species*. New York: Harper and Row, 1972.

Morescalchi, A. "Chromosome evolution in the caudate amphibians." *Evolutionary Biology,* vol. 8, 1975.

———. "Karyology and vertebrate phylogeny." *Bolletin di Zoologia,* vol. 37, no. 1, 1970.

Olmo, E. "Quantitative variations in the nuclear DNA and phylogenesis of the Amphibia." *Caryologia,* vol. 26, no. 1, 1973.

Savage, J. M. "The geographic distribution of frogs: Patterns and predictions." *In* J. Vial (ed.), *Evolutionary Biology of the Anurans.* Columbia: University of Missouri Press, 1973.

Sill, W. D. "The zoogeography of the Crocodilia." *Copeia,* no. 1, 1968.

Tarling, D., and M. Tarling. *Continental Drift: A Study of the Earth's Moving Surfaces.* Garden City: Anchor Press, 1975.

Udvardy, M. D. F. *Dynamic Zoogeography with Special Reference to Land Animals.* New York: Van Nostrand Reinhold, 1969.

Williams, E. E. "The ecology of colonization as seen in the zoogeography of anoline lizards on small islands." *Quarterly Review of Biology,* vol. 44, 1969.

Williams, G. C. *Adaptation and Natural Selection.* Princeton: Princeton University Press, 1966.

CHAPTER TWELVE

CAECILIANS, SIRENS, AND SALAMANDERS

CAECILIANS, SIRENS, AND SALAMANDERS are treated together in this chapter simply as a matter of convenience. They are mostly small, inconspicuous, secretive animals, so it is not surprising that they are the least familiar of the herps. The burrowing caecilians are few in number and restricted to the tropics, so they are little more than names and pictures in a book to most biologists. The sirens have only two modern genera, both confined to the southern United States. The salamanders are somewhat more numerous, and because of their availability in the north temperate zone they have been the subject of many excellent studies on their life history, evolution, and ecology. *Necturus,* of course, is familiar to most comparative anatomy students. Much of our knowledge of how organs are differentiated during development has come from experimental transplantation of tissues on the embryos and early larvae of *Ambystoma* and *Notophthalmus.* Otherwise, the salamanders have been almost ignored by experimental biologists, and most people continually confuse them with lizards.

ORDER GYMNOPHIONA (APODA)

A caecilian is a slim, wormlike creature with no limbs or limb girdles and practically no tail. Most are not more than 240–300 mm long. The body is usually divided into segments by a series of folds in the skin that enhance the wormlike appearance. Frequently, secondary folds are present between the primary ones. The vent is close to the posterior end of the body on the ventral side; the eyes are minute, without lids, and are buried in the skin; the skull is compact; the intestine is not differentiated into large and small portions. An unusual sense organ—the tentacle—is present in all caecilians and only in caecilians. It grows forward along the side of the brain and emerges from the skull either at the eye socket or at a point in front of and below the eye. It may not appear externally until after metamorphosis. Its exact purpose is unknown, although an olfactory function is suspected. Adults lack gills or gill slits. Fertilization is internal, and the cloaca of the male is modified to form a protrusible copulatory organ. The more primitive genera have tiny dermal scales imbedded in the skin; the scales are apparently a heritage from the early, scaled amphibians of the Carboniferous.

Caecilians are primarily forest animals. They have been reported from savanna areas, but only along rivers bordered by patches of forest. Members of one family, Typhlonectidae, are river dwellers. The rest are probably terrestrial and fossorial, living in burrows in damp earth. Snakes, which can hunt caecilians in their burrows, are probably their main predators.

The caecilians are divided into four families, whose relationships remain to be discovered. Their distribution is shown in Figure 12–1.

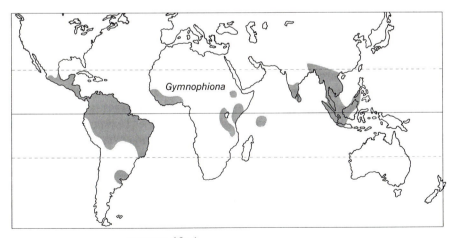

FIGURE 12–1
Distribution of the caecilians (order Gymnophiona).

Family Ichthyophiidae

The ichthyophiid family consists of four genera—*Ichthyophis, Rhinatrema, Caudacaecilia,* and *Epicrionops*—occurring in both Asia and South America. Adults are terrestrial and have distinct tails, at least some scales, and from two to four secondary folds on each body segment. *Ichthyophis,* a native of Asia, breeds in the spring. The female prepares a burrow in moist ground close to running water. She coils her body around the twenty or more relatively large-yolked eggs and guards them zealously from predacious snakes and lizards. The eggs absorb water and gradually swell until they are about double their original size. The larva at hatching weighs approximately four times more than the newly laid egg. The external gills are lost soon after the larva hatches. The young go through a long aquatic stage before they metamorphose into burrowing, terrestrial adults. Members of the genus *Rhinatrema* of northern South America also lay eggs from which hatch aquatic larvae with external gills.

Family Typhlonectidae

Typhlonectids lack tails, scales, and secondary folds. They are the most aquatic of all caecilians. Some species have been taken in the nets of fishermen who were seining streams with gravelly or rocky bottoms. The typhlonectids are ovoviviparous, and since the embryo has no placental or yolk-sac connection to the uterine wall, the source of its energy is an enigma. It has been suggested that it feeds on the soft tissue of the uterine wall. The family is confined to South America.

Family Caeciliidae

By far the largest family of caecilians, and also the most widespread are the caeciliids. The hundred-odd species are divided into two subfamilies: the caeciliines comprise the South American *Caecilia* and *Oscaecilia;* the dermophiines, with 23 genera, are scattered throughout Asia, Africa, and South America. This heterogeneous assemblage is hard to characterize. The tail is indistinct. The larvae may or may not be aquatic, but the adults are always terrestrial. The life histories of most forms are unknown, but *Gymnopis* and *Geotrypetes* retain their eggs in the oviducts until they hatch. The wall of the oviduct has compound oil glands, and the larvae subsist by eating the wall and its oil droplets. They metamorphose before they are born, and thus the newborn young are replicas of the adults.

Family Scolecomorphidae

The small scolecomorphid family includes only a single African genus with six species. These caecilians lack secondary folds, scales, and a tail, but have very large tentacles. In the young the eye is covered by a bone of the skull, but as the tentacle grows forward it may carry the eye forward, so that in some large specimens the eye moves from under the bone. Unlike other caecilians, members of this family lack a stapes. Little is known of their life history, but embryos with curious, large, branching gills have been reported for one species.

ORDER CAUDATA (URODELA)

As the name of the order indicates (*cauda* = tail), a salamander retains its tail throughout life instead of losing it at metamorphosis, as does a frog. The head and trunk regions are distinct. Two pairs of legs and a primitive, poorly developed sternum are present. Fertilization is either external or internal by means of spermatophores. Most salamanders are oviparous. The larvae, which closely resemble the adults, have true teeth in both jaws. Except in the perennibranchs, which are permanent larval types that retain the gills as adults, the lateral line system is lost at metamorphosis.

The salamanders of today do not show the extensive adaptive radiation displayed by many of the other tetrapods. Some are terrestrial, some aquatic, and some may live in either environment. A number are fairly efficient burrowers. Many of the tropical forms have become arboreal, living and reproducing in bromeliad plants. The majority of the species must stay in or on moist ground, or entirely in water.

Living salamanders are divided into seven families comprising about 310 species (Figure 12–2). Many attempts have been made to arrange these families into natural groups; they are here divided into two suborders.

SUBORDER CRYPTOBRANCHOIDEA

The cryptobranchids and hynobiids, known collectively as cryptobranchoids, are the most primitive of the salamanders and the only ones known to have external fertilization. Other salamanders have a complex of glands in the cloaca that contribute to the formation of the spermatophores. The cryptobranchids and hynobiids possess only one type of cloacal gland, and spermatophores are not formed (there is one exception, *Ranodon sibiricus*). The eggs are laid in gelatinous sacs.

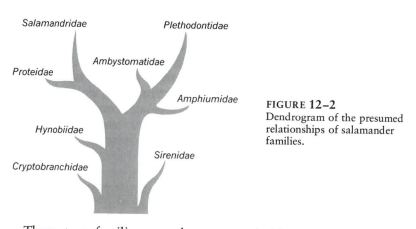

FIGURE 12–2
Dendrogram of the presumed relationships of salamander families.

These two families are also more primitive structurally than other salamanders. The earliest amphibians resembled the fishes in having many bones in the skull. In the evolution of the group, the tendency has been toward a reduction in the number of bones through loss or fusion. In the cryptobranchids and hynobiids, two of the bones of the lower jaw, the angular and prearticular, are still separate. In the higher salamanders they are fused.

Family Cryptobranchidae

The cryptobranchid family contains the giant salamanders and Hellbenders—squat, ungainly water animals that never completely metamorphose. In contrast to the hynobiids, adult cryptobranchids lack eyelids (as do all larval salamanders) and retain larval teeth. Although they never leave the water, they undergo a partial metamorphosis, and adults of both New and Old World forms lose their gills. In the American genus, *Cryptobranchus,* one gill slit remains open, whereas in *Andrias* of eastern Asia all are closed in the adult. The cryptobranchids also differ from the hynobiids in having flattened skulls from which certain bones (lacrimals and septomaxillaries) have disappeared. *Andrias,* which reaches a length of 160 cm, is the largest salamander, and indeed the largest living amphibian. *Cryptobranchus* is about 68 cm long (Figure 12-3).

Although it was apparently more extensive at one time, this family is now represented by only these two genera. The Hellbender of the eastern United States, *Cryptobranchus alleganiensis,* ranges from New York south to Georgia and west to the Ozarks. *Andrias* comprises two species, *A. japonicus* of Japan and *A. davidianus* of China (Figure 12–4).

Like the hynobiids, the cryptobranchids have external fertilization. *Cryptobranchus* mates in the late summer. The male excavates a nest in the stream bottom beneath a sheltering object, usually a flat rock. He allows

FIGURE 12–3
The Hellbender of the eastern United States, *Cryptobranchus alleganiensis.*

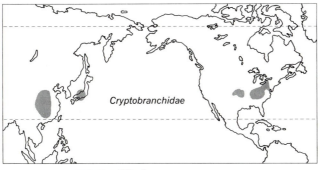

FIGURE 12–4
Distribution of the cryptobranchids.

females that have not deposited their eggs to enter the nest, but drives spent females and other males away. The eggs are laid in long, rosarylike strings, one from each oviduct; these form a tangled mass at the bottom of the nest. As many as 450 eggs may be deposited by a single female, and at times several females may lay in a single nest. As the eggs are deposited, the male discharges a whitish, cloudy mass that consists of the seminal fluid and the secretions of the cloacal glands. While the eggs develop, the male often lies among them with his head guarding the opening of the nest. The incubation period lasts between 10 and 12 weeks. When the young larvae hatch they are about 30 mm long; they lose their gills when they are about 125 mm long and about 18 months old.

Family Hynobiidae

The hynobiids—Asiatic land salamanders—undergo a more complete metamorphosis than the cryptobranchids. They develop eyelids, nonlarval teeth, and other metamorphic characters. A stapes develops in the larva and remains separate in the adult, and an operculum is present in adults of only some species.

The family is concentrated in eastern Asia and the adjacent islands (Figure 12–5). There is one widespread, primitive genus, *Hynobius,* whose range

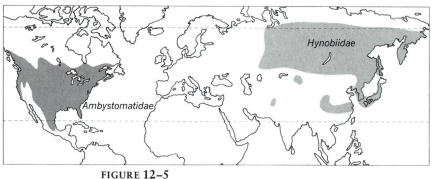

FIGURE 12–5
Distribution of the hynobiids and ambystomatids.

extends from Japan to western Asia. (*H. keyserlingi* is sometimes placed in a separate genus, *Salamandrella.*) The other four genera apparently were all independently derived from *Hynobius,* and are all included within its geographic range. *Pachypalaminus* and *Batrachuperus* have well-developed, horny epidermal pads on the soles, palms, and digits. *Batrachuperus* differs from both *Hynobius* and *Pachypalaminus* in having the vomerine teeth separated into two small, isolated patches, rather than in a V-shaped series. *Ranodon* and *Onychodactylus* are mountain-stream forms. Salamanders that live in mountain torrents must keep themselves from being carried downstream by the current. Air-filled lungs make an animal buoyant and increase the chances of its being swept away. Moreover, the water in mountain streams is cool and highly oxygenated. A salamander living in cool water has a low body temperature, and consequently a low metabolic rate. It does not need as much oxygen as a warm-water form, and is able to get what it needs from the oxygen-rich water through cutaneous respiration. In such an environment, the disadvantages of lungs outweigh their advantages, and they tend to be reduced or absent. Thus lungs are small in *Ranodon,* and *Onychodactylus* has become completely lungless, paralleling the plethodontid salamanders of the Americas.

Hynobiids have not developed the rather elaborate courtship behavior shown by other salamanders. The male is stimulated to sexual activity by the extruding egg sacs, and his sole response to the female is apparently an attempt to push her away as he fertilizes the eggs. *Batrachuperus karlschmidti* is a common salamander of small mountain streams at elevations of about 1.8 to 4 km in western China. The female attaches the egg case under or to the side of a large stone in flowing water. The end attached to the stone is flat and sticky, and the body of the case is a cylindrical tube that is largest in the middle and smaller toward the transparent free end. The free end is covered with a smooth, rather delicate, cuplike cap. This cap is forced off by the movement of the fully developed embryos, which free themselves through the hole thus formed. An individual egg case contains from 7 to 12 eggs or developing embryos. Since as many as 45 eggs in the same stage of development have been taken from a single specimen, each female presumably deposits 5 or 6 separate egg cases. The larvae are fairly typical salamander stream larvae.

Ranodon sibiricus has been reported to deposit a sticky spermatophore, to which the female attaches her eggs.

SUBORDER SALAMANDROIDEA

The five families constituting the salamandroids have internal fertilization. Males have three sets of cloacal glands and produce a spermatophore; females have a spermatheca for storing sperm. Reproductive behavior grades from the deposition of clusters of spermatophores and their eventual discovery and retrieval by females in some ambystomatids to more specialized dancing and clasping in male salamandrids and plethodontids, by which the impregnation of the females is ensured.

Family Ambystomatidae

The ambystomatids are North American salamanders, usually rather sturdily built, broad-headed, and of small to medium size. The majority of the adults are terrestrial, and an aquatic larval stage is always present. The sides of the body are marked with vertical grooves, the costal grooves, which mark the position of the ribs. Three subfamilies are commonly recognized. Their distribution is shown in Figure 12–5.

Subfamily Ambystomatinae.   The ambystomatines comprise the two genera *Ambystoma* and *Rhyacosiredon*. The diverse and widespread *Ambys-*

toma ranges from the Atlantic to the Pacific coast and from Canada to Mexico. Of the roughly twenty species, the ones with consistently terrestrial adults are centered in the eastern United States; the species frequently possessing neotenic—and hence aquatic—adults (axolotls) are centered on the Mexican Plateau. *A. tigrinum* populations occurring in both places show a similar pattern: typically neotenic in Mexico, only sporadically so in the eastern United States (Figure 12–6). *Rhyacosiredon*, with four species, is confined to the high mountains at the southern edge of the Mexican Plateau.

FIGURE 12–6
The Barred Tiger Salamander, *Ambystoma tigrinum mavortium,* a typical ambystomatid.

The breeding habits of *Ambystoma jeffersonianum* are probably typical of the family. Adults migrate to the breeding ponds in early spring. Females usually outnumber males, and must often bid for attention. There is a characteristic Liebespiel before the spermatophore is deposited. The female lays the eggs in small, cylindrical masses that contain an average of 16 eggs. These are deposited in quiet pools, attached to slender twigs or other objects below the surface. Since the female's egg complement may total over 200, it takes a number of such masses to complete egg deposition. The upper part of the egg is dark brown or black, a pigmentation characteristic of amphibian eggs laid in the open. Under field conditions, the incubation period ranges from about 30 to 45 days. The dark-colored hatchling is about 12 mm long and has well-developed balancers; the forelegs are represented by elongate buds directed backwards, but the hindlegs are not yet developed. The tail fin is continuous with the back fin, which extends almost to the base of the head. Metamorphosis usually takes place two to four months after hatching, and the transforming young may be found from July through September.

Most species of *Ambystoma* lay their eggs in water in the early spring, but *Ambystoma opacum* lays them on land in the fall. The female coils around them and protects them. The young, which have all the typical larval characteristics, hatch when winter rains begin and then make their way to the water.

Subfamily Dicamptodontinae. The dicamptodontine subfamily comprises two species confined to the damp forest and associated streams of the Pacific coast from British Columbia to California. *Dicamptodon ensatus* is one of the largest terrestrial salamanders, attaining a reported length of 271 mm. Eggs are usually laid in the spring-fed shallows of creeks and lakes. Larval development is prolonged, and by the time metamorphosis occurs the larva is nearly adult size. Recently, large, sexually mature larvae were discovered that appear to represent a new species, *D. copei*. Presumably it is paedogenetic, but we know too little of its life history to be certain.

Subfamily Rhyacotritoninae. *Rhyacotriton* is a small, mountain-stream genus with vestigial lungs, the only representative of the rhyacotritonine subfamily. Lungs cause buoyancy problems in aquatic salamanders, and have tended to be lost or reduced in mountain-stream forms. Many salamanders have an ypsiloid (Y-shaped) cartilage extending forward from the pelvic girdle on the ventral side, which helps control the shape of the inflated lungs. It is reduced in *Rhyacotriton,* as it is in most lungless salamanders.

Family Amphiumidae

The amphiumids are the smallest family of salamanders. There is only one genus *Amphiuma*. Three species, all familiarly but inappropriately known as Congo Eels, live today. They are restricted in distribution to the southeastern United States (see Figure 12–13). Amphiumas are dark-colored, semilarval animals with long, cylindrical bodies and tiny, useless arms and legs (Figure 12–7). Large specimens of *Amphiuma tridactylum* may be more than 90 cm long. Adults lose their gills, though one pair of gill slits remain open. They also show other larval characteristics, such as the lack of eyelids. They have costal grooves and lungs, but no ypsiloid cartilage. Amphiumas are savage, and a large one can inflict a painful bite.

   Courtship takes place in the water. Unlike most salamanders, in which the female plays a passive role, several Amphiuma females may compete for the attention of a male by rubbing his body with their snouts. The spermatophore is simple, and is apparently transferred directly from the cloaca of the male to the cloaca of the female during a mutual embrace. The eggs are laid in depressions beneath old logs or boards in shallow water, which frequently dries up, so that development is finished out of water. The eggs are guarded by the female. The rosarylike strings contain about 150 eggs,

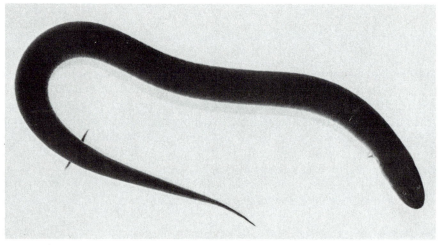

FIGURE 12–7

The Two-toed Amphiuma, *Amphiuma means,* of the southeastern United States. Its eellike form has led to the common name Congo Eel.

each approximately 9 mm in diameter. The newly hatched larvae are from about 60 to 75 mm in total length, of which about 10 mm is tail. They have short, white gills, and are dark brown above and on the sides. Many small, round, light-colored dots are scattered over the skin.

Family Plethodontidae

The plethodontid family is the largest and most successful group of living salamanders. Its members are mostly small to medium in size and, for salamanders, occupy a wide variety of habitats. Some are aquatic, some terrestrial; some are burrowers, some arboreal. *Typhlomolge, Haideotriton,* and *Typhlotriton* are blind, white salamanders found in caves and underground waters. The first two are permanent larvae, but *Typhlotriton* metamorphoses, and only the adults are blind. All plethodontids are lungless and lack the ypsiloid cartilage; costal grooves are present. The vomerine teeth lie in patches on the surface of the parasphenoid bone. A small, gland-lined groove—the nasolabial groove—runs from the nostril to the upper lip on each side; it apparently transfers olfactory sense data from the substrate to Jacobson's organ. The middle-ear bones of plethodontids differ from those of all other salamanders. The operculum develops first, and fills the oval window. The stapes is either absent or present as a slender rod fused to the operculum; it does not enter the oval window. The operculum of other salamanders usually forms at metamorphosis by a pinching-off of part of the otic capsule. In plethodontids, it develops by an ingrowth of cells from the margin of the oval window.

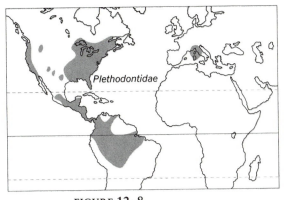

FIGURE 12–8
Distribution of the plethodontids.

The plethodontids appear to have arisen in eastern North America (Figure 12–8). Most of the common land salamanders of the United States belong to this family. One group (*Bolitoglossa* and related forms) extends into Central America and northern South America, where it has undergone an extensive adaptive radiation. Another genus, *Hydromantes,* has species in California and southern Europe.

Life histories of the plethodontids include all gradations from the aquatic to the completely terrestrial. Hedonic glands are well developed and widely distributed over head, body, and tail of most males. Courtship includes rubbing and prodding of the female, and the "tail-walk," in which the pair move along with the female straddling the tail of the male. The Rusty Mud Salamander, *Pseudotriton montanus floridanus,* is an aquatic type. Its eggs are deposited in small groups on tiny rootlets and other submerged objects in cool, muddy springs (Figure 12–9). In a single clutch, found hanging from rootlets at the edge of an undercut bank, there might be 27 eggs in tiny clusters of 2 to 8 eggs. The female stays with the eggs. It is not known how long embryonic development takes, but the young hatch at lengths of about 12 to 14 mm. Balancers are absent in this species.

The Dusky Salamander, *Desmognathus fuscus,* lays its eggs in small clusters on land—in shallow excavations in soft earth, within beds of sphagnum, or beneath stones or logs, usually near water. Egg clusters of the southern race contain 9 to 19 eggs approximately 5 by 7 mm in size, each attached by a short, twisted stalk, about 2 mm long, to a common base along a rootlet. The cluster has the appearance of a bunch of grapes. When the young hatch they are 16 to 20 mm in total length. They do not go to the water at once, but remain in the nest with the mother and show definite terrestrial adaptations. The hindlimbs are longer in proportion to the trunk region than at any time during later development. These young thus are not merely larvae that have not had a chance to reach the water, but are basi-

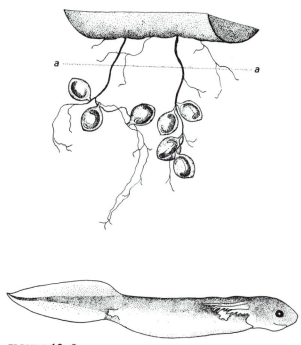

FIGURE 12–9
The eggs and newly hatched larvae of the Rusty Mud Salamander, *Pseudotriton montanus floridanus;* the line *a–a* represents the waterline. [After Goin, *Nat. Hist. Misc.,* 1947.]

cally terrestrial salamanders, able to move about in the damp crannies and crevices that lead from the nest to the nearest pool or stream. After one or two weeks, they enter the water to exist as aquatic larvae until they are 7 to 9 months old. They are about 45 mm long when they metamorphose.

The Red-backed Salamander, *Plethodon cinereus,* of the eastern United States, has a life history typical of terrestrial plethodontids. The female lays 3 to 12 large, unpigmented eggs in crannies and holes in partly decayed logs. Each egg adheres to the one that preceded it, to form a small mass seemingly contained in a single envelope. This egg cluster is usually attached to the roof of a small cavity in the log. The embryos develop rapidly and soon show large, well-developed external gills. The gills are lost at hatching, and the young have the same form as adults. They never take up an aquatic larval existence.

The family Plethodontidae is divided into two subfamilies:

Subfamily Desmognathinae. The desmognathines include three genera, *Desmognathus, Leurognathus,* and *Phaeognathus;* all of them are found in the eastern United States. They differ from the other plethodontids in the way they open their mouth. The lower jaw is held in a relatively rigid position by ligaments joining it to the atlas, and the mouth is opened by raising the upper jaw and the skull proper. The animals are streamlined, with flattened, well-ossified skulls. The hindlimbs are larger than the forelimbs. *Leurognathus* is completely aquatic; *Phaeognathus* is a large, terrestrial, burrowing form. *Desmognathus* includes the majority of the species: some are semiaquatic, others terrestrial.

Subfamily Plethodontinae. The large plethodontine subfamily contains twenty genera and many species. The mouth is opened by dropping the lower jaw. The tribe Hemidactylini of eastern North America primarily includes aquatic forms such as *Gyrinophilus*, although *Eurycea* has some strongly terrestrial species. *Plethodon* and *Aneides* of the tribe Plethodontini are predominantly terrestrial and occur on both coasts (Figure 12–10). The Bolitoglossini are the tropical salamanders of Central and South America; they are largely, but not exclusively, arboreal.

FIGURE 12–10
The Slimy Salamander, *Plethodon glutinosus*, a typical plethodontine salamander.

Family Proteidae

The proteids are two genera of aquatic salamanders. They are permanent larvae that retain the gills and two pairs of gill openings as adults. They develop lungs and have costal grooves, but lack nasolabial grooves, the ypsiloid cartilage, maxillary bones, and eyelids. The angular and prearticular bones are fused.

Proteus, the blind, white cave salamander of Adriatic Europe, has a long, pigmentless body, bright red gills, and skinny appendages that have only three fingers and two toes. Members of the genus *Necturus* have pigmented bodies, stouter limbs, and four toes on each foot; these are the Mud Puppies of the eastern United States. There are four species of *Necturus,* only one of *Proteus* (Figure 12–11).

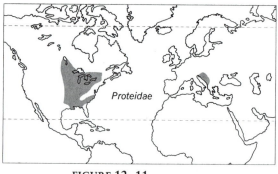

FIGURE 12–11
Distribution of the proteids.

Fertilization in the proteids is internal. Little is known of the actual court-ship, but in the Mud Puppy, *Necturus maculosus*, mating apparently takes place in the fall and the eggs are laid the following spring. The nests are excavations beneath stones, boards, or other objects, lying in water at depths of 10 to 150 cm. In a lake habitat, nests are found 4 to 8 m from shore; they have also been found in streams. The eggs are deposited singly and are attached to the undersurface of the object sheltering the nest. Clutch size varies from 18 to 180; clutches deposited in streams are usually larger. For example, in five lake nests examined, the average number of eggs per nest was 66, whereas in three stream nests it was 107. The eggs, which may be as large as 10 or 11 mm in diameter, hatch after 4 or 5 weeks. The newly hatched larvae are 20 to 25 mm long. There is, of course, no metamor-phosis, since the proteids are perennibranchs.

Proteus usually lays eggs, but sometimes retains them in the oviduct, where one or two of the young develop. They are born as miniature replicas of the adults. No special modifications, either of the larvae or of the oviducts, are known to accompany this change in life history.

Family Salamandridae

The salamandrids are the typical salamanders and newts. They usually metamorphose completely and spend at least part of their lives on land, though occasional neotenic populations are found. Salamandrids have lungs and an ypsiloid cartilage; costal grooves are not evident. The vomerine teeth are in two long rows, one on either side of the parasphenoid. The stapes fuses to the ear capsule and does not appear as a distinct element in adults.

In ancient times, the European *Salamandra* was credited with being able to live in fire. It is easy to see how such a myth arose. Salamanders some-times take shelter in crevices in fallen logs. When such a log is thrown on a fire, the animal is roused by the heat and tries to escape, seeming to the unknowing to emerge from the flame. Even today, objects that can withstand a great amount of heat or that are used in producing heat are sometimes called salamanders.

Considerable confusion often arises about the words *salamander, newt,* and *eft. Salamandra* is a Greek word meaning a lizardlike animal. It is used as the generic name for the common terrestrial tailed amphibians of Europe. As a common noun, the word *salamander* is applied generally to all the caudate amphibians, and especially to the more terrestrial salamandroids. *Newt* and *eft* come from the Anglo-Saxon *efete* or *evete,* a word used for both lizards and salamanders. In medieval English this word became *ewt* and finally "a *newt*" ("an ewt"). Since the only caudates found in England are moderately aquatic members of the family Salamandridae, *newt* has come to be used as the common name of the more aquatic salamandrids. Larvae of some of the American newts *(Notophthalmus)* frequently metamorphose into tiny spotted creatures, often bright red or orange, that leave the water and live on land for as much as three years. These land creatures are called efts, red efts or spring lizards. The eft later returns to the water, develops a tail fin and adult coloration, and changes into a sexually mature, aquatic newt. This condition is not fixed, even in a single species, for the larva may metamorphose directly into the aquatic adult. The eft stage is not known in European newts, but adults frequently spend part of their time on land.

The salamandrids are a widespread family, occupying Europe, eastern Asia, and both eastern and western North America (Figure 12–12). There

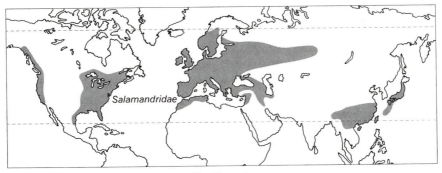

FIGURE 12–12
Distribution of the salamandrids.

are about 40 species, most of them Palearctic. Best known are the European salamanders of the genus *Salamandra* and the European Newts, *Triturus. Pleurodeles, Triturus,* and *Salamandra* extend into northern Africa and are the only salamanders known from that continent. *Cynops, Triturus,* and

Tylototriton are found in Asia. Only two genera of newts, *Notophthalmus* in the east and *Taricha* in the west, occur in North America.

The salamandrids exhibit a wide variety of rather elaborate courtship patterns. Sight, smell, and touch may all play a part in arousing the female to pick up the spermatophore. During the breeding season, the males of some species develop vivid colors and special structures, such as high dorsal crests, which they display before the females. The male may also stimulate the female by the secretions of special hedonic glands, by rubbing, prodding, nipping, or carrying her on a "piggyback" ride. Male Mountain Newts of Europe *(Euproctus)* actually clasp the female, and may place the spermatophore directly in her cloaca. Variation in courtship pattern is probably an important isolating mechanism.

The Red-spotted Newt of eastern North America *(Notophthalmus v. viridescens)* mates in water in the early spring. The male seizes the female and rubs her with his cheeks and chin, smearing her with the odorous secretion of his hedonic glands. He then moves a short distance and deposits a spermatophore. She follows, takes it into her cloaca, and, some hours later, begins to lay. A female may deposit from 200 to 375 eggs; the period of deposition sometimes lasts for several months. The eggs are laid singly, usually fastened to a leaf or the stem of a small plant, in quiet waters; less often they may be attached to the surface of a stone. The egg is pigmented, with the animal pole varying from light to dark brown. The period of incubation varies from about 20 to 35 days. The larva at hatching is about 7 or 8 mm long and has well-developed balancers, one on each side of the head just below the eye. The forelegs are short, blunt buds, and the hindlegs are undeveloped. As in the larvae of ambystomatids and most other salamanders that breed in quiet waters, a well-developed dorsal keel, continuous with the tail keel, extends nearly to the base of the head. Metamorphosis usually takes place in late summer or early fall, after a larval period of two or three months.

The European *Salamandra salamandra* and *S. atra* retain the eggs in the oviducts for at least part of the developmental period. The young *S. atra* are born fully metamorphosed. The female *S. salamandra* goes to water to bear the young, which are usually born as late-stage larvae. If the embryos of these two species are dissected from the oviduct, they are found to have long, filamentous gills and rudimentary balancers, indicating that these forms evolved from salamanders having pond-type larvae.

ORDER MEANTES

The sirens, unique "salamanders" of the southern United States, are permanent aquatic larvae that develop few adult characteristics. The maxillaries are tiny and float in connective tissue, and the coracoid ossifies as a separate element. The eyes are tiny. The forelegs are minute, and the hindlegs are lacking entirely. The expanded haemal arches of the tail vertebrae resemble those of the extinct nectridians. The upper jaw lacks marginal teeth, the lower has patches of teeth only on the splenial bones. The vomerine teeth lack the zone of weakness between crown and pedicel that is found in the teeth of frogs and salamanders. Both jaws are sheathed with horn. A Jacobson's organ, lungs, and three pairs of external gills are present.

Sirens lack glands of Leydig, but have unicellular glands of a different kind scattered through the skin. The ventricle of the heart is almost divided by an interventricular septum. It resembles that of some reptiles, rather than that of the salamanders.

The urogenital system of the sirens differs from that of the salamanders. Glomeruli are well developed in the anterior part of the kidney, the Wolffian duct is relatively straight and enclosed in the kidney, and the ducts from the posterior part of the kidney show no tendency to fuse into a ureterlike structure. The kidneys fuse posteriorly to form a "tail kidney." The male lacks both a Bidder's canal and a vestigial Müllerian duct. Sirens are unusual in the structure of the spermatozoon. It has two axial filaments, each with an undulatory membrane bordered by a flagellum (see Figure 6–2). In addition to these morphological differences, the sirens have striking biochemical differences from the salamanders: the adenosine deaminase enzymes of salamanders are more like those of frogs than of sirens.

The family Sirenidae includes two recent genera: *Siren* with two species and *Pseudobranchus* with one. It is now restricted to the southeastern and south central United States and extreme northeastern Mexico (Figure 12–13). *Pseudobranchus* is a slender little animal with only three toes on each

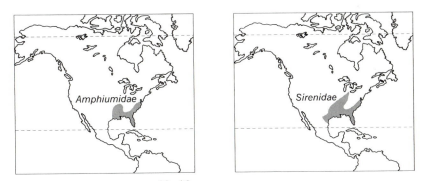

FIGURE 12–13
Distribution of the amphiumids and sirens.

foot and a single pair of gill slits. Members of the genus *Siren* have four toes on each foot and three gill slits. They are much longer and more heavily built than *Pseudobranchus*. Indeed, *S. lacertina,* which reaches a length of 90 cm, is one of the largest living amphibians (Figure 12–14).

FIGURE 12–14
The Greater Siren, *Siren lacertina*, an abundant amphibian in the southeastern United States.

Courtship and egg deposition have not been observed in any of the three species, but there is evidence that the sirenid reproductive pattern differs from that of any known salamander. Males lack the cloacal glands, which contribute to the formation of the spermatophore in salamanders with internal fertilization, and females lack the spermatheca for sperm storage. The absence of these structures has been taken to indicate that sirenids have external fertilization. Among salamanders with external fertilization, the eggs are enclosed together in gelatinous sacs or strings; either the male sheds milt on the egg mass or (in one hynobiid) the eggs are deposited on a sticky spermatophore. The eggs of the siren *Pseudobranchus* are deposited singly and attached to the roots of water plants. In Florida, where mats of water hyacinths are the preferred site of deposition, we have never found more than a single egg on any one plant. The eggs may be so widely scattered that fewer than a dozen can be collected in the course of an afternoon. Since a female may have more than 100 ripe eggs in her oviducts, egg deposition must take a considerable amount of time. All salamanders that deposit single eggs attached to the substrate have internal fertilization. The pair separate after mating, and the female deposits a few eggs each day over a period of a month or so. Either the sirens practice internal fertilization by some means other than spermatophore deposition, or the male accompanies the female during a protracted period and fertilizes each egg singly.

The eggs of *Pseudobranchus* vary from about 7 to 9 mm in diameter (Figure 12–15). Although the exact development time is not known, it must

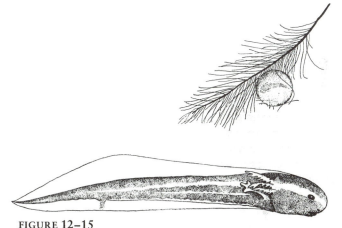

FIGURE 12–15
The egg and newly hatched young of the Narrow-striped Dwarf Siren, *Pseudobranchus striatus axanthus*. The egg is attached to a water hyacinth rootlet. [From Goin, *Nat. Hist. Misc.*, 1947.]

require several weeks, since a series of eggs collected in the neural-groove stage did not hatch until 17 days later. The newly hatched larvae range from about 14 to 16 mm in total length. There are no balancers, the toes are differentiated, and a well-developed dorsal fin extends from the base of the head to the tip of the tail. There is perhaps less evidence of metamorphosis in the sirenids than in any of the perennibranch salamanders, but *Siren* does develop an adult-type skin.

Sirens are able to estivate when their ponds begin to dry up. As in the African Lungfish, the skin secretes a slimy mucous coat at the onset of estivation that hardens on contact with air to form a protective cocoon in the drying mud.

READINGS AND REFERENCES

Arnold, S. J. "The evolution of courtship behavior in salamanders," vol. 2. University of Michigan Ph.D. thesis, 1972.

Bishop, S. C. *Handbook of Salamanders*. Ithaca, N. Y.: Comstock, 1947.

Dunn, E. R. *Salamanders of the Family Plethodontidae*. Northampton, Mass.: Smith College, 1926.

Francis, E. T. B. *The Anatomy of the Salamander*. Oxford University Press, 1934.

Gorham, S. W. *Checklist of World Amphibians*. Saint John: New Brunswick Museum, 1974.

Larsen, J. H., and D. J. Guthrie. "Parallelism in Proteidae reconsidered." *Copeia*, no. 3, 1974.

Leon, P., and J. Kezer. "The chromosomes of *Siren intermedia nettingi* (Goin) and their significance to comparative salamander karyology." *Herpetologica*, vol. 30, no. 1, 1974.

Ma, P. F., and J. R. Fisher. "Multiple adenosine deaminases in the Amphibia and their possible phylogenetic significance." *Comparative Biochemistry and Physiology*, vol. 27, no. 3, 1968.

Salthe, S. N. "Courtship patterns and the phylogeny of the urodeles." *Copeia*, no. 1, 1967.

Taylor, E. H. *The Caecilians of the World*. Lawrence: University of Kansas Press, 1968.

———. "A new family of African Gymnophiona." *University of Kansas Science Bulletin*, vol. 48, no. 10, 1969.

Thorn, R. *Les Salamandres d'Europe, d'Asie et d'Afrique du Nord*. Paris: Lechevalier, 1968.

Twitty, V. C. *Of Scientists and Salamanders*. San Francisco: Freeman, 1966.

Wake, D. B. "Comparative osteology and evolution of the lungless salamanders, family Plethodontidae." *Memoirs of the Southern California Academy of Sciences*, vol. 4, 1966.

———, and N. Ozeti. "Evolutionary relationships in the family Salamandridae." *Copeia*, no. 1, 1969.

CHAPTER THIRTEEN

FROGS
AND TOADS

ALTHOUGH FROGS AND TOADS are seldom of much interest to amateurs, they are, for a number of practical reasons, of prime importance to professional zoologists. Their tendency to congregate in large breeding choruses makes them easy to collect in large numbers. They are easy to maintain alive or to preserve, large enough to work on conveniently, small enough not to raise serious storage problems. Cheap and convenient, they are the laboratory animals *par excellence* for introductory biology courses. They are ideal for studies in experimental embryology—because they lay large numbers of eggs that are easy both to collect and to maintain and hatch in the laboratory, because their eggs are surrounded by a clear jelly rather than an opaque shell, and because most of them pass through a free-swimming larval stage. They have further potential as research animals that are only now beginning to be explored. Ethologists are discovering that frogs exhibit territoriality and social hierarchies. Some of the tropical genera speciate readily and have developed hundreds of distinct forms. These genera may give us clues to mechanisms of speciation that have so far eluded us in studies of more restricted genera.

There is confusion in many people's minds about the use of the words *frog* and *toad*. All members of the order Salienta (or Anura) are collectively and correctly known as *frogs*. Some frogs, however, may bear the common name *toad*—for example, the genus *Bufo*, the Surinam Toad *(Pipa pipa)*, and the Midwife Toad *(Alytes obstetricans)*. The term *tree frog* is generally used only for those anurans in which there is a short intercalary (extra) cartilage between the ultimate and penultimate phalanges of the digits. These include the families Hylidae, Centrolenidae, Rhacophoridae, and Hyperoliidae. The term "tree toad" is best left to the poets.

All anurans are easily recognized: there is never any doubt whether a given animal belongs to this order. Every frog lacks a true tail in the adult stage; the head is not separated from the trunk by a constricted neck; the legs are well developed, the hind ones longer than the front. This characteristic body form is largely an adaptation for jumping. The long, muscular hindlegs propel the body forward, the lack of a neck streamlines the body, and the short, heavy forelegs cushion the shock of landing. Frogs are the most primitive vertebrates to have a middle-ear cavity. The tympanic membrane usually lies flush with the surface of the head, and is often a very large and obvious structure. Correlated with the development of the ear as a hearing organ is the appearance of a true voice box. Movable eyelids protect the eyes, and glands keep the eyes moist.

Although the modern frogs are remarkably uniform in gross structure, they have undergone considerable adaptive radiation in their reproductive habits. Fertilization is almost always external. Usually the eggs are laid in water and hatch into aquatic larvae; but all stages can be found from this reproductive pattern to true terrestrialism, in which the eggs are laid on land and the young hatch as tiny frogs. Several species are ovoviviparous.

Most anurans require moist surroundings and cannot stand prolonged exposure to low humidities, but a few, such as the Marine Toad, *Bufo marinus,* are able to adapt to rather arid conditions. The group as a whole is well adapted to relatively few niches, and in those niches has spread around the world. With the possible exception of the lizards, the frogs are the most widely distributed herps.

The history of anuran classification has been one of continual uncertainty, and remains so today. With the acceptance of Darwin's theory of evolution, systematists attempted to adapt a linear, hierarchical system of classification to the highly branched evolution of frogs. Frog families have been split only to be recombined or further subdivided; suborders or superfamilies established to encompass one set of families have gradually changed their content or have been eliminated. This uncertainty results from the high degree of parallelism and convergence in anuran evolution. The initial development of a saltatory (jumping) body form and a bipartite life history canalized later evolution by restricting the number of paths that adaptations could follow. Thus different lineages have adapted to similar niches by similar morphological and physiological modifications. As a result, many characters that were once thought to indicate close relationship are now believed to show only an adaptation to a similar way of life.

Subordinal classifications have tended to be based on one or two character sets. Historically, frogs have been subdivided by the presence or absence of a tongue, pupil shape and type of amplexus, or the structure of the pectoral girdle. If the epicoracoid cartilages at the ventral ends of the two halves of the girdle are free and overlapping, the girdle is said to be arciferal; if the cartilages are fused, the girdle is firmisternal (Figure 13–1). One

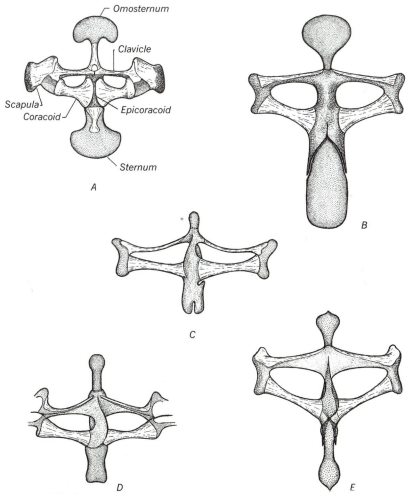

FIGURE 13–1
Pectoral girdle types in frogs: *(A)* the firmisternal girdle in *Rana tigrina,* and variations in the arciferal type, as in *(B) Rhinoderma darwini, (C) Sooglossus seychellensis, (D) Eleutherodactylus bransfordi,* and *(E) Sminthillus limbatus.*

widely followed classification was based largely on the shape and arrangement of the vertebrae: an opisthocoelous (*opistho* = behind, *koilos* = hollow) vertebra has the centrum convex in front and concave behind, a procoelous (*pro* = in front) one is concave in front and convex behind, an amphicoelous (*amphi* = both) one is concave at both ends. Under this system, the frogs were divided into the suborders Opisthocoela, Anomocoela, (*anomo* = irregular) Procoela, and Diplasiocoela (*displasio* = twofold); later Amphicoela was added. The names refer to the structure of the vertebral column. The Amphicoela (Ascaphidae and Leiopelmatidae) have amphicoelous presacral vertebrae and a sacral vertebra that articulates by a single condyle with the urostyle. The Opisthocoela (Discoglossidae, Rhinophrynidae, and Pipidae) have opisthocoelous presacral vertebrae and the sacral vertebra either fused to the urostyle or articulating with it by a biconvex centrum. The Anomocoela (Pelobatidae) have procoelous, rarely amphicoelous, presacral vertebrae and the sacral vertebra either fused to the urostyle or articulating with it by a single or double condyle. The Diplasiocoela (Hyperoliidae, Microhylidae, Ranidae, Rhacophoridae, and Sooglossidae) have procoelous presacral vertebrae except the last, which is amphicoelous; the sacral vertebra articulates by a double condyle with the urostyle. The Procoela (all remaining families in Figure 13–2) have all the presacral vertebrae procoelous and the sacral vertebra articulating with the urostyle by a double condyle.

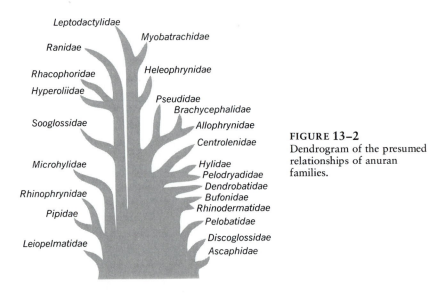

FIGURE 13–2
Dendrogram of the presumed relationships of anuran families.

This classification remained in use for many years, in spite of accumulating evidence that some of the suborders did not constitute natural groups. The critics demonstrated errors in the classification, but were unable or unwilling to propose a better one. Recently, attention has focused on the structure of the tadpoles. Frog larvae can be separated into four groups, each comprising a unique set of families, by the nature of their mouth parts and the position of the spiracle. The four groups have now been given subordinal status: the names of the suborders are those that were originally proposed for the tadpole types. We are using this system here, but recognize that such a classification built on one set of related characters will require further modification.

SUBORDER XENOANURA

Xenoanuran tadpoles are the least specialized. Their jaw cartilages and muscles are simple and cannot be used for biting or scraping, so keratinous denticles and complex lips are absent. The tadpoles are filter feeders. The branchial (gill) chambers are separate, each with its own opening (spiracle) to the exterior. The forelimbs develop posteriorly to the branchial chambers. Adults have an opisthocoelous vertebral column, expanded sacral diapophyses, and either a firmisternal or an arciferal pectoral girdle. In the pipids, ribs are free in juveniles and fused to the diapophyses in adults; in *Rhinophrynus* they are completely absent. Amplexus is pelvic, fertilization external, and embryonic development either direct or indirect.

Family Pipidae

The pipids are stout-bodied, big-footed, aquatic frogs. Three genera, *Xenopus, Hymenochirus,* and *Pseudhymenochirus,* occur in Africa; one genus, *Pipa,* is found in South America (Figure 13–3). Pipids are the most aquatic of all frogs—so aquatic that the adults of some species have retained the larval lateral line system.

Subfamily Pipinae. The South American pipines have starlike appendages on the tips of their fingers and small sterna. Their eggs develop in temporary, individual pits in the soft skin on the back of the female. The method by which the eggs are transferred to the female's back has been described for a pair of Surinam Toads *(Pipa pipa)* that bred in an aquarium (Figure 13–4). The male clasped the female inguinally as the pair rested on the bottom of the tank, the dorsal part of the female's vent pressed against the male's abdomen. The pair then rose to the top, turning over as they did so, and paused momentarily in an upside-down position. At this point, three to five eggs were extruded by the female and caught in transverse skin folds on the belly of the male. The two then returned headfirst to a tilted, resting position on the bottom. Fertilization apparently occurred during the descent, and as the frogs righted themselves, the eggs dropped to the female's back and

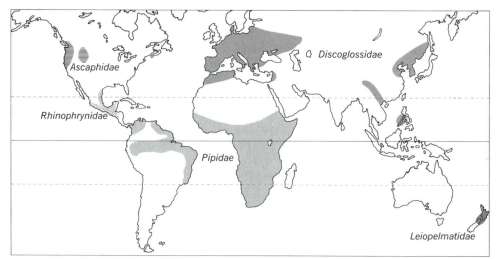

FIGURE 13–3
Distribution of the ascaphids, discoglossids, leiopelmatids, pipids, and rhinophrynids. [After Savage, 1973.]

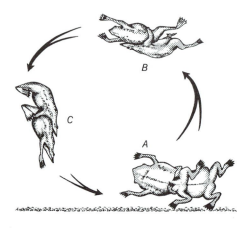

FIGURE 13–4
The egg-laying turnover maneuver (A–C) of breeding Surinam Toads, *Pipa pipa*. The eggs are extruded by the female at B, when both individuals are upside down. [After Rabb and Rabb, *Copeia*, 1960.]

adhered there. A number of turnovers were required to complete the deposition. Fifty-five eggs were thus implanted on the back of the female; eleven more were lost as they were laid. Eggs of some species of *Pipa* hatch into tadpoles, but others hatch directly into tiny frogs.

Subfamily Xenopinae. The African pipids, the xenopines, have undivided fingertips and large sterna. More striking than the differences in structure between the African and South American subfamilies are the differences in their life history. The African pipids deposit their eggs in water. The male, at the time of mating, makes a sudden dash at the female and clasps her by the pelvic region. The female expels each egg separately, holding it for a moment with her cloacal lips. She then grasps a leaf or twig with her hindlimbs. The egg is ejected sharply from her cloaca and propelled along a shallow groove on the venter of the male, past his cloaca for fertilization, and then to the weed, to which it becomes attached.

Family Rhinophrynidae

The Mexican Burrowing Toad *(Rhinophrynus)*, the only member of this family, is an egg-shaped, short-legged, toothless creature with a smooth, blotched skin. The voice of these grotesque frogs is equally grotesque: a chorus of them sounds like a shipload of seasick landlubbers. The body form of *Rhinophrynus* is an adaptation for a burrowing life and ant- and termite-eating habits. A small head and stout body and limbs are also seen in many of the microhylids, which have similar burrowing and feeding habits. The tongue is not attached in front, and the toad is apparently able to protrude it like a mammal, rather than flick it out in the usual toad fashion. The foot is specialized for digging—the prehallux is covered with an enormous cornified "spade," and the single phalanx of the first toe is shovel-like. Eggs are deposited in temporary pools, and the tadpoles are free-swimming.

SUBORDER SCOPTANURA

Only one family of frogs, the microhylids, possess the scoptanuran type of tadpole. The jaw muscles are modified to allow for forward extension of the lower jaw. The tadpoles lack keratinous denticles, although the lips are frequently folded in complex shapes. The branchial chambers are separate, and each opens into an elongated spiracular chamber with a single ventral spiracle to the outside. The forelimbs develop posteriorly to the branchial chambers. Adults have diplasiocoelous (occasionally procoelous) vertebral columns, expanded sacral diapophyses, and firmisternal pectoral girdles occasionally lacking clavicles. The ribs are fused to the diapophyses in juveniles and adults. Amplexus is axillary, fertilization external, and embryonic development either direct or indirect.

Family Microhylidae

The microhylids are an enigmatic group. Although they occur in most zoogeographic regions, they are usually inconspicuous components of anuran faunas (Figure 13–5). Many are small, tiny-headed, stout-bodied, ant-

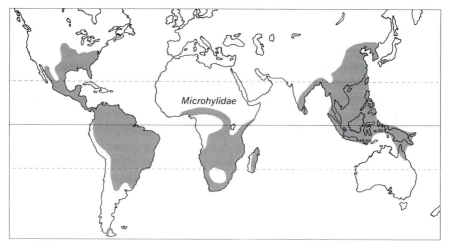

FIGURE 13–5
Distribution of the microhylids. [After Savage, 1973.]

and termite-eating frogs living a semifossorial life. Only in the Old World tropics, especially New Guinea, have they radiated into terrestrial and arboreal habitats.

The unique microhylid tadpole indicates an early derivation from the ancestral stock leading to the advanced frogs. Among the advanced frogs, the ranids share some structural characteristics with the microhylids, but whether the two families are related is uncertain. Adaptive radiation of this group is clearly indicated by the seven subfamilies. Some of these subfamilies will undoubtedly be elevated eventually to familial status.

Subfamily Asterophryninae. The asterophrynine subfamily has its distribution centered in New Guinea, with outliers in Southeast Asia, the East Indies, the Philippines, and northeastern Australia. All members lay their eggs on land and have direct development, with embryos much like those of the leptodactylid *Eleutherodactylus*. They occupy a variety of niches from completely fossorial to arboreal.

Subfamily Brevicipitinae. The brevicipitines are confined to southern and eastern Africa and seem to form a closely allied, compact group of four genera. All possess the stout-bodied, fossorial body form. *Breviceps,* and probably the other members of the subfamily, lay their eggs in burrows on land. The embryo differs from that of the Asterophryninae in having an operculum and a muscular rather than a vascular tail. It hatches at an advanced stage and completes metamorphosis in the nesting site.

Subfamily Cophylinae. All eight cophyline genera are confined to Madagascar. Little is known of their natural history. Breeding habits are unknown, but females with large ovarian eggs have been found in two genera, which suggests that egg laying may be terrestrial and development direct or partly so.

Subfamily Dyscophinae. Like the cophylines, the single dyscophine genus, *Dyscophus,* is confined to Madagascar, and little is known of its ecology and behavior. Eggs are laid in water, and free-swimming tadpoles hatch from them.

Subfamily Hoplophryninae. The hoplophrynines, a subfamily of one genus *(Hoplophryne),* occur only in the mountainous region of Kenya and Tanzania. Females lay their eggs in the internodes of bamboo or between the leaves of wild bananas. The tadpoles are specialized for egg eating, and superficially resemble those of bromeliad-breeding hylas.

Subfamily Microhylinae. The microhylines are the most diverse (in number of genera) and most widespread family of microhylids (Figure 13–6). They range from southern India throughout Southeast Asia to the East Indies, and from central North America to south central South America. A strong relationship between the Asiatic and American forms is assumed, but has not been adequately investigated. *Gastrophryne* has two species in the United States; their breeding choruses can always be recognized by the "lost-sheep" call. All microhylines are aquatic breeders with indirect development.

Subfamily Phrynomerinae. The single phrynomerine genus, *Phrynomerus,* is widely scattered over Africa south of the Sahara. These frogs are arboreal and have intercalary cartilages between the penultimate and ultimate phalanges. For this reason, they are frequently given familial status. Reproduction is aquatic and development indirect.

SUBORDER LEMNANURA

The lemnanuran tadpole possesses extra jaw cartilages and muscles, which produce a flexible and mobile feeding apparatus. The circumoral region is folded into a large labial disc bearing keratinous denticles. The branchial chambers are separate, and each chamber has a short spiracular tube that fuses with its opposite mate to form a single ventral spiracle at midbody. The forelimbs develop close to the branchial chambers. Adults have amphicoelous vertebral columns in ascaphids and leiopelmatids and opis-

FIGURE 13–6
The Mexican Sheep Frog, *Hypopachus cuneus,* a microhyline. The common name refers to the call of the breeding male. In many of this group the call closely resembles that of a forlorn lamb.

thocoelous ones in discoglossids, with expanded sacral diapophyses and arciferal pectoral girdles. Ribs are distinct and free in both juveniles and adults. Amplexus is pelvic, fertilization internal or external, and development direct or indirect.

Family Ascaphidae

The ascaphid family contains only one species, *Ascaphus truei,* living in and along the swift-flowing mountain creeks of the northwestern United States (Figure 13–7). Although primitive in many respects, *Ascaphus* has evolved a number of specializations to allow successful reproduction in swift streams. The male is voiceless. Instead of sitting still and calling for a mate, he swims about on the bottom until he finds a female. He then clasps her with a pelvic embrace, humps his body, and maneuvers his extended cloacal appendage into position to thrust into her cloaca. The cloacal appendage thus transfers the sperm directly into the female's cloaca and prevents them from being swept downstream before they can enter the ova. The eggs are deposited in

FIGURE 13–7
The Tailed Frog, *Ascaphus truei*, of the Pacific coast of North America. The tail is really the everted cloaca, which is used as an intromittent organ.

coils of rosarylike strings, which adhere to rocks on the bottom of the stream. In the cold water of the mountain stream, embryonic development is slow and metamorphosis does not occur until the following summer. The tadpoles are streamlined for living in the swift-flowing streams, and possess enlarged oral suckers for clinging to rocks while feeding. Convergent adaptations occur in stream-dwelling members of other frog families—for example, the ranids and microhylids.

Family Discoglossidae

The four discoglossid genera are all Old World forms. *Discoglossus* is ranalike in appearance and habits and is found in southwestern Europe and northwestern Africa. The four species of *Bombina* are flattened, highly aquatic toads of Europe, Asia Minor, China, and Korea (Figure 13–8). They are brilliantly marked on the underside—black mottled with bright red, orange, yellow, or white. If highly disturbed, *Bombina bombina* will bend its legs and head over its back, exposing its bright belly. This "unken" reflex may act to startle a predator or to display warning coloration. *Alytes* of

FIGURE 13–8
The Korean Fire-bellied Toad, *Bombina orientalis*.

western Europe is the most terresterial member of the family. The completely aquatic *Barbourula* is known only from the Philippines.

Bombina maxima, the Yellow-bellied Toad, breeds in water in the spring and summer. The male clasps the female with a pelvic embrace. Eggs are laid in small masses that sink to the bottom or come to rest suspended on vegetation, and hatch in about a week.

The breeding habits of *Bombina* are probably fairly typical of most other members of the family, but the genus *Alytes* shows one of the most remarkable modifications of any frog. The male Midwife Toad, *Alytes obstetricans,* calls from a small hole in the ground. Mating takes place on the ground and is apt to last most of the night. The male clasps the female tightly around the head. Just before she extrudes the string of eggs, he moves his hindlegs forward so that his heels are together, in front of and above her cloaca. As the eggs are emitted, he catches the rosarylike strings in his feet and, by stretching his legs backward, delivers from 20 to 60 eggs. He then moves his legs so as to twist the strings around them. He carries the eggs in this way for several weeks, until the tadpoles are ready to hatch. Then he makes a brief visit to a pool where no other tadpoles are present. Here the eggs hatch and the tadpoles finish their development.

Family Leiopelmatidae

Inclusion of the leiopelmatids with the lemnanurans is a matter of convenience, rather than of known relationships. *Leiopelma* comprises three species, *L. archeyi*, *L. hamiltoni*, and *L. hochstetteri*, which occur only in New Zealand. They share many primitive characters with *Ascaphus* and are, for this reason, frequently included in the family Ascaphidae. Sharing primitive characters is not necessarily an indication of close relationship. The widely separated distribution and different embryonic stages suggests only a distant relationship with *Ascaphus*. Leiopelmas are small frogs (20–50 mm) that live on mountainsides, where streams and pools of water are scarce. They lay small clutches (2–8) of large-yolked eggs in damp crevices under logs and stones. The eggs hatch in about six weeks directly into tiny froglets, bypassing the need of tadpoles for free-standing water.

SUBORDER ACOSMANURA

Acosmanurans are the most abundant and diverse group of frogs. Their tadpoles possess complex jaw cartilages and muscles like those of the lemnanurans. Similarly, they have complex labial discs with keratinous denticles. However, acosmanurans have a single branchial chamber with a single spiracle on the left side. Their forelimbs develop in the branchial chambers. Three types of adult vertebral columns are found: anomocoelous in the pelobatids, diplasiocoelous in the ranids and related families, and procoelous in the bufonids and their relatives. Sacral diapophyses and pectoral girdles also show a variety of modifications. Ribs are fused to the diapophyses in both juveniles and adults. Amplexus is usually axillary, fertilization almost always external, and development usually indirect.

Family Pelobatidae

In most classifications, the pelobatids couple the supposedly primitive groups, the xenoanurans and lemnanurans, with the more advanced acosmanurans. This intermediacy reflects the characters used for classification, and says only that, in these characters, pelobatids possess roughly half ancestral and half derived traits. In many ways, pelobatids are highly specialized. They have developed a specialized larval physiology, burrowing apparatus, and locomotor modifications. Figure 13–9 shows the distribution of the family. Three subfamilies are usually recognized.

Subfamily Megophryinae. The half-dozen megophryine genera occur in southeastern Asia from China through the East Indies. There are more than three dozen species, and some of them are quite large and spectacular. The

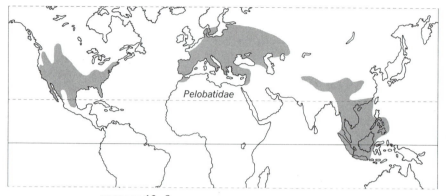

FIGURE 13-9
Distribution of the pelobatids. [After Savage, 1973.]

quaint Nose-horned Frog of the East Indies *(Megophrys monticola)* has each upper eyelid extended to form a large, thin, pointed "horn," along with a flexible projection from the snout, so that the animal looks as though it had three horns. In *Megophrys longipes* of Malaya, approximately a dozen large eggs are laid in humid moss, with the young probably metamorphosing before hatching.

Subfamily Pelobatinae. The pelobatines are the spadefoot toads of the Palearctic *(Pelobates)* and Nearctic *(Scaphiopus)*. Both bear a broad, sharp-edged, keratinous tubercle on the inside of each hindfoot. With these "spades," the toad rapidly digs itself backward into the ground. Members of both genera are explosive breeders. Living in dry areas, they require extremely heavy rainfall following several warm days. The rainfall fills the temporary pools and ponds; the toads appear almost immediately, and in two days have completed the egg laying. The eggs develop rapidly and may hatch in as little as one and a half days, and the tadpoles may metamorphose in fifteen to twenty days. Such speed is necessary in the arid grasslands and semideserts, where evaporation is intense and pools short-lived.

Subfamily Pelodytinae. The pelodytine subfamily includes only one living genus, the slender *Pelodytes* of southwestern Europe and Caucasia. Its natural history is poorly known, although its unique fusion of the calcaneum and astragalus (ankle bones) into a single bone suggests some specialized locomotor behavior. Surprisingly, it shares a number of similarities with the discoglossids, particularly *Bombina:* as yet, it is not known whether these similarities reflect convergence or relationship.

Family Hyperoliidae

The hyperoliids are small to medium-sized frogs, primarily of mainland Africa, but with one representative in Madagascar and one in the Seychelles (Figure 13–10). They are unquestionably related to the ranids, but the group has shown little systematic stability as various zoologists have rearranged and reinterpreted the status of its members. We recognize four subfamilies.

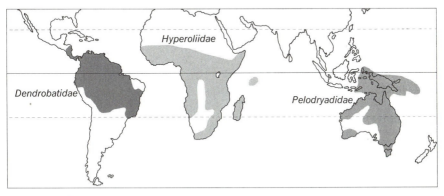

FIGURE 13–10
Distribution of the dendrobatids, hyperoliids, and pelodryadids. [After Savage, 1973.]

Subfamily Arthroleptinae. Four genera of African frogs constitute the arthroleptines: *Arthroleptis, Coracodichus, Schoutedenella,* and *Cardioglossa.* All are little frogs of the forest floor and brushy country. Males of a few species of *Arthroleptis* and of the genera *Coracodichus* and *Cardioglossa* have the third finger greatly elongated, sometimes as much as two or three times as long as the rest of the hand—a striking secondary sex character unknown among other amphibians.Development is direct in all except *Cardioglossa,* which retains a free-swimming tadpole.

Subfamily Astylosterninae. The astylosternines are West African forest frogs of four closely related genera. The terminal phalanges of both fingers and toes of *Nyctibates* are only slightly curved; those of two or more toes of *Scotobleps, Astylosternus,* and *Gampsosteonyx* are bent sharply downward and may pierce the integument to form little bony spines protruding below the tips of the digits. The functional significance of these modified toes is not known, although they may give a firmer grip for jumping. The male of *Astylosternus robustus,* the Hairy Frog, has a peculiar cutaneous growth of hairlike processes on the thighs and flanks. These are most fully developed during the breeding season. They are not true hairs, but vascular papillae that presumably aid in respiration.

Subfamily Hyperoliinae.   The small- to large-bodied hyperoliines of Africa, Madagascar, and the Seychelles are the most diverse group of hyperoliids. *Hylambates* and *Kassina* are terrestrial; most of the other hyperoliines are arboreal, although *Cryptothylax* is frequently found on aquatic vegetation. *Kassina,* which refuses to hop but instead walks or runs, lays its eggs in the water. *Leptopelis* buries large eggs in the ground near water; the tadpoles later migrate to the water. Most arboreal members, *Hyperolius* included, lay their eggs in a gelatinous mass on vegetation hanging over water. As the tadpoles hatch, they drop into the water and complete their development. The centrolenids of Central and South America have adopted similar arboreal egg-laying habits.

Subfamily Scaphiophryninae.   The three scaphiophrynine genera of Madagascar (*Dyscophus, Pseudohemisus* and *Scaphiophryne*) range from semifossorial to arboreal. Body form varies from stout-bodied and short-limbed to elongate-bodied and long-limbed. The natural history of this group is poorly known.

Family Ranidae

The ranids are the so-called true frogs. The classroom *Rana* is cosmopolitan in distribution, occurring on all major continents, and largely encompasses the range of all the other ranids (Figure 13–11). Body form in most members of the family tends to mimic that of *Rana.* Most lay aquatic eggs and have free-swimming tadpoles.

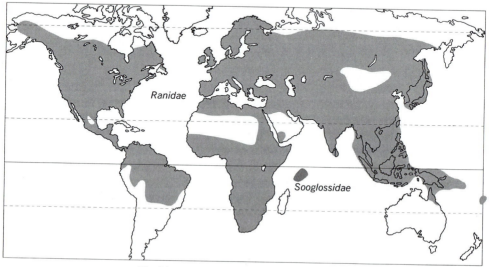

FIGURE 13–11
Distribution of the ranids and sooglossids. [After Savage, 1973.]

Subfamily Hemisinae. The hemisines are small fossorial frogs, primarily of East Africa. They have small, pointed heads projecting from bloated, spherical bodies. Eggs are laid in a pit close to water; when the tadpoles hatch, the guarding female digs them a tunnel to the water.

Subfamily Phrynobatrachinae. The phrynobatrachines, a widespread group of small to medium-sized African frogs, lack vomerine teeth. They generally live by the edge of standing water, and escape from their predators by a series of long, erratic jumps along the shore.

Subfamily Platymantinae. The platymantines occur from Southeast Asia through the Indoaustralian Archipelago to the Fijis. They are very diverse in appearance and habits. *Staurois* of Southeast Asia lays eggs in mountain streams. The tadpole has an abdominal sucking disc behind the mouth for clinging to rocks in swift water. *Platymantis* of the Pacific Islands lays large-yolked terrestrial eggs. Development is direct; there is no free-swimming tadpole stage.

Subfamily Raninae. The ranine subfamily includes the genus *Rana* and its close allies (Figure 13–12). About 200 species of *Rana* have been described. Of the major land masses, only Greenland, New Zealand, central and southern Australia, and southern South America lack at least one representative of this vigorous genus. Many are large, and most are colored with soft, muted browns, greens, and yellows. Most of the other ranine genera contain only a few forms and have a limited distribution in Africa or southern Asia.

The largest known frog, *Conraua goliath* of Africa, belongs to this subfamily. Some say that this giant of all frogs is really nothing but a large *Rana,* but it differs from that genus in the weak calcification of its epicoracoid cartilages.

Family Rhacophoridae

The rhacophorids are small to large frogs, primarily of the Oriental region, but with several members on Madagascar and a single genus, *Chiromantis,* in Africa (Figure 13–13). Like the hyperoliids, they are closely allied to the ranids and have often been regarded as a subfamily of that group.

Subfamily Mantellinae. The five rhacophorids (*Boophis, Gephyromantis, Mantellus, Mantidactylus, Trachymantis*) found on Madagascar constitute the mantelline subfamily. *Mantidactylus* is the most primitive member of this group and apparently the most primitive of all rhacophorids. Little is known of the natural history of the mantellines, although they probably have free-swimming larvae.

FIGURE 13–12
A New Jersey specimen of the Bullfrog of the eastern United States, *Rana catesbeiana*. This frog has now been introduced into various regions of the world as a potential source of food.

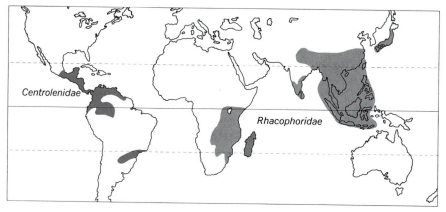

FIGURE 13–13
Distribution of the centrolenids and rhacophorids. [After Savage, 1973.]

Subfamily Rhacophorinae. Only *Chiromantis* among the rhacophorines lives in tropical Africa; all other members of this subfamily are found from India through Indochina to the East Indies. Most are medium- to large-bodied arboreal frogs possessing digital pads and intercalary cartilages between the penultimate and ultimate phalanges.

Rhacophorines typically lay their eggs in masses of foam on the leaves of plants or other structures above water. The female does most of the work in producing the foam mass. Before the eggs appear she ejects a small amount of fluid, which she beats into a froth by moving her feet mesially and laterally, turning them as she crosses them on the midline. When the foam has been prepared, the eggs and fluid are ejected together. During the egg-laying process the male is passive, grasping the female under her armpits and simply holding his body closely applied to her back, his eyes half-closed. His pelvic region is bent downward, with the cloacal opening near that of the female, and the eggs are apparently fertilized as they leave the female's cloaca. When the egg-laying process is concluded, the female stands on her forelimbs and the male tries to get away from the foam in which the distal ends of his hindlegs are buried. The female usually disentangles herself later by moving her legs and body sideways with the help of large, sticky finger discs.

After egg laying, the foam changes color from white to light brown. The eggs are scattered singly or in small groups in the foam mass, but are mostly concentrated near the basal part, where the foam is attached to the substrate. Just before the eggs hatch, the foam begins to liquefy and the active movement of the fully developed embryos or tadpoles drops them into the water. Sometimes the whole egg-foam mass may be washed into the pool by rain. Tadpoles of a few species sometimes go through their entire development in the nest.

Rhacophorus microtympanum of Ceylon lays about 20 large-yolked eggs on land. The female does not produce a foam nest, but remains with the clutch. Development is direct. The embryos bear a striking resemblance to embryos of *Eleutherodactylus,* which also develop directly on land.

Family Sooglossidae

The two genera of the sooglossid family occur only on the Seychelles Islands. *Sooglossus* resembles a small arthroleptine externally, whereas *Nesomantis* is more toadlike in build. Males transport the tadpoles from egg-laying sites on land to water, a behavioral trait shared with the neotropical dendrobatids. The larvae of *S. seychellensis* lack gills and hatch with the hindleg rudiments already developed.

Family Allophrynidae

A single genus, *Allophryne,* of northeastern South America, constitutes the bufo-like allophrynid family (Figure 13–14). It is toothless, with peculiar, scalelike patches of roughened skin on the head and back. This small tree frog breeds in small ponds in the usual anuran fashion.

FIGURE 13–14
Distribution of the allophrynids, brachycephalids, pseudids, and rhinodermatids. [After Savage, 1973.]

Family Brachycephalidae

The brachycephalid family includes only the single genus *Brachycephalus,* from southeastern Brazil. *Brachycephalus* looks like a small *Atelopus* with a broad, bony dorsal shield, which is confluent with the diapophyses of the second to seventh vertebrae. Eggs are laid in water, and development passes through a tadpole stage.

Family Bufonidae

The bufonids are the true toads, including the familiar squat-bodied, short-legged *Bufo* of which most of us think when we hear the word "toad" (Figure 13–15). The genus *Bufo* occurs natually on all the major land masses except Greenland, Australia, New Guinea, and New Zealand, and largely encompasses the ranges of all the other bufonid genera (Figure 13–16). *Atelopus, Dendrophryniscus,* and *Oreophrynella* are several of the genera found in South America (Figure 13–17). Twelve other genera of bufonids occur in the tropics and south temperate zone of Africa, southern Asia, and the East Indies.

Bufonids typically breed in open water. The small-yolked eggs are characteristically laid in long strings on the bottom. As with other frogs that lay in open water, the eggs are numerous. The tadpole is short and plump-bodied—the typical polliwog. The tadpole of the Philippine toad *Ansonia,* which breeds in mountain torrents, is depressed, with a strong, muscular tail, reduced fins, and a suckerlike oral disc. The African *Nectophryne* lays

FIGURE 13–15
"Which like the toad, ugly and venomous. . . ." The American Toad, *Bufo a. americanus.*

its eggs on land and lacks a free-swimming tadpole stage. Several of the genera (for example, *Pelophryne* of the Philippines and *Laurentophryne* of Africa) produce relatively few unpigmented, large-yolked eggs. Little is known of the life histories of these forms; it is possible that they also lay their eggs away from water. The Brazilian *Dendrophryniscus brevipollicatus,* a rough-skinned little toad of the forest floor, is reported to breed in bromeliads.

The African Live-bearing Toads, *Nectophrynoides,* diverge most remarkably from the normal anuran reproductive pattern. Breeding takes place on land, and fertilization is internal. The male lacks a copulatory organ, but the opening of his cloaca is more ventral in position than that of the female. He grasps her in the axillary region and brings his cloacal opening into apposi-

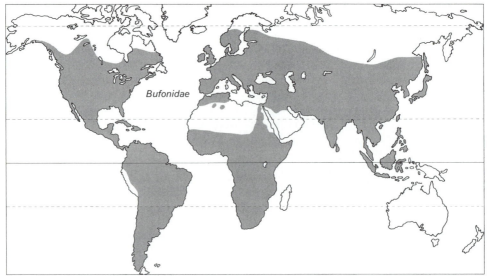

FIGURE 13–16
Distribution of the bufonids. [After Savage, 1973.]

tion with hers, transmitting the sperm directly to her cloaca. The young go through their larval development in the oviduct of the mother. The number of eggs is much smaller than in the water-breeding bufonids, but even so, more than 100 young have been taken from a single female *N. vivipara*. Although the young are born as fully formed frogs, the embryos still retain some typical tadpole characteristics. However, they lack gills, adhesive organs, labial denticles, and horny beaks. They seem to represent a transitional stage in the evolution toward direct development.

Family Centrolenidae

The small centrolenid family of arboreal frogs (*Centrolene, Centrolenella,* and *Teratohyla*) are apparently closely related to the hylids. Most centrolenids are small, bright green frogs, usually possessing some degree of transparency, so that bones, muscles, or viscera can be seen through the skin. Their distribution is principally in Central America and northern South America, but with a small range in southeastern Brazil and Paraguay. Most species deposit their eggs on large leaves overhanging running water. The placement of the eggs is species-specific: they may be deposited on the upper or lower leaf surface, or may hang from the leaf-tip like a large teardrop. The eggs are guarded by a parent, usually the male, to protect them from insect predators. When the tadpoles hatch, they fall into the stream and complete their development there.

FIGURE 13–17
Atelopus varius zeteki, the Golden Frog of El Valle de Antón, Panama. El Valle is now a tourist resort, and this frog is one of the popular attractions there. Lapel pins modeled after it can be purchased in Panama.

Family Dendrobatidae

The dendrobatids comprise three genera of small frogs, *Colostethus, Dendrobates,* and *Phyllobates.* They are found only in tropical America from Costa Rica to southern Brazil. Their affinities appear to be with hylids and bufonids, rather than with the ranids as commonly believed. *Dendrobates,* Poison Frogs, are often brightly colored, apparently to serve as a warning to would-be predators that their skin secretions are highly toxic (Figure 13–18). Males of all three genera transport the tadpoles on their backs from the egg-laying site on land to water. The female of *D. auratus* lays from one to six large-yolked eggs surrounded by an irregular, sticky, gelatinous material with no definite external film. The male either guards or periodically visits the clutch. The eggs hatch in about two weeks, and the tadpoles wriggle onto the male's back. He then visits water, and the young leave his back and finish their development as free-swimming tadpoles. They may not metamorphose until six weeks later.

FIGURE 13–18
The Gold and Black Poison Frog, *Dendrobates auratus,* of Central America.

Family Hylidae

The hylids are small- to large-bodied frogs of the Neotropic and Holarctic (Figure 13–19). Most have arboreal or semiarboreal habits and possess dilated digit tips. In all, intercalary cartilages lie between the ultimate and penultimate phalanges. Four subfamilies are commonly recognized, and more will undoubtedly be recognized in the future.

Subfamily Amphignathodontinae. The eight genera of amphignathodontines share similar morphological features, and the females of all except *Anotheca* and *Nyctimantis* carry the eggs on their backs. Development may be direct, or tadpoles may be released into the water. A dorsal pouch is present in *Flectonotus, Amphignathodon,* and *Gastrotheca.* In *Stefania,* the eggs simply adhere to the back; they lie in a single large depression in *Fritziana,* and in individual depressions in *Cryptobatrachus. Anotheca* deposits its eggs in bromeliads, and the tadpoles are often cannibals. The distribution of the subfamily is primarily in northern South America.

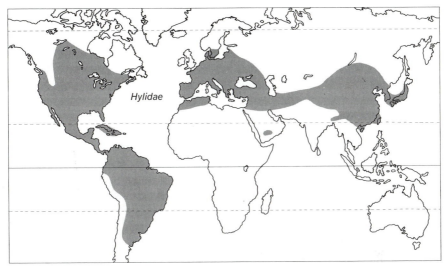

FIGURE 13–19
Distribution of the hylids. [After Savage, 1973.]

Subfamily Hemiphractinae. The casque-headed *Hemiphractus* is the only member of the hemiphractine subfamily. It is confined to lower Central America and northwestern South America. As in the amphignathodontines, *Hemiphractus* females carry their eggs on their backs and development is direct.

Subfamily Hylinae. The hyline subfamily is the hylid waste can. Its members are unquestionably hylids, but their relationships to one another and to other subfamilies are still a matter of conjecture. The Hylinae and the genus *Hyla* have the same distribution as the entire hylid family. There are approximately twenty genera. Most hylines are arboreal, with enlarged digital discs for climbing; (Figure 13–20) but a few, such as the Cricket Frogs *(Acris)* of the United States, have become secondarily terrestrial and possess reduced discs. Some *(Pternohyla, Triprion)* have the skin of the head fused to the skull and have developed grotesque, bony casques. Many are spectacularly colored. *Hyla heilprini,* for example, has been described as "pea green . . . brightly variegated with gold and sky blue."

The hylines show a variety of life-history patterns. Most of them, including all those found in the United States, lay large numbers of eggs in open water. Some species breed in a still pond or lake, others in streams, and the tadpoles are modified accordingly. Others have more specialized habits. *Hyla rosenbergi* build basins of mud in or at the edges of pools for the eggs. The tadpoles have enormous gills, with which they cling to the surface film of the small amount of water contained in these mud basins. They metamorphose before leaving the nest.

FIGURE 13–20
The Mexican Tree Frog, *Smilisca baudini*.

Subfamily Phyllomedusinae. The phyllomedusines are three genera of medium- to large-bodied arboreal frogs, *Agalychnis, Pachymedusa,* and *Phyllomedusa*. They occur from Mexico through tropical South America.

Females of these leaf frogs select leaves over water on which to deposit their eggs. The spawning pair move slowly forward from the tip to the stalk of the leaf, folding it into a nest, which is open at both ends. When the tadpoles hatch, they fall through the opening into the water below.

Family Heleophrynidae

The heleophrynid family contains only the single genus *Heleophryne*, from South Africa (Figure 13–21). Little is known of the breeding habits of these frogs. They live in the vicinity of mountain streams, and the tadpoles are modified for survival in swift-flowing water. *Heleophryne* has been considered a leptodactylid, but its isolation in southern Africa indicates a long separation from the leptodactylid stock of the Neotropic.

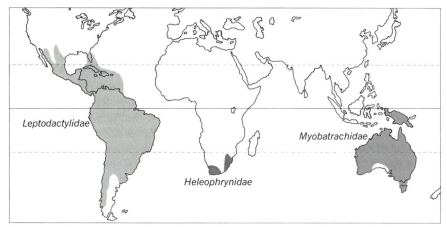

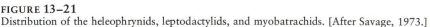

FIGURE 13–21
Distribution of the heleophrynids, leptodactylids, and myobatrachids. [After Savage, 1973.]

Family Leptodactylidae

The New World leptodactylids comprise four subfamilies. Of all the neo-tropic frogs, they show the greatest adaptive radiation and species diversity. Some are completely aquatic, others entirely arboreal. Reproduction ranges from aquatic egg laying through foam nest to terrestrial eggs with direct development.

Subfamily Ceratophryinae. Five genera of toadlike, wide-mouthed, rather bizarre forms are included in the small ceratophryine subfamily, restricted to South America. Their toadlike mien and enormous mouths distinguish them from their close relatives, the leptodactylines.

Not much is known of the life histories of most ceratophryines, although all appear to lay their eggs in water and have free-swimming tadpoles. Frogs of the genus *Ceratophrys* are extremely belligerent animals that snap and bite viciously at anything that approaches (Figure 13–22). *Ceratophrys* and *Lepidobatrachus* are known to feed on frogs, and it is likely that members of the other genera do also.

Subfamily Hylodinae. The hylodine subfamily encompasses the majority of the leptodactylid genera and species, and tends to show the most ad-vanced characteristics. *Hylodes* and its close relatives lay large clutches of eggs in water or moist places, and these hatch into free-swimming tadpoles. *Cycloramphus* and its relatives show an intermediate evolutionary step be-tween indirect and direct development. They lay their eggs in wet areas on the forest floor. When the tadpoles hatch, they live in rotting jelly masses, but do not feed. *Eleutherodactylus* has developed completely terrestrial

FIGURE 13–22
The Brazilian Horned Toad, *Ceratophrys cornuta*. These frogs use their huge mouths for catching other frogs, on which they feed.

eggs, although moist egg-laying sites are required. Since the larvae lack a free-swimming and feeding stage, eggs are large and heavy-yolked. The increased size means the female can produce and carry fewer eggs, but this lower reproductive capacity is compensated by fewer deaths in the developing larvae. One species is ovoviviparous. *Eleutherodactylus* occurs from southern North America to southern South America.

Subfamily Leptodactylinae. The leptodactylines include many South American frogs; a few species reach the extreme southern United States. The group includes the so-called South American Bullfrog (*Leptotdactylus pentadactylus*).

Leptodactylus, Physalaemus, and *Pleurodema* deposit their eggs in frothy nests, usually constructed in or near bodies of water. The larvae have very slim bodies and, after hatching, wriggle through the foam to reach the water. A few species have become more terrestrial. *L. marmoratus* scoops out a small basin in the ground away from water. The eggs and frothy mass are deposited in this basin, which is then covered with mud. The young hatch and pass through the tadpole stage in the nest, escaping after metamorphosis through a tiny hole left in the roof.

Subfamily Telmatobiinae. The telmatobiines are largely confined to the Andean highlands and appear to be the most primitive leptodactylid lineage. Most are either aquatic or semiaquatic, and all lay their eggs in water and have free-swimming tadpoles. *Telmatobus culeus,* the famous Lake Titicaca frog, is completely aquatic throughout its life, yet other *Telmatobus* species are highly terrestrial.

Family Myobatrachidae

The myobatrachids are the Australian "leptodactylids." They probably share a common origin with the South American leptodactylids. However, the two groups have been separated for such a long time that the myobatrachid genera are more closely related to one another than to any American genera, and this compact relationship is obscured by lumping them with the American leptodactylids. Like their neotropic counterparts, the myobatrachids show the greatest adaptive radiation and species diversity of all the frogs in their region. Some have adapted to the arid interior, others to cold, moist mountains. Three subfamilies are now recognized.

Subfamily Limnodynastinae. The ten limnodynastine genera occupy almost all of Australia. *Notaden* is toadlike, and has evolved behavioral and physiological adaptations to enable it to survive in the arid interior. Some of the limnodynastines produce foam nests similar to those of *Leptodactylus.* The Marsh Frog, *Limnodynastes tasmaniensis,* lays small eggs in a foam nest floating on any available water supply. The eggs hatch in about 48 hours. *Kyarranus* lays its eggs in sphagnum or moist earth, and apparently lacks a free-swimming tadpole.

Subfamily Myobatrachinae. The myobatrachines are not as widely distributed as the limnodynastines, and tend to be confined to moister habitats. Some, such as *Glauertia,* lay their eggs in water and show the typical anuran developmental patterns. In *Crinia* and *Myobatrachus,* the eggs are laid in damp areas on land, and part or all of the development may occur within the original egg mass.

Subfamily Rheobatrachinae. The recently discovered *Rheobatrachus* has the strangest parental care of all the frogs—gastric brooding. The female swallows the eggs, and the larvae develop in the female's stomach. This strange creature, discovered in the mountain streams of southeastern

Queensland, is as peculiar structurally as reproductively. It has been placed in its own subfamily, the rheobatrachines, and temporarily assigned to the myobatrachid family while its true affinities are being determined.

Family Pelodryadidae

The Australian tree frogs, *Litoria* and *Nyctimystes,* constituting the pelodryadid family, are similar to the hylids in habits and morphology. Indeed, the species of *Litoria* were formerly placed in the genus *Hyla.* Most of these frogs are arboreal, with strongly expanded digital discs. Eggs are laid in water, and development includes a free-swimming tadpole stage. *Cyclorana* has long occupied an enigmatic position in Australian frog classification, because it appears to be intermediate in morphology between the myobatrachids and pelodryadids. Its reproductive behavior is more similar to the latter.

Family Pseudidae

The pseudid family includes two small genera of highly aquatic South American frogs. An extra phalanx is present in each of the digits. This increase in phalangeal formula is apparently an adaptation for swimming, similar to that of aquatic mammals. The extra element is a different structure, and serves a different function than the one found in the foot of the various groups of tree frogs. Reproduction is aquatic: the eggs are laid in a frothy mass. *Pseudis paradoxa* is remarkable for the large size of the tadpoles, which may be more than 250 mm long (the adults are only about 65 mm long).

Family Rhinodermatidae

The rhinodermatid family includes only the peculiar-looking *Rhinoderma* of the Chilean and Argentinian Andes. Egg laying occurs on land, and amplexus follows the usual anuran pattern. After the embryos have begun to develop, the male picks up the eggs and places them in his vocal sac. Here they hatch, and the tadpoles remain in the sac until they metamorphose.

READINGS AND REFERENCES

Duellman, W. E. *The Hylid Frogs of Middle America.* Monograph of the Museum of Natural History, the University of Kansas, no. 1, 1970.

Gorham, S. W. *"Checklist of World Amphibians."* Saint John: New Brunswick Museum, 1974.

Griffiths, I. "The phylogeny of the Salienta." *Biological Reviews,* vol. 38, 1963.

Heyer, W. R. "A preliminary analysis of the intergeneric relationships of the frog family Leptodactylidae." *Smithsonian Contributions to Zoology,* no. 199, 1975.

————, and D. S. Liem. "Analysis of the intergeneric relationships of the Australian frog family Myobatrachidae." *Smithsonian Contributions to Zoology,* no. 233, 1976.

Inger, R. F. "The development of a phylogeny of frogs." *Evolution,* vol. 21, no. 2, 1968.

Kluge, A. G., and J. S. Farris. "Quantitative phyletics and the evolution of anurans." *Systematic Zoology,* vol. 18, no. 1, 1969.

Laurent, R. F. "Review of Liems' morphology, systematics, and evolution of the Old World tree frogs (Rhacophoridae and Hyperoliidae)." *Copeia,* no. 1, 1972.

————. "The natural classification of the Arthroleptinae (Amphibia, Hyperoliidae)." *Revue Zoologique et Botanique Africaine,* vol. 87, 1973.

Liem, S. S. "The morphology, systematics, and evolution of the Old World tree frogs (Rhacophoridae and Hyperoliidae)." *Fieldiana: Zoology,* vol. 57, 1970.

Lynch, J. D. "Evolutionary relationships, osteology, and zoogeography of leptodactyloid frogs." *University of Kansas Museum of Natural History Miscellaneous Publication,* no. 53, 1971.

————. "The transition from archaic to advanced frogs." *In* J. L. Vial (ed.), *Evolutionary Biology of the Anurans.* Columbia: University of Missouri Press, 1973.

McDiarmid, R. W. "Comparative morphology and evolution of frogs of the Neotropical genera *Atelopus, Dendrophryniscus, Melanophryniscus,* and *Oreophrynella." Bulletin of the Los Angeles County Museum of Natural History,* no. 12, 1971.

Reig, O. A. "Proposiciones para una nueva macrosystematica de los anuros." *Physis,* vol. 21, no. 60, 1958.

Savage, J. M. "The geographic distribution of frogs: Patterns and predictions." *In* J.L. Vial (ed.), *op. cit.*

Starrett, P. H. "Evolutionary patterns in larval morphology." *In* J. L. Vial (ed.), *op. cit.*

Tihen, J. "Evolutionary trends in frogs." *American Zoologist,* vol. 5, 1965.

CHAPTER FOURTEEN

TURTLES

SALAMANDERS ARE SOMETIMES mistaken for lizards, and lizards for snakes, but a turtle can never be mistaken for anything but a turtle. The short, wide body is encased in a protective armor, the shell, which is composed of a dorsal carapace and a ventral plastron. The carapace is formed of dermal bones that are usually fused to each other and to the underlying vertebrae and ribs, and typically covered with large epidermal scales, the laminae (sing. lamina) or scutes. The bones of the plastron apparently evolved from parts of the shoulder girdle, gastralia, and possibly other dermal bones, and are also covered with scutes. The scutes do not correspond in shape or size to the underlying dermal bones. Usually a bony bridge on either side, formed by an extension of the plastron, connects the upper and lower shell (Figure 14–1).

All turtles lack teeth. Instead, each jaw is usually covered with a horny sheath, the beak, which has a sharp cutting edge, the tomium. The skull is anapsid. The shoulder girdle is unique in that it lies beneath the ribs. The vent is a longitudinal slit, and the glans penis is single except in the families Trionychidae and Carettochelyidae, in which it is multilobed. The tail is short.

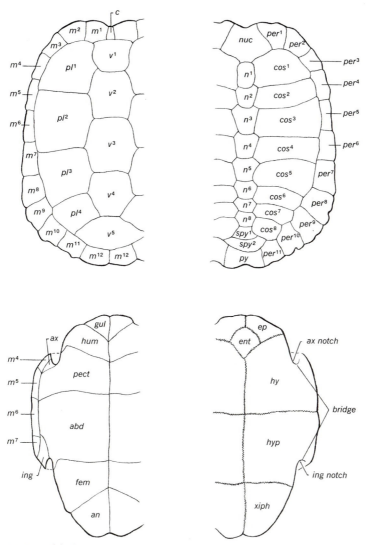

FIGURE 14–1

Epidermal shell *(left)* and bony shell *(right)* of an emydid turtle, *Chrysemys scripta*. Epidermal scutes: *abd*, abdominal; *an*, anal; *ax*, axillary; *c*, cervical; *fem*, femoral; *gul*, gular; *hum*, humeral; *ing*, inguinal; *m*, marginal; *pect*, pectoral; *pl*, pleural; *v*, vertebral. Shell bones: *ax notch*, axillary notch; *cos*, costal; *ent*, entoplastron; *ep*, epiplastron; *hy*, hyoplastron; *hyp*, hypoplastron; *ing notch*, inguinal notch; *n*, neural; *nuc*, nuchal; *per*, peripheral; *py*, pygal; *spy*, suprapygal; *xiph*, xiphiplastron. Some turtles have a series of scutes, inframarginals, lying on the bridge between the marginal and plastral scutes. [Terminology follows Zangerl, *The Biology of the Reptilia*, 1969.]

The confusion arising from the varied usages of the words *turtle, tortoise,* and *terrapin* is even worse than that between "frog" and "toad." Thus, although the terrestrial forms are usually called tortoises, the members of the terrestrial genus *Terrapene* are called neither terrapins nor tortoises, but Box Turtles. Similarly, tortoiseshell is not obtained from the terrestrial tortoises, but from the marine Hawksbill Turtle. All members of the order may properly be called "turtles." We shall restrict the name *tortoise* to the members of the family Testudinidae—the giant tortoises and their allies—and use *terrapin,* not for any systematic group, but for such small to medium-sized, more or less aquatic, hard-shelled turtles as are commonly used for food.

Many modern turtles are semiaquatic marsh dwellers, and it seems probable that this has been their characteristic mode of existence throughout their evolutionary history. There have been three significant adaptive trends away from this mode: the tortoises have become terrestrial, other turtles truly aquatic or even marine, and still others have adopted a bottom-dwelling existence. It is a curious anomaly that, although the primary turtle adaptation was the development of a heavy, bony box, these three adaptive trends have all entailed a reduction of the shell.

Many fossil turtles are noteworthy for the thickness of the carapace bones, but such a heavy shell would be too much of a burden for a creature moving overland. Modern turtles, and particularly the larger terrestrial tortoises, have usually met the problem of weight-to-volume ratio, not through a reduction of the number of bones in the shell, but through a reduction in the thickness of the individual bones. Many land turtles also developed a high, domed shell, perhaps in part as a protection against gnawing predators that might more easily crack a flat one. The African Pancake Tortoise, *Malacochersus tornieri,* which lives in rocky terrain, has such thin bones in the flattened carapace that the shell is flexible (Figure 14–2). The animal

FIGURE 14–2
The African Pancake Tortoise, *Malacochersus tornieri,* which is adapted for living in rocky crevices.

can squeeze into narrow crevices or between boulders, and when it is in place it inflates its body with air, making it almost impossible to remove.

Except perhaps the sea snakes, the most aquatic of all living reptiles are the sea turtles, which are adapted for swift movement through the water. The Leatherback, *Dermochelys,* has undergone an extensive loss of shell bones; its carapace is a mosaic of small dermal elements. Other marine turtles have a reduced plastron, and the embryonic gaps between the lower ends of the ribs sometimes persist even to maturity. The sea turtles have also developed an efficient swimming stroke. The forefeet have become flippers, and are moved with a figure-eight beat similar to the motion of a bird's wing in flight. Tortoises are traditionally "Slow-and-Solid," but marine turtles are among the swifter of the tetrapods with a probable top swimming speed of 32 kilometers per hour.

Other turtles, such as the snappers, the softshells, and the bizarre South American *Chelus fimbriatus,* are adapted for bottom dwelling. The plastron of the snappers is greatly reduced. The softshells lack scutes entirely. The shell is covered instead with an undivided, leathery skin whose edges conform to the bottom contour as the animal lies hidden under silt or sand.

Of all the reptiles, turtles are of the greatest direct economic benefit to man. They are probably eaten everywhere they occur. Not only do they provide the gourmet with green turtle soup and terrapin stew, but they and their eggs are a major source of protein in many parts of the world where meat is hard to get. The dorsal scutes of the marine Hawksbill Turtle are the source of the beautifully marked tortoiseshell, used for many centuries to make jewelry and other *objets d'art.*

Turtles are conservative in their breeding habits. All species are oviparous, and even the marine forms must come to shore to bury their eggs above the high-tide mark.

The order Testudines, today a small group of about 230 species, is divided into two suborders, Cryptodira and Pleurodira. The pleurodires consist of two families restricted to the southern hemisphere. The cryptodires include nine families, and are worldwide in distribution. Their relationships are indicated in Figure 14–3.

SUBORDER CRYPTODIRA

The cryptodires includes almost all the modern turtles. They are distributed on all continents, though only marine forms reach the shores of Australia. The skull roof is usually very emarginate (indented) from behind. There are never any mesoplastra (small bones between the hyo- and hypoplastral bones), and the pelvis is never fused to the plastron. The cervical vertebrae are distinctive: the postzygapophyses are set wide apart, the central articulations well developed, broad, and typically double on the posterior cervicals.

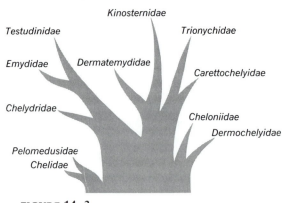

FIGURE 14–3
Dendrogram of the presumed relationships of turtle families.

The posterior cervical spines are low, and the transverse processes are almost absent. The neck is more or less retractable in a vertical position, bending in a sigmoid curve as the head is drawn into the shell. An intergular scute is absent from the plastron, and a cervical scute is usually present on the carapace. There are three superfamilies within the suborder Cryptodira.

SUPERFAMILY TESTUDINOIDEA

The three testudinoid families include most of the modern amphibious and terrestrial turtles. The shell is always complete, although the bridge joining the plastron and carapace may be cartilaginous rather than bony. The shell is covered by epidermal scutes. The forelimbs are not paddlelike. The seminal groove of the penis is undivided, except for a bifurcate condition in tortoises. The fourth cervical vertebra is usually biconvex.

Family Chelydridae

The small chelydrid family contains only three genera, the Snapping Turtle (*Chelydra*) and Alligator Snapping Turtle (*Macroclemys*) of Central and North America (subfamily Chelydrinae) and the Big-headed Turtle (*Platysternon*) of Southeast Asia (subfamily Platysterninae) (Figure 14–4). All chelydrids have big heads, powerful jaws, long tails, and reduced cruciform plastrons. The plastron has a narrow bridge and only a ligamentous attachment to the carapace. The carapace is broad and relatively flat; the nuchal bone is also broad and anteriorly concave. In the skull, the premaxillae form a distinct hooklike projection, and the alveolar surfaces of the upper and lower jaw tend to be moderately expanded and smooth.

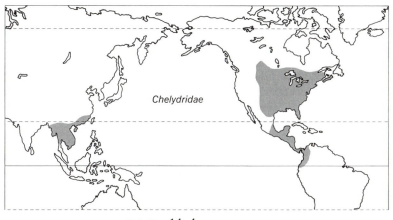

FIGURE 14–4
Distribution of the chelydrids.

Subfamily Chelydrinae. Although primarily aquatic, *Chelydra* sometimes wanders overland. The massive *Macroclemys* (weight to 80 kg) is the largest fresh-water turtle, and leaves the water only to lay its eggs. Both genera can swim, but most of their aquatic locomotion is by bottom-walking. They eat both plants and animals. *Chelydra* actively hunts its prey; *Macroclemys* lies on the bottom with its mouth open and lures fish into its mouth with a wormlike appendage (Figure 14–5).

FIGURE 14–5
The Alligator Snapping Turtle, *Macroclemys temmincki,* of the southeastern United States. Note the bait on the floor of the open mouth.

The breeding habits of the Snapping Turtle *(Chelydra serpentina)* are highly variable. Mating occurs between late April and November. During copulation, the male holds his position atop the female by clinging to the under-edge of her shell with the claws of all four feet. He then twists his tail upward and manipulates it until it establishes contact between the vents. The eggs are most commonly laid in June, but nesting has been reported from May to October. The nest is dug at various distances from the water, its site apparently determined not only by the nature of the soil but also by the whim of the female. Nests have been found at distances of 1 to 25 meters from water; they are about 10 to 18 centimeters deep, and vary widely in form and in manner of excavation. Sometimes, at least, the cavity is definitely flask-shaped, and the excavation slants downward. The female digs by alternately working her hindfeet. She guides the eggs to the bottom with one foot and covers the nest before leaving. The number of eggs deposited varies from 8 to 77 (the latter in a single nest in Manitoba). The eggs may vary from about 25 to 33 millimeters in diameter, and although often spherical, they are sometimes slightly elongated. They hatch in about three months.

Subfamily Platysterninae.   The Big-headed Turtle *(Platysternon)* has a head so massive it cannot be withdrawn within the shell, covered with a shield that forms a continuum with the carapace when the head is retracted as far as possible. *Platysternon* inhabits swift-flowing streams and rivers, but sometimes travels overland and can even climb trees. It is strictly carnivorous. Only two eggs are laid at a time.

Family Emydidae

The emydids are the most diverse group of modern turtles. By and large, most species are aquatic and well deserve the group name, pond turtles, but some are semiaquatic and a few terrestrial. All have small heads, short tails, and well-developed shells. The plastron and carapace are usually united by a broad, bony bridge, and the carapace is usually domed. The premaxillae do not form a hooklike process, and although the alveolar surface varies from narrow to broad, there is a tendency toward the development of a secondary palate. Figure 14–6 shows the distribution of the emydids. The family is divided into two subfamilies.

Subfamily Batagurinae.   Batagurines are pond turtles with a distribution centered in southeast Asia. Of the 21 genera, only *Mauremys* has a European species; *Callopsis,* with six species, is confined to Central America and northern South America. Until recently, the batagurines were placed in the emydine subfamily. Although closely related to the emydines, they can be

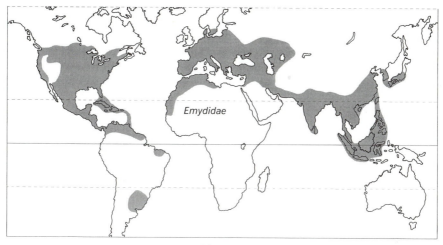

FIGURE 14–6
Distribution of the emydids.

distinguished from them by several characters, including the suture between the twelfth marginal and last vertebral scute, which lies on the suprapygal bone.

The batagurines show convergence in habits and structure with the New World emydines; the semiaquatic *Cuora* has a hinged plastron like that of the terrestrial Box Turtles, *Terrapene*. The batagurines have more semiaquatic and terrestrial representatives than the emydines. *Mauremys, Geoclemys, Cyclemys* and *Callopsis* have terrestrial members. *Orlita* and *Batagur* are the largest aquatic emydids. *Batagur baska* is found in estuaries, slow-flowing rivers, and canals of southeastern Asia. Around the mouth of the Irrawaddy River, egg laying takes place during January and February. Every afternoon the turtles come ashore, gather in large herds, and sun themselves on the sand. At night the females dig holes 45 to 60 centimeters deep in the sand above the high-tide mark and deposit 10 to 30 eggs, each about 75 millimeters long. Each female lays between 50 and 60 eggs, in three batches, over a period of about six weeks. The incubation period is about 70 days.

Subfamily Emydinae. The emydines are New World pond turtles, with only one exception, *Emys,* in Europe. The other six genera are predominantly North American, although the distribution of *Chrysemys* extends into northern South America. The suture between the twelfth marginal and last vertebral scute lies on the pygal bone.

Terrapene carolina and *T. ornata* are fully terrestrial species. Genera such as *Clemmys* and *Emydoidea* are semiaquatic; *Chrysemys* and *Graptemys* are aquatic, and leave the water only to bask and lay eggs; *Malaclemys* is a brackish-water inhabitant. Diets range from strictly herbivorous to carnivorous. In all species, the juveniles appear to be insectivorous.

Elaborate courtship behavior has developed in *Chrysemys* and *Graptemys*. The toenails of the male's forefeet become greatly elongated. The male courts the female by vibrating his toenails against her head. His position and the speed of vibration seem to be species-specific. In forms with such elaborate courtship, the male is usually considerably smaller than the female, and thus requires her full cooperation when mating.

Family Testudinidae

The testudinids are the true land tortoises, including the lumbering giants of the Galápagos and Aldabra Islands. The feet are club-shaped, short, and broad, and do not have more than two phalanges in any digit. The toes are completely unwebbed. The cylindrical and columnar hindlegs are like those of an elephant. The skull roof is incomplete posteriorly; the inframarginal series is incomplete; there are ten or eleven marginals on each side. Usually the shell is high and dome-shaped. Not all tortoises are large—the little *Testudo kleinmanni* reaches a shell length of only 120 millimeters. The largest tortoise recorded, a specimen from Aldabra Island, measured 1400 millimeters in straight-line shell length and weighed 254 kilograms.

Nine recent genera are recognized in this family. By far the largest and most widely distributed is *Geochelone,* found in South America, Africa, Asia, and the oceanic island groups of the Galápagos and the Seychelles. North America has only one genus, the Gopher Tortoises, *Gopherus,* of the southern United States and Mexico. *Testudo* occurs in southern Europe, Asia, and Africa. The remaining genera are all confined to Africa and Madagascar (Figure 14–7).

In general, each tortoise has an individual resting place—a burrow it has dug, a cranny in a rock, or a sheltered nook under a plant—from which it emerges to graze during the day or, in hot weather, in the cool of the evening. Tortoises are largely herbivorous, but *Testudo graeca* has been reported to take insects, mollusks, and worms. One captive Galápagos tortoise was seen to catch two rats and a pigeon, and a number of zoo specimens from the same area have eaten raw meat greedily. Tortoises that live in desert areas, or on rocky islets where no fresh water is available, are able to get along without drinking; other species drink copiously and enter water freely.

Courtship of tortoises seems to consist mainly of the male pursuing the female and butting her with his shell. Copulating males are surprisingly vocal—the "voice" has been described for various species as "a muffled, whistling cry," "a peculiar grunting noise," and "the intermittent winding up of a metal spring" (Figure 14–8). In spite of their relatively large size,

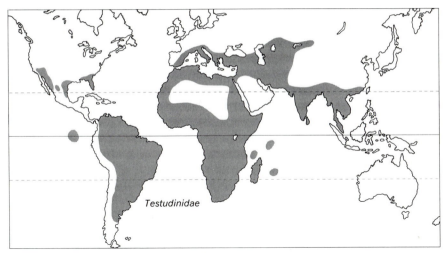

FIGURE 14–7
Distribution of the testudinids.

FIGURE 14–8
Hermann's Tortoise, *Testudo hermanni*, of southern Europe. Breeding males of this and other species of *Testudo* exercise their voices in the spring.

most female tortoises lay few eggs in a single clutch, usually fewer than seven. Some species may lay only a single egg per clutch. As with most other turtles, the eggs are buried in a hole that the female digs with her hindlegs.

SUPERFAMILY TRIONYCHOIDEA

The trionychoids includes two pairs of families, the mud turtles (Dermatemydidae and Kinosternidae) and the soft-shelled turtles (Carettochelyidae and Trionychidae), which are overtly quite distinct. In spite of their superficial dissimilarities, the two groups share many characters that are not found in any other turtles. The stapedial artery is small, and the medial branch of the internal carotid artery is large. The seminal groove of the penis is furcate.

All four families are highly aquatic. The mud turtles tend to walk on the bottom rather than swim; the softshells are excellent swimmers. The mud turtles are New World forms, with their greatest diversity in Central America. They have complete shells, although the plastron may be reduced in size or possess one or more hinges. The shell is covered with epidermal scutes. The softshells lack epidermal scutes, and the shell tends to be reduced. They are widely distributed except in Australia and the American tropics; their greatest diversity occurs in the Old World tropics.

Family Dermatemydidae

The apparently primitive dermatemydid family is monotypic. Only a single species, the Jicotea *(Dermatemys mawi)*, is extant today; it is found from Veracruz, Mexico to Guatemala and Honduras (Figure 14–9). It is a large

FIGURE 14–9
The Jicotea of Central America, *Dermatemys mawi.*

(45 cm) aquatic species with a well-developed and streamlined shell. The plastron is large, covering most of the ventral surface, and firmly linked to the carapace by a bony bridge. In superficial appearance, the Jicotea resembles the large aquatic emydines of the southeastern United States. Like them, it is predominantly herbivorous.

In countries where these turtles live, natives fish for them in muddy backwaters and oxbow lakes, and sometimes sell them in the markets. The species is extremely aquatic and has difficulty moving on land. In fact, on land it seems unable or unwilling even to hold its head off the ground, a phenomenon that can be noted in most photographs. Little is known of its breeding habits other than that it lays about 20 eggs in the fall when the rivers are flooded; the eggs are then covered with decaying vegetation.

Family Kinosternidae

The four kinosternid genera of the New World differ consistently from other turtles in having ten rather than eleven marginal scutes on each side. The four genera are clearly divided into two subfamilies (Figure 14–10).

FIGURE 14–10
Distribution of the dermatemydids and kinosternids.

Subfamily Kinosterninae. *Kinosternon* and *Sternotherus* constitute the kinosternine subfamily. Both lack an entoplastron and have a hinged plastron. Most species are small, seldom exceeding 150 millimeters. Although aquatic, they seldom swim, but rather bottom-walk. They are largely carnivorous.

Sternotherus has three species *(S. odoratus, S. minor,* and *S. carinatus)*, with distribution centered in the southeastern United States. Their plastron is small; its anterior lobe is shorter than the posterior lobe and scarcely movable. *Kinosternon* has at least ten species, centered in Central America. Their plastron is large; the anterior and posterior lobes are of nearly equal size and usually movable.

The Eastern Mud Turtle *(Kinosternon s. subrubrum)* usually nests in the early summer, but nesting sometimes lasts until September. The female searches until she finds a suitable site. She digs with her forefeet, thrusting out the dirt laterally until she is almost concealed, then turns around and completes the nest with her hindfeet. While she is digging with the hindfeet and while she is laying, only her head is visible above the ground. After two to five eggs have been deposited, the turtle crawls out and may return directly to the water, or may make a slight effort to conceal the nest cavity by leveling and scratching the earth around the site. The completed nest is a semicircular cavity 75 to 130 millimeters deep and inclined at an angle of about 30 degrees. The cavity extends slightly beyond the eggs; the soil immediately around them is firmly packed, indicating that the turtle covers them carefully even though she makes no effort to conceal the entrance to the nest. Rains, pounding on the loose, sandy soil, soon obliterate all sign of it. Apparently, the eggs sometimes over-winter and hatch the next spring, for hatching eggs have been plowed up in April. In the laboratory, hatchlings have emerged in late September.

Subfamily Staurotypinae. The two genera of the staurotypine subfamily, *Staurotypus* and *Claudius,* occur from southern Mexico to Honduras. Their plastron contains an entoplastron and lacks a hinge. *Claudius* is small, like most kinosternines. *Staurotypus,* in contrast, is large (to 375 mm) and aggressive. Both are aquatic and carnivorous.

Family Carettochelyidae

The sole surviving member of the carettochelyid family is *Carettochelys insculpta,* the Pitted-shell Turtle of the rivers of northern Australia and southern New Guinea (Figure 14–11). Scutes are absent, and the pitted shell (about 50 cm long) is covered only by a thin layer of soft skin. The forelimbs approach those of the sea turtles in their flipperlike appearance and are the principal locomotor force. Like sea turtles or penguins, *Carettochelys* flies through the water. This little known fresh-water turtle is apparently herbivorous as an adult, depending largely on plant material falling into the river from bank vegetation.

Family Trionychidae

Unlike *Carettochelys,* trionychid turtles have an incomplete or reduced shell. The carapace is usually flattened, the peripheral bones lost, and the distal ends of the ribs extended beyond the pleural bones. The plastron is

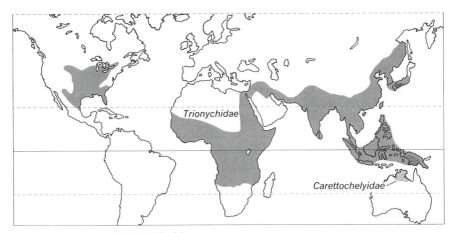

FIGURE 14–11
Distribution of the carettochelyids and trionychids.

similarly reduced and joined to the carapace through a wide ligamentous and cartilaginous connection. The forelimbs are less paddlelike than in *Carettochelys,* and each forefoot bears three claws. These extremely aquatic turtles are excellent swimmers, but spend much of their time partially buried on the bottoms of rivers, lakes, and swamps. Their length varies from 23 to 80 centimeters. Most species are carnivorous; they wait for their prey while concealed in mud or sand, or they actively forage. Occasionally plants are eaten. Soft-shelled turtles are widely esteemed as food. They are probably also useful as scavengers. Like other turtles with reduced shells, many, though not all, of the species are short-tempered and quick to bite. The living forms are divided into two subfamilies.

Subfamily Lissemyinae.   Lissemyines have the hyoplastron and hypoplastron fused. A pair of strong, hinged, cutaneous flaps at the rear of the plastron close over the hindlimbs when they are withdrawn (Figure 14–12). Of the three genera in the subfamily, *Lissemys* is found in India and Burma, *Cyclanorbis* and *Cycloderma* in Africa. *Lissemys* is not closely related to the two African genera and differs from all other soft-shelled turtles in the presence of peripherals along the posterior margin of the carapace (Figure 14–13). These peripherals are probably not homologous to the peripherals of the other turtles. The breeding habits of *Lissemys* probably do not differ greatly from those of the trionychines. Nesting has been observed in the fall; the young appear from July to September, early in the monsoon season.

Subfamily Trionychinae.   The typical softshells of the trionychine subfamily have the hyoplastron distinct from the hypoplastron and are without cutaneous femoral flaps on the plastron. There are three genera. The wide-

FIGURE 14–12
Ventral view of *Lissemys punctata,* showing the pelvic flaps characteristic of the lissemyines.

FIGURE 14–13
The Hinged Softshell, *Lissemys punctata,* of India.

spread *Trionyx* is found in North America, Africa, and Asia. *Chitra* and *Pelochelys* occur in southeastern Asia, with the latter extending to New Guinea.

The breeding habits of the Eastern Spiny Softshell, *Trionyx s. spiniferus* of North America, are probably typical for the family. Nesting occurs in June and July. The female digs a flask-shaped hole, 100 to 250 millimeters deep and 75 to 125 millimeters in diameter, with a narrow neck. It takes her about 40 minutes to dig the nest. She deposits a few eggs at a time, arranges them with her feet, rakes some earth down into the hole with them, gently packing it in, then lays more eggs. This process is continued until the nest is finally completed. She apparently makes no effort to conceal the nest after she has covered the last of the eggs. The 10 to 25 eggs are 25 to 27 millimeters in diameter but are not quite spherical. The shell is thick and not very brittle. The incubation period in this subspecies is not known, but eggs removed from an adult female of the Florida Softshell *(T. ferox)* hatched in 64 days.

SUPERFAMILY CHELONIOIDEA

The two chelonioid families are entirely marine. The shell may be complete or incomplete, and may or may not be covered with epidermal scutes. The forelimbs are distinctly paddle- or flipperlike and are the principal locomotor elements. The seminal groove is undivided. The fourth cervical vertebra is biconvex, separating the anterior opisthocoelous vertebrae from the posterior procoelous ones.

Family Cheloniidae

The cheloniid family includes four living genera: *Caretta,* the massive Loggerhead, which may weigh as much as 400 kilograms; *Chelonia,* the succulent Green Turtles, so called from the color of their fat; *Eretmochelys,* the Hawksbill, the source of the beautiful tortoiseshell (Figure 14–14); and *Lepidochelys,* the Ridleys, the smallest of the marine turtles, with a maximum shell length of 790 millimeters. These turtles are circumtropical in distribution; stragglers are sometimes carried by warm currents as far north as Newfoundland and Scotland (Figure 14–15). Two subfamilies are usually recognized.

Subfamily Carettinae. The carettines, *Caretta* and *Lepidochelys,* are most easily differentiated from the cheloniines by the presence of five pairs of costal scutes and by the size and shape of the first vertebral scute. The members of this subfamily are largely carnivorous. *Caretta* is the only sea turtle that nests on temperate beaches, regularly as far north on the Atlantic coast as Virginia. *Lepidochelys* has synchronized herd nesting (arribada).

FIGURE 14–14
The Atlantic Hawksbill, *Eretmochelys imbricata,* the source of the tortoiseshell of commerce.

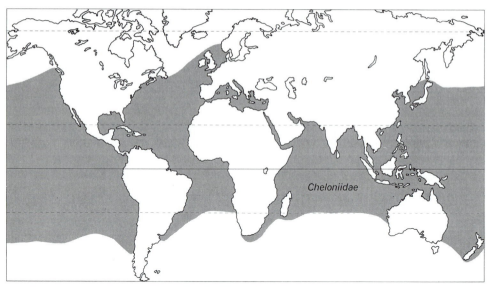

Cheloniidae

FIGURE 14–15
Distribution of the cheloniids.

All the females of one population arrive and begin their nesting nearly simultaneously: hundreds of nesting turtles can be found concentrated on a single stretch of beach.

Subfamily Cheloniinae. *Chelonia* and *Eretmochelys,* of the cheloniine subfamily, normally have four pairs of costal scutes. Adult *Chelonia* are largely herbivorous, grazing on the submerged grasses in shallow offshore waters; *Eretmochelys* tends to be carnivorous. *Chelonia* and most other sea turtles nest every two to four years. During each nesting season, a female will lay three to four clutches. Nesting is an arduous duty for the female *Chelonia*. She will normally crawl onto the beach after twilight. Her initial approach is quite cautious and she is easily frightened away. Once on the beach, she crawls to the foredunes above the maximum high-tide mark. Exactly how a nesting site is selected is unclear, but apparently sand texture and dampness are determined by touch and smell. First the female digs a shallow cavity for her body by alternate swipes of the forelimbs and body movement. Then the hindlimbs begin to dig the flask-shaped egg chamber with a regular scooping movement. The depth of the egg chamber is determined by the reach of the hindlimbs. When the hindfeet can no longer dig, egg laying begins. Usually a hundred or so eggs are dropped gently into the nest. The chamber is carefully filled with sand, and as the female begins to return to the sea, a short period of apparently unorganized kicking and thrashing effectively hides the actual nesting site. On entering the sea, the female is met by males, who will fertilize the next clutch of eggs, to be deposited in about two weeks.

The clutches of marine turtles are concentrated in a small area above high-tide mark on the laying beaches. Hordes of predators gather to dig them up, the most destructive being man and feral dogs. After hatching, the baby turtles claw upward through the sand and head for the water. On their journey across the beach, they must run the gauntlet of hungry crabs, mammals, and hovering sea gulls. Those that survive are met in the water by swarms of predatory fish. Yet marine turtles have persisted for millions of years—cheloniid sea turtles are known from the upper Cretaceous. The marked decline in turtle populations in recent years has probably resulted largely from the activity of human hunters waylaying the females on the beaches before they deposit their eggs.

Family Dermochelyidae

The dermochelyids are marine turtles that lack epidermal scutes and have the dermal bones of the carapace and plastron largely replaced by a mosaic of small platelets set in a leathery skin. The limbs are paddle-shaped and without claws: the anterior ones are very large, and the posterior ones of the adults are broadly connected to the tail by a web.

A single species, *Dermochelys coriacea*, the Leatherback, is the only living member of the family. This extraordinary creature, one of the most remarkable of all living reptiles, can be confused with nothing else. It is distinguished from all other sea turtles by the scaleless black skin of its back and by the seven narrow ridges, formed by enlarged platelets of the dermal mosaic, that extend down the length of its back. Five similar keels occur on the ventral surface. There is a strongly marked cusp on each side of the upper jaw. This largest living turtle may reach a weight of 680 kilograms or more, but such giants are rare. The large specimens encountered from time to time along the coasts today probably weigh around 360 kilograms, about the size of a large Loggerhead.

A remarkably strong and rapid swimmer, the pelagic Leatherback is widely distributed, though usually scarce, in tropical seas, and occurs sporadically in temperate waters (Figure 14–16). It feeds on crustaceans, mollusks, and small fishes, as well as on marine plants. Jellyfish also appear to be an important food in some areas.

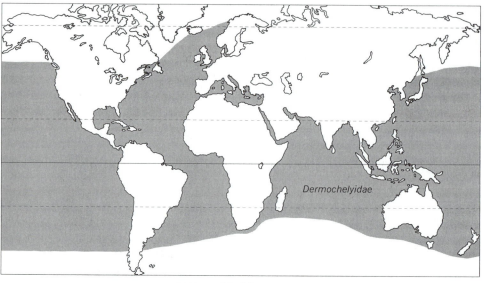

FIGURE 14–16
Distribution of the dermochelyids.

Although the nesting of Leatherbacks is confined to tropical beaches, the adults wander into temperate and subarctic waters. Their occurrence in such cold water was once thought to be accidental, but recent evidence of temperature regulation and body temperatures higher than ambient indicates an adaptation to permit normal behavior in colder areas. The high oil content of the shell would also act as insulation, like blubber on whales.

SUBORDER PLEURODIRA

The pleurodires are known as side-necked turtles, because they withdraw their heads by bending their necks laterally instead of vertically, as do the cryptodires. The cervical vertebrae have rather high spines posteriorly and well-developed transverse processes for the insertion of the muscles that bend the neck. Their central articulations are well developed, but are always single. The pelvis is fused to the plastron and sutured to the carapace. The temporal roof may be emarginate (notched) from behind, and is usually more or less emarginate from below. A pair of mesoplastra are sometimes present. An intergular scute is present on the plastron, and the cervical scute is absent from the carapace.

These aquatic turtles are found today only in the southern continents—South America, Africa including Madagascar, and Australia. The suborder is divided into two families.

Family Pelomedusidae

The pelomedusids are able to tuck the head and neck into the shell so that the neck is concealed; hence they are sometimes called hidden-necked turtles. The skull is emarginate behind, nasal bones are absent, and the second cervical is biconvex. Mesoplastra are present. The family includes three Recent genera: the African *Pelomedusa* and *Pelusios* are moderate-sized forms, with maximum shell lengths of around 30 centimeters (Figure 14–17); *Podocnemis* of South America and Madagascar is large, attaining a shell length of 80 centimeters. The distribution is shown in Figure 14–18.

All of these turtles are aquatic, although *Pelomedusa* may wander freely overland. It is reported to estivate during the African dry season. *Pelomedusa* is said to be carnivorous; *Podocnemis* is largely herbivorous.

FIGURE 14–17
The Marsh Side-necked Turtle of Africa, *Pelomedusa subrufa.*

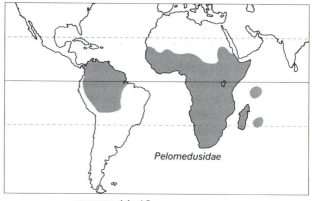

FIGURE 14–18
Distribution of the pelomedusids.

Podocnemis expansa breeds on sandy islands midstream in the Orinoco and Amazon rivers. The females crowd together on the tiny islands, each digging a hole 50 to 60 centimeters deep and depositing 80 to 200 eggs. For years, natives of these regions have conducted highly organized egg hunts, gathering them by the millions and mashing them to extract the oil. The natives also prey on the hatchlings as they emerge, and hunt for the adults in the rivers. Only a species with a very high biotic potential could long survive such heavy predation, and the turtles are no longer as numerous as they once were. The population will probably continue to decline until it is no longer profitable to hunt them.

Family Chelidae

The head of any of the chelid turtles can be more or less withdrawn under the margin of the carapace, but the neck remains exposed. The skull is little emarginate from behind, nasal bones are usually present, the fifth and eighth cervical vertebrae are biconvex, and mesoplastra are lacking. The jaw is usually long, slender, and weak. Because the neck is often longer than the carapace, these turtles are sometimes known as snake-necked turtles (Figure 14–19). The carapace is about 15 to 40 centimeters long.

There are ten genera of chelids: four in Australia and New Guinea, six in South America (Figure 14–20). Although the Australian and South American genera are related, they have been isolated for a long time and show a high degree of structural divergence. For this reason, some herpetologists believe they should be placed in separate families. A similar division has also been suggested for the African and South American pelomedusids.

One of the most bizarre creatures that has ever lived is the South American Matamata *(Chelus fimbriatus)*. Its snout is drawn out into a snorkellike

FIGURE 14–19
A snake-necked South American chelid, *Platemys platycephala.*

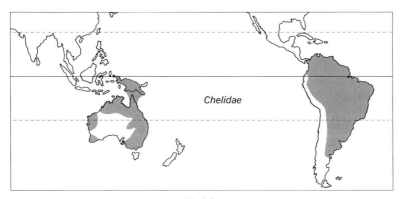

FIGURE 14–20
Distribution of the chelids.

proboscis with the nostrils at the tip, and the tiny eyes are placed far forward in the head. The mouth is very large, reaching back to the region of the ears, but the jaws are weak. The skin of the sides, the lower part of the head, and the long, thick neck is fringed and frayed. The carapace is rough and ridged and has three broad keels.

The female of the Long-necked Turtle of Australia *(Chelodina longicollis)* scoops a circular hole in the ground and deposits as many as 20 elongate, white eggs. They are laid in November or December and the young hatch in February or March.

READINGS AND REFERENCES

Agassiz, L. *Contributions to the Natural History of the United States,* vols. 1–2. Boston: Little, Brown, 1857.

Boulenger, G. A. *Catalogue of the Chelonians, Rhynchocephalians and Crocodiles in the British Museum.* London: British Museum (Natural History), 1889.

Brongersma, L. D. *European Atlantic Turtles.* Leiden: E. J. Brill, 1972.

Bustard, R. *Sea Turtles.* New York: Taplinger, 1973.

Carr, A. F. *Handbook of Turtles.* Ithaca, N.Y.: Comstock, 1952.

———. *The Turtle: A Natural History of Sea Turtles.* London: Cassell, 1967.

Ernst, C. H., and R. W. Barbour. *Turtles of the United States.* Lexington: University Press of Kentucky, 1972.

Gaffney, E. S. "A phylogeny and classification of the higher categories of turtles." *Bulletin of the American Museum of Natural History,* vol. 155, art. 5, 1975.

Loveridge, A., and E. E. Williams. "Revision of the African tortoises and turtles of the suborder Cryptodira." *Bulletin of the Museum of Comparative Zoology,* vol. 115, no. 6, 1957.

McDowell, S. B. "Partition of the genus *Clemmys* and related problems in the taxonomy of the aquatic Testudinidae." *Proceedings of the Zoological Society of London,* vol. 143, no. 2, 1964.

Pritchard, P. *Living Turtles of the World.* New York: T.F.H. Publications, 1967.

Smith, M. A. *Fauna of British India: Reptilia and Amphibia,* vol. I, *Loricata, Testudines.* London: Taylor and Francis, 1931.

Wermuth, H., and R. Mertens. *Schildkröten, Krokodile, Brückenechsen.* Jena: Gustav Fischer, 1961.

Williams, E. E. "Variation and selection in the cervical central articulation of living turtles." *Bulletin of the American Museum of Natural History,* vol. 94, no. 9, 1954.

Zangerl, R. "The turtle shell." *In* C. Gans (ed.), *Biology of the Reptilia,* vol. 1. London: Academic Press, 1969.

Zug, G. R. "The penial morphology and the relationships of cryptodiran turtles." *Occasional Papers of the Museum of Zoology, University of Michigan,* no. 647, 1966.

———. "Buoyancy, locomotion, morphology of the pelvic girdle and hindlimb, and systematics of cryptodiran turtles." *Miscellaneous Publications, Museum of Zoology, University of Michigan,* no. 142, 1971.

CHAPTER FIFTEEN

LIZARDS
AND AMPHISBAENIANS

SNAKES, LIZARDS, AND AMPHISBAENIANS form the most diverse and successful group of modern herps. Members of the order Squamata occur on all
continents and most islands within a latitude range of 52° south to 64°
north. They have adapted to a myriad of environments, from the tree lines
on the mountains to the middle of the ocean, and have adopted a full range
of habits, from fossorial to arboreal to aquatic.

The three suborders of squamates—Amphisbaenia, Sauria, and
Serpentes—share a number of characters that differentiate them from other
reptiles. They have diapsid skulls with paired openings in the temporal
region; frequently one or both of the temporal arches that separate the
temporal openings are missing. Males have uniquely paired copulatory organs, the hemipenes. The abdominal wall contains no gastralia. The teeth
are either acrodont or pleurodont, the vertebrae normally procoelous. Like
all reptiles, squamates have epidermal scales, which commonly overlap one
another, unlike the abutting scales of other reptiles.

SUBORDER AMPHISBAENIA

Amphisbaenians, also called ringed lizards or worm lizards, are small, highly specialized burrowers that do not seem to be closely related to either lizards or snakes. Their bodies are elongate and nearly uniform in diameter, and their tails are short. Except for one genus, *Bipes,* which has short forelegs, external limbs are lacking and the girdles are vestigial. There are no external ear openings, and the eyes of adults are hidden under the skin. Osteoderms are absent. The soft skin of the body is folded into numerous rings divided into quadrangular areas representing the flattened and reduced scales. These rings, combined with the absence of limbs and the cylindrical body form, give the amphisbaenians a remarkable resemblance to earthworms.

The skull of an amphisbaenian is highly specialized for digging: the bones are closely united, the snout region is short and usually expanded, the brain case is almost completely enclosed. There are no skull arches. The teeth are large but few in number, and are absent from the palate. There is only one functional lung, the left. Most species are about 300 millimeters long. The largest, *Monopeltis* of central Africa, attains lengths of about 675 millimeters. There are two families.

Family Amphisbaenidae

Of the two worm lizard families, the amphisbaenids are the most successful, with 19 genera and about 35 species in the tropics and warm temperate regions of the Americas, Africa, and Europe (Figure 15-1). The amphis-

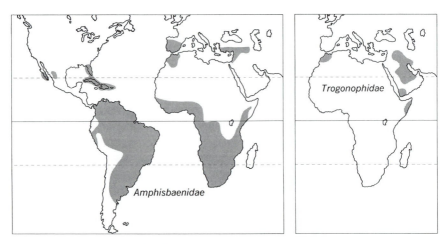

FIGURE 15–1
Distribution of the amphisbaenians. [After Gans, 1968.]

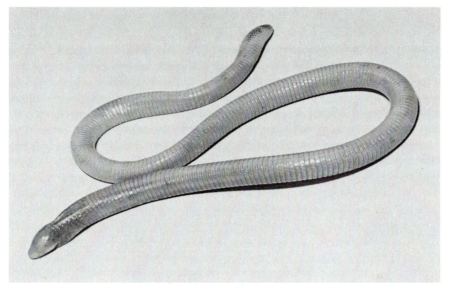

FIGURE 15–2
The Florida Worm Lizard, *Rhineura floridana*, known locally as the "Graveyard Snake."

baenids have either an unpigmented skin, appearing pink in life because of the color of the blood, or a skin lightly pigmented on the dorsal surface. The skin is highly permeable to water, and the animal can absorb moisture directly from the surrounding soil.

The name of the family comes from two Greek words that, translated literally, mean "walk at both ends," a reference to the animals' ability to move backward as well as forward in their underground tunnels. Although some of the amphisbaenids have rounded heads and burrow with a random head movement, many have specialized burrowing movements with associated modifications of the head. The keel-headed species have laterally compressed heads, and burrow by ramming the head forward and then to the side, to create a tunnel. The shovel-headed species have dorsoventrally depressed heads, and burrow by ramming the head forward and then lifting it up, to compact the soil onto the roof of the tunnel.

One species occurs in the U.S., *Rhineura floridana* of Florida (Figure 15-2). It feeds on worms and small insects, especially ants and termites. So far as is known, most amphisbaenids lay eggs, though one form has been reported to be ovoviviparous.

Family Trogonophidae

The small trogonophid family consists of four genera and about six species from North Africa and Socotra Island in the Indian Ocean. All trogonophids have acrodont dentition, in contrast to the pleurodont dentition of the amphisbaenids. Their heads tend to be rounded, with no sharp edges or with low rostral ridges. Most are strongly pigmented, with a distinct dorsal pattern. Their burrowing habits are fairly simple—the head is waved randomly or swung regularly from side to side in association with forward motion.

SUBORDER SAURIA

It is not always easy to determine whether a given animal is a lizard or a snake by looking at the intact specimen. Most lizards have two pairs of legs, but some have lost one or both pairs, although traces of the girdles remain in all. Usually lizards have visible external ear openings and movable eyelids with a nictitating membrane (third eyelid). The two halves of the lower jaw are firmly united at a mandibular symphysis, so that the size of the mouth opening is restricted; the tongue is well developed. In many lizards, each caudal vertebra is divided by a transverse septum at which the lizard is able to break off the tail by muscular contractions (autotomy). This capacity is a defense mechanism: the shed tail part twitches vigorously, thereby distracting the predator while the lizard escapes. The lizard then grows a new tail that is lighter in color and usually smaller than the original, and is supported by a cartilaginous rod rather than by vertebrae.

Lizards can be readily distinguished from snakes on the basis of internal characters: legless lizards retain traces of a pectoral girdle and sternum, but no snake has either; the last one or two movably attached ribs in snakes are forked, but lizards never have forked ribs. There are also many differences in the soft anatomy: the kidneys lie far forward in the snake, with the right further forward than the left, but the kidneys of the lizard are more posterior, extending beyond the level of the cloaca, and are symmetrically placed.

As a group, the lizards have prospered and now number about 3,000 species. Although most numerous in the tropics, they have successfully invaded all continents except Antarctica, and in Europe they are found as far north as the Arctic Circle. Lizards have undergone extensive adaptive radiation and have developed a bewildering variety of habits. Some spend much of their time in water (either fresh or salt), whereas others have become adapted to life in arid deserts.

The desert-dwelling forms usually have depressed bodies that make it easy for them to hide beneath stones and creep into narrow fissures. Many rock-haunting species have developed special pads on their toes that enable them to run swiftly over smooth surfaces. With the advent of man, some of these shifted their quarters to the walls of houses, where an abundance of insects ensures a steady food supply. Lizards usually do not have webs between the toes; the lizard with the most strongly webbed feet, *Palmatogecko rangei,* is found in the arid deserts of southwestern Africa, where the webbing assists it in maneuvering in the shifting sands.

Some of the ground-inhabiting lizards are very large: the giant Komodo Dragon *(Varanus komodoensis)* probably grows to 3 meters in length. Others are small and agile; some run rapidly on two legs (much as small dinosaurs did in the Mesozoic), others have lost their limbs and glide through the grass like snakes.

Some lizards have become burrowers: their limbs are not used in digging through the soil, but are held close to the sides and are usually greatly reduced. A few have lost their limbs entirely.

Some arboreal lizards have compressed bodies that make them inconspicuous when they bask on the branches of trees. Most spectacular perhaps are the many species of Flying Dragons *(Draco)* of Southeast Asia and the East Indies. A Dragon may be recognized immediately by the winglike expansions of skin on either side of the flank, which are supported by half a dozen of the hindmost ribs. These ribs function in much the same way as the ribs of a parasol, for when the lizard is at rest it presses them against its body, so that the skin is folded and is scarcely noticeable. These expansions are simply gliding surfaces, and cannot be used for true flight. The lizard rests on a tree in a vertical position, and when it is ready to move it leaps into the air, soars down, and lands with an action so rapid that the opening and closing of the "wings" is scarcely perceptible. In this way a Dragon can glide 30 meters or more.

Lizard classification has been fairly stable since 1923. In that year, Charles L. Camp published his thesis, "Classification of the Lizards," which recognized the natural phylogenetic lineages. There have been, of course, subsequent modifications of Camp's classification, but its basic framework stands.

Camp recognized two major groups of lizards, the Ascalabota and the Autarchoglossa. The ascalabotans include the geckos, iguanids, agamids, xantusiids, and chameleons; all other lizards are autarchoglossans. The names refer to the structure of the tongue and indicate a difference in feeding habits (Figure 15-3). Ascalabotans are visual predators who feed largely on moving prey. Taste and smell (chemoreception) are secondary, and are used mainly to reject obnoxious or toxic prey. The tongue is usually thick and fleshy and aids in swallowing. Autarchoglossans, in contrast, depend on chemoreception to recognize their prey. Although the tongue is fleshy in some, it plays an active role in carrying particles to the Jacobson's organ for chemoreception. As in the ascalabotans, the tongue also aids in swallowing.

The ascalabotans are generally considered more primitive than the autarchoglossans. Although this may be true, it is important to remember that

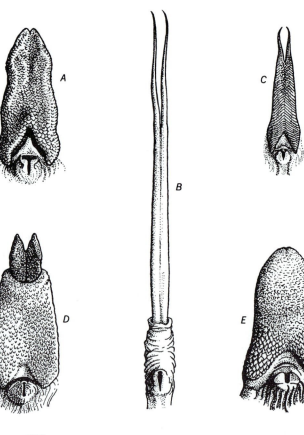
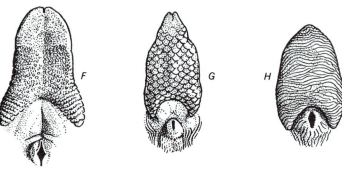

FIGURE 15–3

Tongues of lizards: *(A) Mabuya carinata* (Scincidae); *(B) Varanus niloticus* (Varanidae); *(C) Tachydromus sexlineatus* (Lacertidae); *(D) Ophisaurus harti* (Anguinidae); *(E) Calotes versicolor* (Agamidae); *(F) Gekko gecko* (Gekkonidae); *(G) Nessia monodactyla* (Scincidae); *(H) Dibamus novaeguineae* (Dibamidae). [After M. A. Smith, 1931.]

lizards of both groups are highly specialized in one way or another, and cannot be considered primitive in the sense of having remained unchanged since their origin.

The relationships of the two groups are shown in Figure 15–4. The following infraorders are those differentiated by Camp, with the exception of Dibamia, which has only recently been established.

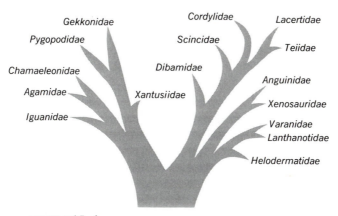

FIGURE 15–4
Dendrogram of the presumed relationships of lizard families.

INFRAORDER GEKKOTA

The geckos and their allies are mostly either stoutly built, short-tailed little lizards, largely nocturnal and well adapted for climbing, or snakelike forms with very reduced limbs. Most geckos are less than 150 millimeters long. They lack an upper temporal arch, a postorbital arch, and lachrymal, squamosal, and postorbital bones in the skull. The jugal bone is small and sometimes absent. The frontal bones are usually united and surround the forebrain. Pleurodont teeth are usually present on the marginal bones (premaxillaries, maxillaries, and dentaries), but palatal teeth are absent. The eyes have either eyelids or "spectacles," which are simply eyelids that have fused and developed transparent areas through which the lizard sees. The tongue is only very slightly cleft, if it is cleft at all. Most species have a postanal sac opening to the outside on each side of the base of the tail, just behind the vent. In the male, a bone lies free just below the skin in front of the opening of the sac on each side. The function of these sacs is unknown. There are usually four or more transverse rows of belly scales per body segment. This infraorder includes those lizards that have well-developed voices.

The Gekkota include three families. The largest of them (Gekkonidae) comprises those lizards with well-developed limbs that are commonly known as geckos. In one family, the scalefoots (Pygopodidae), the hindlegs are represented by scarcely noticeable flaps on either side of the vent, and the forelegs are lacking. The Xantusiidae are known as night lizards.

Family Gekkonidae

The gekkonids form a large, tropicopolitan family of about 75 genera and hundreds of species (Figure 15–5). Most are stoutly built little lizards, nocturnal and arboreal, with big, spectacled, catlike eyes, the pupils contracting to slits in daylight and opening wide at night. The vertebrae are procoelous or amphicoelous. The digits frequently have both claws and friction pads for climbing. Males usually have preanal and femoral pores, as well as postanal sacs and bones.

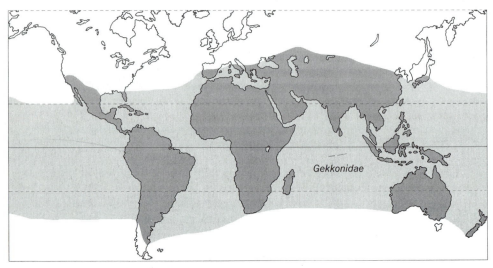

FIGURE 15–5
Distribution of the gekkonids. [After Kluge, 1967.]

Many geckos have attached themselves to human dwellings. This is really a type of mutualism, since the insects that are attracted to houses provide food for the geckos, and the geckos, by feeding on the insects, help keep these pests under control. Man has been slow to recognize his debt. Indeed, superstitions about the geckos are as widespread as the geckos themselves. It is said that they are highly venomous, that their bite is fatal, that they poison man's food and drink, that they can cause leprosy by running over the face of a sleeper. As with the rat and the mouse, man has unwittingly transported geckos around the globe, so that the original distribution of many of the genera may never be known.

Few lizards can compete with the geckos in their ability to discard the tail. This defense mechanism is used so often by some species that it is nearly impossible to find an adult specimen with its original tail.

The name *gecko* probably arose as an attempt to imitate the call of some species of these lizards. Their ability to vocalize is remarkable, for most lizards are silent creatures. The sound is perhaps produced by clicking the broad tongue against the roof of the mouth. It has been variously transcribed as "checko," "tocktoo," "toki," "tok," or "chick chick."

Perhaps most spectacular to people seeing geckos for the first time is their ability to run over a windowpane or up a vertical wall and across the ceiling (Figure 15–6). The geckos that are capable of this sort of climbing have

FIGURE 15–6
The underside of the Moorish Gecko, *Tarentola mauritanica,* seen as it clings to a sheet of glass.

some part of their digits dilated to form adhesive discs. In the most arboreal forms, the underside of the disc is made of a transverse or fan-shaped series of narrow plates bearing minute, hairlike processes or papillae, which can be pressed into tiny irregularities of the surface. Because the pad does not function well on a smooth surface, a convenient way to collect geckos is with a water pistol. If one squirts the wall on which a gecko is climbing, it will often fall alive and uninjured to the ground.

Most geckos are gentle, but not the Tokay Gecko *(Gekko gecko)* of Indonesia and the Philippines. When annoyed, the Tokay inflates its body and hisses and puffs loudly, holding its jaws wide open in readiness to attack. If the provocation continues, the lizard rushes forward and seizes some part of its annoyer in its powerful jaws, hanging on with bulldog tenacity. The Tokay is the largest of the geckos: its head and body are about 175 millimeters long and its total length is over 300 millimeters.

The gekkonids fall clearly into four natural lineages, which are recognized as distinct subfamilies. The Diplodactylinae, Gekkoninae, and Sphaerodactylinae are closely related; the Eublepharinae are a much earlier divergence from the gekkonid stock. Some systematists recognize the latter as a separate family, and it has even been suggested that they diverged from the gekkonoid stock before the origin of the pygopodids.

Subfamily Diplodactylinae.   The diplodactylines are found in Australia, New Zealand, New Caledonia, and the Loyalty Islands. Each area has its own endemic genera. Diplodactylines tend to be arboreal or semiarboreal. Like most geckos, they lay two eggs, except for the three endemic New Zealand genera, which are ovoviviparous. The eggs are covered by flexible, parchmentlike shells, as in most other lizards.

Subfamily Eublepharinae.   The ground geckos differ from all other geckos in having true eyelids that can be opened and closed. All eublepharines have straight digits without friction pads. These two primitive characteristics and many others point to this subfamily as the most generalized of all the geckos and the earliest derivatives of the original gekkonid stock.

The eublepharines' peculiar distribution also suggests an earlier radiation that has been fragmented by the radiation of the more specialized gekkonines. Eublepharines are widely scattered in desert regions throughout the world. The Banded Geckos, *Coleonyx,* are found from the southwestern United States to Central America, *Holodactylus* in Somalia, *Hemitheconyx* in Somalia and West Africa, *Eublepharis* in southern Asia and on islands off the east coast of Asia, and *Aeluroscalabates* in Malaya, Sumatra, and Borneo.

Like the true geckos, and unlike most other lizards, the ground geckos are nocturnal, hiding by day under rocks or in burrows in the sand, and coming forth at night to forage for insects. Some have loud voices. Females lay two to four eggs at a clutch, with two- or three-egg clutches being most common. The eggshell is parchmentlike.

Subfamily Gekkoninae.   The gekkonines are the most successful of the geckos, with a circumtropical distribution and more than 50 genera. They range in adult size from 5 to over 30 centimeters, and in habits from terrestrial to arboreal. Many possess highly modified digital pads and digital tips. They lack true movable eyelids; instead, each eye is covered by a transparent scale, the spectacle.

Hemidactylus is the most widespread genus, its range nearly coinciding with that of the entire subfamily. It is found on all major land masses in the tropics, and almost all oceanic islands. At least one species, *H. garnoti,* is parthenogenetic, a trait that makes it a particularly good colonizer of islands.

Females characteristically lay two eggs. The eggs tend to be spherical, and possess a hard, calcareous shell. This shell appears to be an adaptation to prevent excessive water loss, and the eggs do have a higher resistance to desiccation than the parchment-shelled eggs of the diplodactylines and eublepharines.

Subfamily Sphaerodactylinae.   The diminutive sphaerodactylines are confined to the New World tropics. Few species exceed 8 centimeters in total length, and many are much smaller. Unlike the other geckos, sphaerodactylines tend to be more active during the day and are highly terrestrial. They also apparently lack a voice. In keeping with their small size, females lay only one calcareous-shelled egg at a time. Larger clutches are often found, but these result from communal nesting.

Sphaerodactylus has its primary radiation centered in the Caribbean. The other genera have radiated on the mainland.

Family Pygopodidae

At first glance, the pygopodids, or snake lizards, bear little physical resemblance to the geckos, but they are very similar to them in many structural characteristics and are probably derived from a gekkonid ancestor. They are slender and snakelike, but are clearly derived from limbed lizards, and show many of the morphological stages in the evolutionary sequence of limblessness. Although none possess forelegs, vestiges of the pectoral girdle remain, as well as remnants of hindlegs in the form of a pair of small, scaly flaps. Members of this family are between 150 and 750 millimeters long. The tail is considerably longer than the body and is easily shed.

Pygopodids are uniquely Australian; in fact they are the only family of lizards endemic to Australia and New Guinea (Figure 15–7). There are 8

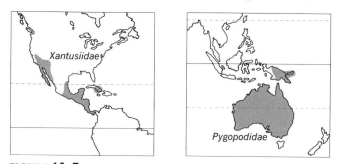

FIGURE 15–7
Distribution of the xantusiids and pygopodids. [After Savage, 1963; and Kluge, 1974, respectively.]

genera and 30 species. *Lialis burtonis* has the largest range—most of the Australian continent and the south coast of New Guinea. It is an active predator, feeding largely on other lizards, whereas other pygopodid species are principally insectivorous. All appear to be oviparous.

Family Xantusiidae

The diminutive night lizards of the xantusiid family seldom exceed 75 millimeters in total length. Superficially, they resemble a small Old World lacertid, with granular dorsal scales and enlarged rectangular belly scales. Their actual relationships are with the geckos, with which they share such features as elliptical pupils, fused eyelids with the enlarged lower lid bearing a transparent window, and procoelous vertebrae with small articular condyles.

Like the geckos, the night lizards are secretive, and some are nocturnal. All are terrestrial. They occupy four types of habitat: rock outcrops, deserts, pebble beaches, and tropical forests. Temperature and humidity appear to regulate much of their activity. When it is too hot and dry or too cool and wet, they remain hidden deep in the rock crevices or beneath a fallen tree.

Xantusiids are truly viviparous. Mating of *Xantusia vigilis,* the Desert Night Lizard, takes place in May, and ovulation occurs one to four weeks later. Usually two eggs are formed at one time (this is another way in which the night lizards resemble the geckos), but rarely only one embryo is found (usually in the right oviduct). Occasionally, three embryos are present. Early in embryonic development, a simple, cellophanelike shell is formed, but it soon disintegrates, and a chorioallantoic placenta develops. Gestation takes about three months. Surprisingly enough, these lizards have developed the typically mammalian custom of eating the fetal membrane. The membrane ruptures before the young is born and remains in the cloaca. The female grasps the protruding edge in her mouth, gradually draws it out, and swallows it.

The family contains four genera, which have been grouped into two subfamilies.

Subfamily Cricosaurinae.   *Cricosaura typica,* the single cricosaurine representative, is confined to the coastal areas of Cabo Cruz in eastern Cuba. It differs from the other xantusiids in possessing two frontonasal scales, one frontal scale, no parietal scale, and a fourth finger with four phalanges.

Subfamily Xantusiinae.   The remaining three xantusiid genera have one frontonasal, two frontal, and two parietal scales. The fourth finger of the hand has five phalanges. *Xantusia,* with three species, is found in the southwestern United States; *Klauberina,* with one species, on the Channel Islands off southern California; and *Lepidophyma,* with six or more species, in Central America.

INFRAORDER IGUANIA

The infraorders Iguania and Scincomorpha include the majority of lizards. As a general rule, but with exceptions, iguanians tend to have robust bodies, short necks, distinct heads, and overlapping and noniridescent scales. In contrast, scincomorphs tend to have slim bodies, with heads not clearly demarcated from the neck; if the scales overlap, they are iridescent. On a more technical level, iguanians have fleshy tongues, calyculate hemipenes, and four or more transverse belly scales per body segment.

The iguanians are numerous and varied—some large, others quite small. They are frequently brightly colored, and often ornamented with crests, spines, frills, and throat fans. They are diurnal and may be either arboreal or terrestrial. *Amblyrhynchus* of the Galápagos Islands is semimarine. No iguanian shows any tendency toward the development of a snakelike body or a reduction of the limbs.

The temporal arch is present, the skull is high, the teeth are either pleurodont or acrodont, the parietals are fused to form a single bone. There are six cervical vertebrae, and four or more rows of transverse belly scales per body segment. The tongue is simple, rather than divided into anterior and posterior portions. The eyelids are well developed, and the pupils are round. Femoral pores are occasionally present. Breaking-point septa are present in the caudal vertebrae, and some forms have rather fragile tails.

This infraorder contains two very similar families, Iguanidae and Agamidae, along with the strange-looking Chamaeleonidae.

Family Agamidae

The large agamid family is the Old World counterpart of the New World iguanids (Figure 15–8). The agamids differ from the iguanids in the structure of their jaws and teeth. The teeth are acrodont and the dentition is heterodont. Many species have well-developed ornamental crests, frills, or throat pouches, frequently brilliantly colored. The "wings" of the Flying Dragons are as bright as those of some butterflies. Most agamids are of moderate length, but *Hydrosaurus,* the Water Lizards of the East Indies and New Guinea, may reach 900 millimeters, and the Toad-headed Agamids, *Phrynocephalus,* may be less than 125 millimeters.

The agamids have undergone extensive adaptive radiation in terrestrial and arboreal habits. The majority run on four legs (quadrupedal locomotion); a few, like *Physignathus,* are bipedal and run rapidly on only two legs.

288

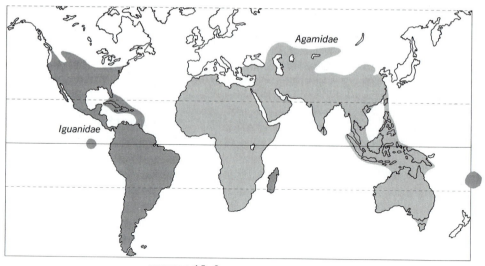

FIGURE 15–8
Distribution of the iguanids and agamids.

Draco, the Flying Dragons, are arboreal gliders. They have modified rib cages that can be depressed and expanded into pseudowings, with which *Draco* can jump from the tops of trees and glide for thirty meters or more.

Being visually-oriented, many agamids have developed ornamental crests, frills, throat pouches, or brightly colored patches on the head and body. These ornamentations serve as behavioral signals in establishing and maintaining territories or in courtship. Most agamids are insectivorous, although a few (some *Agama* and *Uromastyx*) are herbivorous as adults (Figure 15–9). Herbivory appears to be restricted to adult reptiles; perhaps a vegetable diet lacks sufficient nourishment for young, growing reptiles.

Agamids are predominantly oviparous. Only two genera, *Phrynocephalus* of Central Asia and *Cophotes* of Ceylon, are known to give birth to living young.

Family Chamaeleonidae

The bizarre chamaeleonids owe their strange appearance to their successful adaptation to arboreal life. Their most apparent arboreal adaptations are the laterally compressed body, prehensile tail, and zygodactylous feet. The zygodactylous (= "yoked-toed") feet permit a strong grip on branches: the two inner digits of the forefeet are bound together in a bundle to oppose the similarly joined three outer digits; the hindfeet reverse this grouping. The prehensile tail acts as an extra limb. The laterally compressed body provides stability for walking on narrow branches, since the center of gravity always lies directly over the branch.

FIGURE 15–9
This North African Mastigure, *Uromastyx acanthinurus werneri*, refused food until offered the zinnia, whereupon it bit off the petals as fast as it could swallow them.

The chamaeleonid skull often bears horns, crests, and tubercles, and these may extend backward to form helmetlike casques. These ornamentations serve as visual signals in territorial and courtship behavior. The most striking feature of the head is the protruding eyes covered with mufflerlike lids, which leave only a narrow opening for vision. The restricted opening and the ability to rotate each eye independently of the other are important for capturing prey. Once an insect is sighted with one eye, the head is moved so that both eyes can see the prey. This movement aims the head so that the extensile tongue can be accurately shot out to catch and pull the insect into the mouth.

Chameleons are famous for their ability to change color, from white through shades of yellow, green, and brown to black, variously spotted and blotched with contrasting colors. Some iguanids, such as *Anolis,* are also able to change color, though with less virtuosity, and so the Anoles are often called chameleons. The color changes are responses to changes in light, heat, and the emotional state of the animal, and not, as is popularly believed, to the color of the background.

The chameleons are clearly derived from the agamids. There are more than 80 species, not all of them arboreal: some live on grasses and sedges, and a few are terrestrial. Their distribution is centered in Africa, with an extension into Asia as far east as India. The greatest diversity is found in central Africa and Madagascar (Figure 15–10).

Reproductive habits vary: *Brookesia* and *Rhampholeon* are completely oviparous, the small South African *Microsaura* is completely ovoviviparous, and *Chamaeleo* shows both patterns. In *Chamaeleo,* the ovoviviparous species tend to occupy cooler environments. Clutch size is larger in oviparous species.

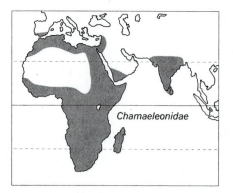

FIGURE 15–10
Distribution of the chamaeleonids.

Family Iguanidae

The iguanids are the largest family of New World lizards. There are two genera *(Chalarodon* and *Oplurus)* in Madagascar and one *(Brachylophus)* on the Fiji and Tonga Islands, but there are more than 50 genera in the western hemisphere. The iguanids have a pleurodont dentition. The splenial bone of the lower jaw is well developed, in contrast to its reduced condition in the sister family, Agamidae.

The iguanids range from smaller forms less than 125 millimeters long to the giant iguanas, which may be 1800 millimeters long. The latter are used as food in some tropical countries. The smaller iguanids are mostly insectivorous or carnivorous, but many of the larger ones are herbivorous. The Marine Iguanas of the Galápagos enter salt water to forage for seaweed, which seems to be their favorite food.

A characteristic habit of the family is head bobbing, which, in such forms as *Anolis,* is correlated with the distension of the throat fan. The head is raised above the surface and bobbed backwards so that the brightly colored fan is spread, but the characteristic bobbing motion occurs also in iguanids that do not have a throat fan.

Iguanids are usually oviparous, but some species of Horned Lizard *(Phrynosoma)* and some Spiny Lizards *(Sceloporus)* are ovoviviparous. The eggs are soft-shelled and are frequently buried in the ground, even by species that are largely arboreal.

Six groups of iguanids can be recognized. However, they have not been adequately defined and so cannot be designated as formal taxonomic categories.

The anolines are small to medium-sized lizards, largely arboreal or semiarboreal, that characteristically have calcified endolymphatic glands lacking in the other iguanids. The group apparently evolved in Central America and northern South America. *Anolis, Polychrus,* and *Phena-*

cosaurus are among the better-known genera: *Anolis* has been the most successful, with well over a hundred species.

The medium-sized basiliscines of Central America tend toward arboreality—*Laemanctus* and *Corythophanes* strictly so, *Basiliscus* at least when resting. All possess cranial modifications that give them helmeted appearances, and unique clavicular fenestras in their pectoral girdles. Their closest relatives are probably the anolines.

The iguanines include all the large iguanids—*Amblyrhynchus, Conolophus, Ctenosaura, Cyclura, Iguana,* and *Sauromalus*—and a couple of medium-sized ones, *Brachylophus* and *Dipsosaurus* (Figure 15–11). All have compact heads, small, granular body scales, and distinctive doubled transverse processes on their caudal vertebrae. They appear to be most closely related to the tropidurines.

FIGURE 15–11
The Iguana, *Iguana iguana.*

The sceloporines and tropidurines are sister groups of small to moderate-sized, rough-scaled lizards. The sceloporines *(Sceloporus, Uta, Phrynosoma, Holbrookia, Crotaphytus)* have evolved on the Mexican Plateau and in the southwestern United States. The tropidurines *(Leiocephalus, Liolaemus, Stenocercus, Tropidurus)* have their greatest diversity in South America, notably in the Andes. The former usually possess femoral pores, which are generally absent in the latter. The two Malagasian genera, *Chalarodon* and *Oplurus,* resemble the tropidurines and sceloporines in body form and behavior. The resemblance is in part convergent, for they have long been isolated from the New World stocks.

INFRAORDER ANGUINOMORPHA

The anguinomorphs include a heterogeneous assemblage of animals ranging from the bulky, three-meter-long Komodo Dragon, *Varanus komodoensis,* to the wormlike California Legless Lizard, *Anniella pulchra,* no bigger than a lead pencil. At first glance, they seem to have nothing in common, but underlying structural similarities justify their being placed together. The tongue is divided into two parts, with a notched, inelastic forepart separated by a transverse fold from the elastic hindpart, which serves as a sheath when the tongue is withdrawn. The teeth are nearly solid, not hollowed at the base as are those of other lizards, and are replaced alternately (that is, the new tooth comes up behind, not beneath, the older tooth). There are relatively few anguinomorphs living today, but the group includes a number of large, heavily armored extinct forms, notably the huge, aquatic mosasaurs, the most spectacular lizards that have ever lived. The infraorder is divided into two superfamilies.

Superfamily Anguinoidea

The anguinoids, a group of small to moderate-sized lizards, are also known as diploglossans in reference to the dual structure of their tongues. Most are more or less armored, with osteoderms lying beneath the epidermal scales. Many have reduced limbs and snakelike bodies.

Family Anguinidae

The anguinid family includes some lizards with well-developed limbs, such as the Alligator Lizards *(Gerrhonotus),* and some that lack limbs, such as the Glass Lizards *(Ophisaurus)* and Slowworms *(Anguis).* The tropical American genera known as galliwasps show intermediate stages in limb reduction. The temporal arches are present; the temporal openings are long and narrow, and in some forms are roofed. Palatal teeth may be present or absent. Osteoderms are well developed. Most anguinids are medium-sized; the largest, *Ophisaurus apodus,* may attain a length of nearly 1200 millimeters.

Most of the anguinids are terrestrial, a few are somewhat arboreal. Some species hide in burrows during the day and go forth at night to hunt for insects and other small invertebrates.

Both oviparous and ovoviviparous forms occur, not only within the family, but within a single genus. The San Francisco Alligator Lizard, *Gerrhonotus c. coeruleus,* which is ovoviviparous, copulates in April. The copulatory process is lengthy, sometimes lasting for many hours. The number of young varies from 2 to 15, but usually about 7 are born in late

August or September. When first born, they have a snout-to-vent length of 25 to 30 millimeters. In contrast, the Oregon Alligator Lizard, *G. multicarinatus scincicauda,* is oviparous. Mating occurs from the middle of May to the middle of June. The eggs, which may number more than a dozen in a single clutch, are laid in late July and early August, in burrows dug by mammals. The young hatch in September.

The anguinid family includes eight genera, grouped into four subfamilies. The family has a Holarctic distribution with an extension into the northern Neotropics (Figure 15–12).

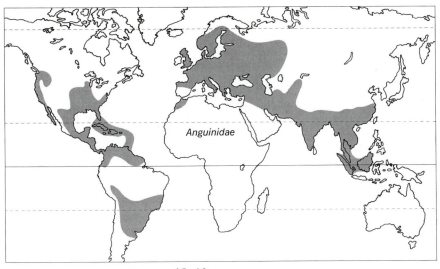

FIGURE 15–12
Distribution of the family Anguinidae.

Subfamily Anguininae. The anguinines comprise two genera: the monotypic *Anguis,* the Slowworm, which is strictly European, and the Glass Lizards, *Ophisaurus,* with about ten species in Eurasia and North America. Both genera are legless and have a fragile tail more than twice as long as the body. They are primarily terrestrial and largely associated with grassy habitats. In many features they are intermediate between the gerrhonotines and diploglossines. They share a lateral body fold and other scutellation and osteoderm characters with the former, and a premaxillary foramen and divided frontals with the latter.

Subfamily Anniellinae. Anniellines are early derivatives of the anguinid stock, represented solely by *Anniella,* the little, shovel-snouted Legless Lizard of California and Baja California. The bones of the skull are closely knit, there are no temporal arches, the snout is short, and the brain case is expanded. There are no palatal teeth, and the osteoderms are reduced. *Anniella* is about 250 millimeters long and lacks external ear openings, but has lidded eyes. The body is covered on all sides with smooth, rounded, uniform, overlapping scales like those of a skink.

Legless lizards lead almost exclusively underground lives, but are sometimes found on the surface of the sand under rocks and logs. The species is ovoviviparous; one to four young are born in late summer and fall.

Subfamily Diploglossinae. The diploglossines include three genera *(Diploglossus, Ophiodes, Wetmorena)* occurring in the West Indies and Central and South America. All have an elongate body, and the legs, when present, are small. They are mainly terrestrial lizards, although a few species occur commonly in the epiphytic bromeliad community.

Subfamily Gerrhonotinae. Three living genera, *Abronia, Coloptychon,* and *Gerrhonotus,* constitute the gerrhonotines. They range from the southwestern United States through Central America. All possess limbs and use them in locomotion. They are usually slow, methodical creatures: many have adapted to the cool, damp forest along the cloud line in the mountains.

Family Xenosauridae

Two poorly known genera, widely separated geographically and of very limited distribution, are included in the xenosaurid family. *Xenosaurus* has three species in eastern Mexico, and *Shinisaurus* one species found in southern China (Figure 15–13). They are medium-sized (about 150 to 375 mm), with normally developed limbs. Two clearly defined longitudinal crests, formed by series of enlarged scales, run down the midline of the back. The rest of the dorsal surface is covered with a scattering of smaller scales

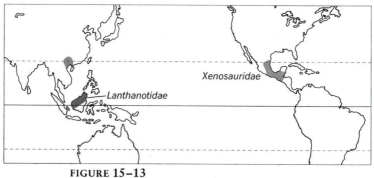

FIGURE 15–13
Distribution of the xenosaurids and lanthanotids.

interspersed with minute granules. The temporal arches are strongly developed, and the temporal openings are large and not roofed by skull bones. The bones of the skull are roughened by the fusion to them of the cranial osteoderms.

Subfamily Shinisaurinae. *Shinisaurus* differs from the xenosaurines in having large osteoderms, a compressed skull, and long, lightly pointed teeth. It lives along streams and feeds partly on tadpoles and fish. It also basks on trees overhanging the water.

Subfamily Xenosaurinae. *Xenosaurus* has tiny osteoderms, a depressed skull, and blunt teeth. Members of this genus generally occur in limestone or volcanic areas in mountainous regions. Their habitat ranges from dry scrub forest to wet cloud forest. They are secretive and apparently crepuscular or nocturnal.

Superfamily Varanoidea

The varanoids include medium-sized to very large predaceous lizards, with jaws adapted more for grasping large prey than for chewing small invertebrates. Thus the bones of the mandible are rather loosely joined, and there is a tendency toward the development of a hinge within the jaw. The maxillary bone barely reaches the level of the orbit, so that the marginal teeth are in front of the eye. The slitlike external nares extend far back into the skull. The tail is not autotomous, and the limbs are never reduced. Three families are included in the superfamily, each with only one genus.

Family Helodermatidae

The Gila Monster *(Heloderma suspectum)* and the Mexican Beaded Lizard *(H. horridum),* found only in the southwestern United States and Mexico, are the only venomous lizards known (Figure 15–14). Unlike those of the

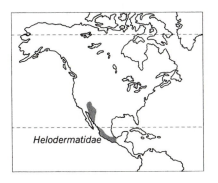

FIGURE 15–14
Distribution of the helodermatids.

poisonous snakes, the venom glands of *Heloderma* are in the lower jaw and are not connected with the teeth. The venom empties into the mouth through several ducts that open between the teeth and the lips. Grooves on the teeth help draw the venom into the wound by capillary action as the lizard hangs on and chews. The bite is painful, but seldom fatal to man. There are a few palatine and pterygoid teeth, no temporal arches, and eight cervical vertebrae. The back and the outer surfaces of the limbs are covered with large osteoderms.

The helodermatids are heavy-bodied, short-tailed, clumsy-looking lizards, gaudily marked with dark reticulations on a yellow or orange background, or vice versa. The Beaded Lizard may reach a length of 900 millimeters. The Gila Monster is shorter, at most about 500 millimeters long (Figure 15–15). They eat a wide variety of animal food—small mammals, other lizards, and bird and reptile eggs.

FIGURE 15–15
The Gila Monster, *Heloderma suspectum.*

Heloderma suspectum has been seen in copulation in mid-July. What little evidence there is indicates that three to seven eggs are laid during a period from late July to mid-August. They are buried to a depth of about 125 millimeters in an open place that is exposed to the sun, but usually near a stream or dry wash. The thin-shelled, white, rather rough eggs are 67 to 75 millimeters long and 33 to 39 millimeters wide. The incubation period is apparently 28 to 30 days.

Family Lanthanotidae

Lanthanotus borneensis, the Earless Monitor, is the sole representative of the lanthanotid family. It is a dull-colored, short-legged lizard about 400 mm long, known only from Borneo. The back and tail are protected by ridges of raised tubercles. The head is covered with small, grainlike nodules. There are a few teeth on the palatine and pterygoid bones, nine cervical vertebrae, and no dorsal temporal arch. *Lanthanotus* bears a superficial resemblance to *Heloderma* and was long classified with that genus, but it has no venom apparatus and is anatomically much closer to *Varanus.*

Only a few specimens have been captured, so our knowledge of this lizard derives from observations of captive individuals. In habits, *Lanthanotus* is lethargic and appears to be most active at night. Its metabolism is low, and it tends to desiccate rapidly. Movement on land is slow and appears to be difficult. In contrast, it swims well by lateral undulations. These behavioral features and its preference for a fish diet suggest that it is naturally aquatic. The lack of ear openings, eyelids with transparent scales, and nares on dorsal mounds somewhat behind the snout also indicate aquatic habits.

Family Varanidae

All living varanids are included in a single genus, *Varanus,* the Monitor Lizards, or Goannas. The smallest Monitor, *Varanus brevicauda* of Australia, is only about 20 centimeters long, but most are large, so much so that they have been mistaken for crocodiles. Indeed, the native name for the Komodo Dragon, *buaja darat,* means "land crocodile." This species may reach a length of three meters. Monitors lack the venom apparatus of the helodermatids. The dorsal temporal arch is complete, there are no palatine or pterygoid teeth, and the osteoderms are reduced and sometimes lacking. The tail is long and muscular and, like that of the crocodile, is a formidable weapon. There are nine cervical vertebrae. A monitor holding its head erect on its long neck has a very alert appearance.

About 30 species of *Varanus* are known in southern Asia, Africa, the East Indies, and Australia (Figure 15–16). All are carnivorous; the larger forms have been reported capable of tackling such prey as pigs and small deer. Although largely terrestrial, they are surprisingly agile climbers for such bulky animals. Many are quite aquatic, and have been seen swimming far out at sea. This undoubtedly explains their distribution throughout the East Indies.

The monitors are oviparous. In Thailand, the Common Water Monitor, *Varanus salvator,* lays 15 to 31 eggs at the beginning of the rainy season in June. They are deposited in holes in riverbanks or in trees beside the water. They measure about 70 by 40 millimeters, have rather soft shells, and are said to taste like turtle eggs.

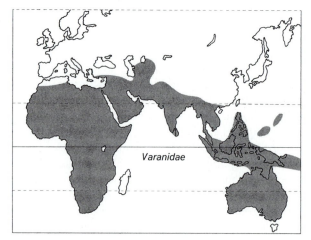

FIGURE 15–16
Distribution of the varanids.

INFRAORDER DIBAMIA

The dibamian infraorder includes only the dibamids—limbless, snakelike lizards that have been volleyed back and forth between the Gekkota and Scincomorpha. They share a few characters with each, but also possess unique characters of cranial and vertebral morphology that suggest they are a distinct lineage of long separation.

Family Dibamidae

Two genera, *Dibamus* of the East Indies and *Anelytropsis* of central Mexico, form the dibamid family (Figure 15–17). Both are limbless and

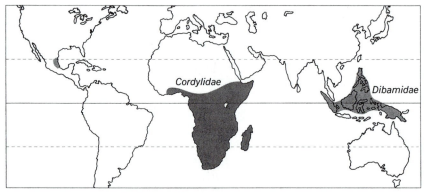

FIGURE 15–17
Distribution of the dibamids and cordylids.

blind, the eyes reduced and hidden beneath the skin. Pectoral girdles are vestigial or absent. The temporal arch of the skull has been lost. Few individuals have been seen, so their life history is speculative. They are apparently oviparous, as calcareous-shelled eggs have been found upon dissection.

INFRAORDER SCINCOMORPHA

The scincomorphs are a large, cosmopolitan group of lizards, medium to small in size, with a strong tendency toward reduction of the limbs and development of a snakelike body habitus. The degenerate forms are usually burrowers. The dentition is pleurodont, the dorsal temporal arch is usually present, and the parietals are fused. The tongue is simple. There are six cervical vertebrae and fewer than four rows of transverse belly scales per body segment. The infraorder is cosmopolitan in distribution and comprises four families.

Family Cordylidae

The cordylids are small to medium-sized African lizards, the largest only a little more than 600 millimeters long, with the tail making up most of the length. The dorsal body scales are usually keeled and overlapping. Osteoderms are present on the head and body. There is a tendency to elongate the body and tail with a corresponding reduction of limbs, but the limbs are never completely lost. Femoral pores are present in most. A lateral body fold, as in some anguinids, is invariably present. The ten genera are partitioned into two subfamilies.

Subfamily Cordylinae. Cordylines are ovoviviparous. They are called girdle-tailed lizards, because their tails are encircled by whorls of large heavy-keeled scales. In many *Cordylus* species these whorls of keeled scales extend onto the body. The spiny scales act to protect the lizard: the Armadillo Lizard, *Cordylus cataphractus,* forms a spiny loop by biting its tail and holding its legs over its belly. Most girdle-tailed lizards have well-developed legs. In *Chamaesaurus* there is a tendency toward reduced legs. The cordylines are predominantly South African.

Subfamily Zonurinae. The plated lizards are oviparous. Most have stout bodies covered with platelike scales, and well-developed limbs. *Tetradactylus* has a tendency toward reduced limbs. Zonurines occur throughout Africa, but not in the rain forest of central Africa. They are terrestrial and insectivorous.

Family Lacertidae

The lacertids are usually small, agile, long-tailed lizards. Alone among the scincomorphs, they show no tendency toward reduction of the limbs. The dorsal temporal arch is complete, but bony dermal plates (osteoderms) cover the temporal fenestra and fuse with the cranial bones when in contact with them, obscuring the structure of the dorsal arch. Osteoderms are lacking on the body, though present on the head. The lateral teeth are often bicuspid or tricuspid. The tongue is moderately elongated, deeply notched anteriorly, and covered with scalelike papillae or with transverse folds. Femoral glands are usually present. Some forms have windows in the lower eyelids. The largest species is the Jeweled Lizard *(Lacerta lepida),* which reaches a length of 750 millimeters.

The lacertids are Old World lizards, occurring in Europe, Asia, and Africa, but not in Madagascar or in the Australian region (Figure 15–18).

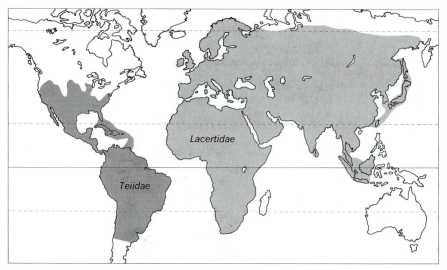

FIGURE 15–18
Distribution of the lacertids and teiids.

They are most abundant in Africa, and comparatively rare in the Oriental region. The common lizard of Europe, *Lacerta vivipara,* exists above the Arctic Circle; no other lizard is found this far north. The family includes about 20 genera.

Lacertids are predominantly terrestrial, often living in grassy or sandy places. Some, such as the European Wall Lizard *(Lacerta muralis),* are agile climbers. They are carnivorous, feeding chiefly on insects and other small invertebrates.

Lacerta vivipara, as its name indicates, usually bears living young, but the other lacertids are oviparous. The Sand Lizard, *Lacerta a. agilis,* breeds during May and early June; the same male and female may mate many times during this period. The eggs, which are laid in June and July, vary from 6 to 13 in number, depending in part on the size of the female. When first laid, the eggs are 12 to 15 millimeters long and 8 to 9 millimeters wide; shortly before hatching they are 15 to 20 millimeters long. The eggs are hidden under stones or in shallow holes dug in the earth, and covered over by the mother. They hatch in 7 to 12 weeks, and the young, at hatching, are 56 to 63 millimeters in combined head and body length.

Parthenogenesis in lizards was first reported in the Rock Lizard, *Lacerta saxicola,* in Armenia.

Family Scincidae

The abundant and ubiquitous scincids, the true skinks, are usually small, secretive semiburrowers with highly polished scales. Most of the scincids are less than 200 millimeters long, and the largest, *Corucia zebrata* of the Solomon Islands, is only about 600 millimeters in length. The temporal openings of the skull are more or less roofed by backward growths of the postfrontals. Pterygoid teeth are often present. The limbs may be present or absent; those species that have lost their limbs always possess some traces of the pectoral and pelvic girdles. Abdominal and parasternal ribs are sometimes present, chiefly among the burrowing forms. The head, body, limbs, and tail are protected by osteoderms. The head is covered with symmetrical shields, the pupil is round, femoral pores are absent. The tail is easily broken, but another is quickly regenerated. Some of the skinks have developed a transparent window in the lower eyelid that enables the lizard to see while the eyelids are shut. The most extreme development along this line is found in the little, active, Snake-eyed Skinks *(Ablepharus)* whose eyelids are wholly transparent and immovable, permanently covering the eyes like a pair of watch glasses. Not all skinks are smooth and shiny: a few bizarre-looking forms have big, spiny scales on their backs, sides, and tails. The stout-bodied, stubby-tailed Shingleback of Australia *(Trachydosaurus)* looks like an animated pine cone (Figure 15–19).

Although cosmopolitan in distribution, skinks are most numerous in Australia, the islands of the western Pacific, the Oriental region, and Africa; they are poorly represented in the Americas (Figure 15–20). There are nearly 50 genera and more than 600 species.

The vast majority of skinks are terrestrial and are usually extremely active; a few have arboreal tendencies, but none is so highly adapted for climbing as the members of the preceding infraorders. Some have become rock dwellers, and a few are good swimmers. A few, like *Neoseps* and

FIGURE 15–19
The bizarre Shingleback of Australia, *Trachydosaurus rugosus*. The short tail is responsible for the common name "Stump-Tail," often applied to these skinks.

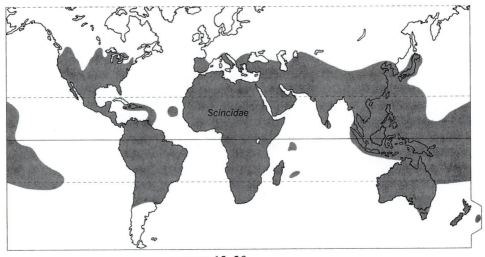

Scincidae

FIGURE 15–20
Distribution of the scincids.

Ophiomorus, have become specialized for living in loose sand, and move through it with a swimming motion. Skinks are often found under piles of dead leaves, coconut husks, and rotting vegetation or decaying logs. The terrestrial skinks are diurnal, whereas the burrowers seem to be largely crepuscular or nocturnal. Most skinks are insectivorous, but some of the larger ones also consume small vertebrates, and a few are partly herbivorous.

Skinks may be either oviparous or ovoviviparous, and a few, such as the European skink *Chalcides ocellatus,* approach true viviparity by having a placenta. *Eumeces fasciatus* of the eastern United States is a typical oviparous skink. In Maryland, courting and copulation take place shortly after emergence from hibernation, generally during early May; the eggs are laid 6 or 7 weeks afterward. Clutch size ranges from 2 to 18 eggs, the smaller clutches being laid by the smaller females. There is also an indication that the size of the clutch decreases toward the northern part of the range. The eggs are deposited in rotten wood or loose soil 50 to 75 millimeters below the surface, and are brooded for the entire incubation period by the mother. Shortly after deposition the eggs are approximately 13 by 7 millimeters, but like the eggs of most lizards (except the geckos), they increase in size during incubation and shortly before hatching may be as much as 20 millimeters long. They hatch 4 to 7 weeks after deposition. At hatching the young are 24 to 28 millimeters in head and body length.

The Australian Blue-tongued Lizard, *Tiliqua scincoides,* is an example of a viviparous skink. The young are born in the middle of the summer (January); litters of from 5 to 18 have been recorded. The newborn lizard is still wrapped in the fetal membrane, but in a few seconds it breaks the membrane and immediately devours it. At birth the young are from 130 to 152 millimeters in total length.

The skinks are divided into four subfamilies. These subfamilies are largely characterized by the morphology of the secondary palate, which occurs among lizards only in skinks and dibamids.

Subfamily Acontinae.   The acontines include three genera of South African skinks. The palatine bones are separated on the midline of the secondary palate and exclude the pterygoids from the infraorbital foramina. All three—*Acontias, Acontophiops,* and *Typhlosaurus*—are limbless, burrowing forms. The ear openings are covered with skin; eyes remain well developed and apparent. All appear to be ovoviviparous.

Subfamily Feylininae.   The feylinines comprise two limbless genera, *Chabanaudia* and *Feylinia* of central Africa. In the secondary palate, the palatines are separated on the midline and are excluded from the infraorbital foramina by the pterygoids. Being burrowers, they also have ear openings covered with skin, and the eyes are somewhat reduced. Like the acontines, they appear to be live-bearers, but unlike the latter, they live in forests.

Subfamily Lygosominae.   These skinks occur on all continents and many oceanic islands. In the secondary palate, the palatine bones meet on the midline and form the medial borders of the infraorbital foramina with the pterygoids. The nearly forty genera range from limbless, burrowing forms to arboreal and terrestrial ones with well-developed legs. Most are oviparous, a few ovoviviparous. Some of the better known genera are *Sphenomorphus, Emoia, Ablepharus, Leiolopisma, Mabuya,* and *Tiliqua.*

Subfamily Scincinae. The twenty-odd genera of scincines are primarily in Africa. Of the two genera found in the United States, the tiny burrowing skink *(Neoseps)* is restricted to Florida, but *Eumeces* is more cosmopolitan, occurring also in Asia and northern Africa. The secondary palate has the palatines separated on the midline, and the medial border of the infraorbital foramen is formed by both the palatine and pterygoid bones. Both limbed and limbless scincines are known. Oviparity and ovoviviparity occur with equal frequency. In addition to *Eumeces* and *Neoseps,* the genera *Chalcides, Scincus,* and *Scelotes* are included in this subfamily.

Family Teiidae

The teiid family occupies a position in the New World similar to the position filled in the Old World by the closely related lacertids. The largest of the teiids *(Tupinambis)* is about 900 millimeters long, but most are small. Some are degenerate burrowers with reduced limbs. Osteoderms are absent from the head and the body. Skull arches are present, and the temporal fenestra are open. The tail is quite long. The tongue, like that of the lacertids, is long and narrow, deeply forked anteriorly, and covered with papillae. The front teeth are always conical, but the lateral teeth on both jaws may be conical, bicuspid, tricuspid, molariform, or even enormous, oval crushers (as in the snail-eating Caiman Lizard, *Dracaena guianensis*) (Figure 15–21).

FIGURE 15–21
The snail-eating Caiman Lizard, *Dracaena guianensis.*

The family is found only in the western hemisphere, where it is represented by 38 genera and about 200 species. Practically all are restricted to South America; only one, *Cnemidophorus,* reaches the United States.

The teiids have long been recognized as containing several distinct evolutionary lineages. The most apparent separation is between the medium to large genera and the small, frequently semifossorial genera. These are often informally referred to as the macroteiids and microteiids, respectively. Recently two subfamilies have been established for them.

Subfamily Gymnophthalminae. The microteiids all possess one or two frontonasal scales separating the anterior nasal scales. These small, secretive lizards show a trend from well-developed to reduced or absent limbs. The majority of them are confined to South America.

Subfamily Teiinae. The macroteiids have the anterior nasals in contact with one another. All have well-developed limbs. Most are terrestrial, but *Dracaena* and *Crocodilurus* are semiaquatic. *Ameiva* and *Cnemidophorus* are the most common members of the subfamily: the former are forest-savanna lizards of the tropics, the latter primarily scrub and desert lizards of the subtropics.

READINGS AND REFERENCES

Boulenger, G. A. *Catalogue of the Lizards in the British Museum (Natural History)*. London: British Museum (Natural History), 1885–1887.

Bogert, C. M., and R. Martín del Campo. "The Gila monster and its allies." *Bulletin of the American Museum of Natural History,* vol. 109, art. 1, 1956.

Camp, C. L. "Classification of the lizards." *Bulletin of the American Museum of Natural History,* vol. 48, art. 11, 1923. (Reprinted in 1971 by Society for the Study of Amphibians and Reptiles; new introduction by G. Underwood summarized lizard classification.)

Etheridge, R. "The skeletal morphology and systematic relationships of sceloperine lizards." *Copeia,* no. 4, 1964.

Gans, C. "A Check List of Recent Amphisbaenians." *Bulletin of the American Museum of Natural History,* vol. 135, art. 2, 1967.

―――. "Relative success of divergent pathways in amphisbaenian specialization." *American Naturalist,* vol. 102, no. 926, 1968.

Greer, A. E. "A subfamilial classification of scincid lizards." *Bulletin of the Museum of Comparative Zoology,* vol. 139, no. 3, 1970.

Hoffstetter, R. "Revue des recentes acquisitions concernant l'histoire et la systematique des squamates." *Colloques Internationaux du Centre National de la Recherche Scientifique,* no. 104, 1961.

King, W., and F. G. Thompson. "A review of the American lizards of the genus *Xenosaurus* Peters." *Bulletin of the Florida State Museum, Biol. Sci.,* vol. 12, no. 2, 1968.

Kluge, A. G. "Higher taxonomic categories of gekkonid lizards and their evolution." *Bulletin of the American Museum of Natural History,* vol. 135, art. 1, 1967.

_____. "A taxonomic revision of the lizard family Pygopodidae." *Miscellaneous Publications, Museum of Zoology, University of Michigan,* no. 147, 1974.

McDowell, S. B., Jr., and C. M. Bogert. "The systematic position of *Lanthanotus* and the affinities of the anguinomorphan lizards." *Bulletin of the American Museum of Natural History,* vol. 105, art. 1, 1954.

Meszoely, C. A. M. "North American fossil anguid lizards." *Bulletin of the Museum of Comparative Zoology,* vol. 139, no. 2, 1970.

Miller, M. R. "The cochlear ducts of *Lanthanotus* and *Anelytropsis* with remarks on the familial relationships between *Anelytropsis* and *Dibamus*." *Occasional Papers of the California Academy of Science,* no. 60, 1966.

_____. "The cochlear duct of lizards." *Proceedings of the California Academy of Science,* vol. 33, no. 11, 1966.

Moffat, L. A. "The concept of primitiveness and its bearing on the phylogenetic classification of the Gekkota." *Proceedings of the Linnean Society of New South Wales,* vol. 97, pt. 4, 1973.

Savage, J. M. "Studies on the lizard family Xantusiidae. IV: The genera." *Contributions in Science,* no. 71, 1963.

Smith, H. M. *Handbook of Lizards.* Ithaca, N.Y.: Comstock, 1946.

Werner, Y. L. "Observations on eggs of eublepharid lizards, with comments on the evolution of the Gekkonoidea." *Zoologische Mededelingen,* vol. 47, no. 17, 1972.

Zug, G. R. "The distribution and patterns of the major arteries of the iguanids and comments on the intergeneric relationships of iguanids (Reptilia: Lacertilia)." *Smithsonian Contributions to Zoology,* no. 83, 1971.

CHAPTER SIXTEEN

SNAKES

AN INTEREST IN SNAKES has probably led more people to study herpetology than an interest in all other groups of herps combined. Although it is true that many of these students have later turned to more prosaic forms, such as the salamanders or turtles, it is still the snakes that first attracted their attention.

The approximately 2,700 kinds of snakes in the world form the suborder Serpentes of the order Squamata. Like the lizards (suborder Sauria), they have undergone extensive adaptive radiation and have come to occupy most of the major habitats of the world. Some, the burrowers, are usually small and have eyes that are hidden beneath the scales of the head. Others have taken to the trees and seldom come to the ground. One group is entirely marine: these include the only modern reptiles to abandon the land completely (the sea turtles come to shore to lay their eggs). In size, the snakes range from tiny forms only about 100 millimeters long to the Anaconda and the Reticulated Python, which reach lengths of more than 9 meters.

Snakes are not as important economically as turtles. Some, probably most, are edible, but they are seldom used as food, although in the Orient not only the Pythons but also the poisonous sea snakes are esteemed by some peoples. Snakeskins make a leather suitable for fancy belts, pocket-books, and shoes, but not much else. Structurally, the snakes are so highly modified that they are not popular specimens in anatomical laboratories. Their main contribution to mankind is the control of pests, particularly the destructive rats and mice.

Snakes are elongate animals either with no girdles or limbs, or occasionally with vestigial pelvic girdles and hindlimbs. They lack a sternum, external ear opening, tympanic membrane, middle ear, and eustachian tube. Except in some burrowing forms, the immovably fused and transparent eyelids form a protective window, the brille, beneath which the eye moves. The viscera are elongated, and the left lung is smaller than the right or altogether absent. The tongue is long, forked, and protractile. There is no urinary bladder. Like the lizard, the snake has a transverse vent, paired copulatory organs, and a body covered with scales.

The skull of a snake is more specialized than that of most lizards. (Figure 16–1). The brain cavity is completely enclosed anteriorly by dermal bones.

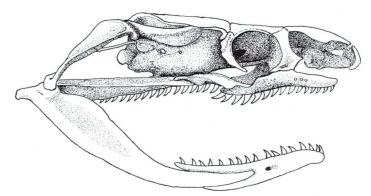

FIGURE 16–1
Skull of the colubrid snake *Drymarchon corais couperi,* the Indigo Snake of the southeastern United States.

Higher forms have the bones of the facial region and jaws loosely joined to each other and to the cranium, so that they can be spread apart. The two halves of the lower jaw are not fused, but are connected by a ligament. Each half of both the upper and the lower jaw can be moved independently of the other half. This allows the snake to engulf objects that look impossibly large. It seems to "walk" its mouth around its food by a forward movement of first one side of the mouth and then the other.

The elongation of the vertebral column results from an increase in the number of vertebrae, rather than from a lengthening of the individual vertebrae. The earliest reptiles seem to have had about 25 presacral vertebrae. Because of the absence of the girdles, it is difficult to classify the snake's vertebral column into regions. The total number of vertebrae reported for different species ranges from 141 to 435. Most of these are dorsals. The caudals (i.e., those located behind the region of the vent) average about 50 to 60 in number, although in the short-tailed burrowing forms the number is much lower, and in some long-tailed snakes the ratio of dorsals to caudals may be only 1:1.

The vertebrae have a complicated structure. In addition to the usual articulating facets—the prezygapophyses and postzygapophyses—each has a pair of zygantra and a pair of zygosphenes (see Figure 3–7). Thus each vertebra has five points of contact with the one in front: the centrum, the two prezygapophyses, which fit against the postzygapophyses, and, dorsal to the zygapophyses on the neural arch, the zygosphenes, which articulate with the zygantra. The vertebrae of the anterior part of the trunk have hypapophyses—ventrally directed processes from the centra—to which muscles are attached. In many groups these are present on the posterior dorsals as well. In place of the hypapophyses, the caudal vertebrae frequently have haemapophyses—paired ventral projections from the centra that surround the caudal blood vessels.

Snake classification is in a period of reorganization and testing. Three different classifications (Underwood's, McDowell's, and Dowling's) have been proposed within the last decade. All three recognize the same basic groups of genera, but differ over how these generic groups are related and to what classification level the groups are assigned. The classification followed here is largely that of Underwood, but contains modifications suggested by other workers (Figure 16–2).

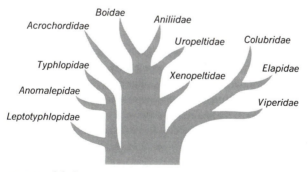

FIGURE 16–2
Dendrogram of the presumed relationships of snake families.

INFRAORDER SCOLECOPHIDIA

The scolecophidians, or blind snakes, are small (most less than 200 millimeters long) burrowing snakes with shiny, equal-sized scales. Their eyes are visible only as darkly pigmented spots beneath the enlarged head scales. Although they possess some primitive characteristics, they are highly specialized for burrowing. Their vertebrae lack neural spines. Usually traces of the pelvic girdle remain, though there are no external spurs representing the hindlimbs. The liver is composed of great numbers of lobes, whereas in most other snakes it is almost unilobed. There is only one pair of thymus glands (two in other snakes). Although previously suggested as relatives of the lizards, the scolecophidians show a clear relationship to other snakes in the structure of the brain case, the bony encasement of Jacobson's organ, the lack of a refracting organelle in the visual cells, and other characteristics found only in snakes. Three families are recognized.

Family Typhlopidae

The typhlopids are the most abundant and widespread of the blind snakes. They occur on all the tropical continents and many islands (Figure 16–3).

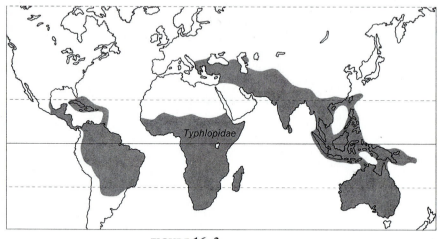

FIGURE 16–3
Distribution of the typhlopids.

The largest species may be 750 millimeters long, but most are less than 200 millimeters. The eyes are more or less distinct, but are covered by head shields. The transversely placed maxilla is loosely attached to the skull and bears teeth that are directed backward. The premaxillary, palatine, and pterygoid bones lack teeth. Some species have a single tooth at the tip of each mandible, but there is never a row of teeth on the lower jaw. Zygosphenes and zygantra are present on the vertebrae. The pelvic girdle may be represented by pubic, ischial, and iliac elements, with traces of pubic and ischial symphyses, or by a single rodlike bone on each side, or it may be absent entirely. Only the right oviduct is present.

The family has at least three genera. *Typhlops* is widely distributed throughout tropical America and the Old World tropics (Figure 16–4). *Typhlina* replaces it in Australia and coexists with it in the East Indies. In central and southern Africa, *Rhinotyphlops* coexists with *Typhlops*.

FIGURE 16–4
Dominican Blind Snakes, *Typhlops dominicana.*

Very little is known of the breeding habits of these small and secretive creatures. The Braminy Blind Snake, *Typhlops braminus,* lays two to seven tiny, elongate eggs, each about 12 millimeters long and 4 millimeters in diameter. But apparently not all species in the genus are oviparous, for a specimen of *Typhlops diardi* was found to contain 14 well-developed embryos.

Family Anomalepidae

Like other scolecophidians, the anomalepids have cylindrical bodies and short tails. The four genera, *Anomalepis, Helminthophis, Liotyphlops,* and *Typhlophis,* are confined to Central and South America (Figure 16–5). They appear to be related most closely to the typhlopids, differing from both the latter and the leptotyphlopids in having teeth in both jaws, the pelvic girdle usually absent, and the left oviduct either present, vestigial, or absent.

FIGURE 16–5
Distribution of the anomalepids.

Family Leptotyphlopidae

The Slender Blind Snakes, or Thread Snakes, of the leptotyphlopid family bear a close resemblance to the typhlopids, but differ from them in many structural features. No teeth are present on the upper jaw or roof of the mouth, and the maxilla borders the mouth instead of being placed transversely. Rows of teeth appear on the mandible. The pelvis consists of the ilium, ischium, and pubis, but is not attached to the vertebral column. A vestigial femur is present and may project through the skin in the anal region. Only the right oviduct is present. The largest species is only about 300 millimeters long; the smaller forms are slightly more than 100 millimeters long.

The species in this family are all included in a single, widely distributed genus, *Leptotyphlops,* found in Africa, southwestern Asia, the southwestern United States, and tropical America (Figures 16–6, 16–7).

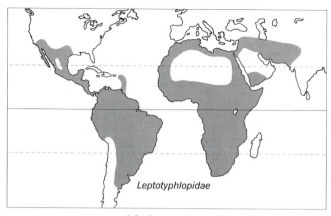

FIGURE 16–6
Distribution of the leptotyphlopids.

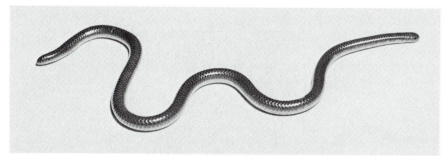

FIGURE 16–7
The Texas Blind Snake, *Leptotyphlops dulcis*. In the Old World, members of the genus *Leptotyphlops* are often called Thread Snakes.

As with many subterranean creatures, the life history of the Thread Snakes is poorly known. They live beneath the surface of the ground, but may come out during the early evening to wander for a short time. They feed largely on termites, adroitly sucking the contents from the termite abdomen. To coexist with the termites, they have evolved a pheromone mechanism. A pheromone is a chemical substance secreted by one animal that affects the behavior of another. The odor of the *Leptotyphlops* pheromone misleads the termites into accepting them as nestmates and not attacking them. *Leptotyphlops* can also recognize a termite's pheromone trail and follow it back to the nest. There are usually four long and slender eggs.

INFRAORDER HENOPHIDIA

The henophidians are a diverse group, seemingly linked more by the retention of primitive characters than by close relationship. They are predominantly tropical in distribution and possess a variety of unusual specializations. They share such primitive characteristics as vestiges of the pelvic girdle, a duplex retina with both rod and cone visual cells, the usual presence of a large left lung, left and right common carotid arteries, a coronoid bone in the lower jaw, and many others.

Family Acrochordidae

The weird-looking wart snakes are blunt-headed, small-eyed, ungainly creatures with unusually stout bodies for snakes (Figure 16–8). The longest females may reach lengths of 180 centimeters. The skin is loose, and the head and body are covered with small, granular or tuberculate, juxtaposed scales. The ventral scales are not enlarged, and those of the head are minute and sometimes pointed in the region of the nostrils; this is the source of the

314

FIGURE 16–8
Acrochordus javanicus, the Javanese Wart Snake.

common name, wart snake. (They are also known as elephant's-trunk snakes.) Hypapophyses are well developed on all the trunk vertebrae. Pelvic girdle and hindlimbs are lacking. The tail is short and compressed.

Wart snakes are aquatic fish eaters; they are found in estuaries and enter the sea quite freely. Lacking enlarged ventral shields, they are unable to glide normally on land, but progress by a slow, clumsy heaving of the body. Since they are ovoviviparous, they do not need to come ashore as do egg-laying species. One female has been reported as giving birth to 27 young. The family includes only two Indoaustralian genera, *Acrochordus* and *Chersydrus* (Figure 16–9).

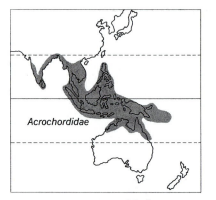

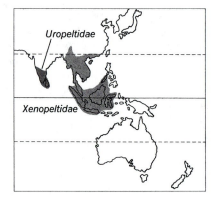

FIGURE 16–9
Distribution of the acrochordids, uropeltids, and xenopeltids.

Family Aniliidae

The aniliids are stout-bodied, short-tailed, cylindrical snakes that may grow as long as 900 millimeters. The scales are small and smooth, those on the ventral side being slightly enlarged. The bones of the skull are solidly united. Teeth are present on the maxillary, palatine, and pterygoid bones, on the dentary bone of the lower jaw, and, in one genus *(Anilius)*, on the premaxillary as well. Hypapophyses and haemapophyses are absent. A vestigial pelvis and rudimentary hindlimbs are present, the latter projecting as clawlike spurs on either side of the vent.

The family includes four genera of burrowing snakes: *Anilius,* the beautiful red and black False Coral Snake of northern South America; *Anomochilus* of Sumatra; *Cylindrophis,* the Pipe Snakes or Two-headed Snakes of southeastern Asia; and *Loxocemus,* the Dwarf Boa of southern Mexico and northern Central America (Figure 16–10).

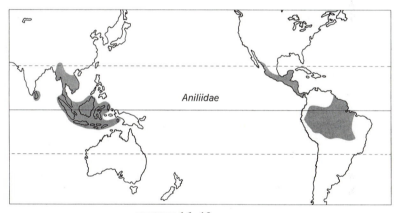

FIGURE 16–10
Distribution of the aniliids.

Snakes in which the bones of the skull are solidly fused cannot open their mouths as widely as can most snakes, and hence they feed largely on insects and worms. However, *Cylindrophis rufus* is reported to feed on other snakes and eels, and to be able to dispose of prey even longer than itself. Aniliids are ovoviviparous: *Cylindrophis maculatus* produces two or three young that may be more than 125 millimeters long at birth.

Family Boidae

The boids are the most diverse and widespread of the henophidians. They range in size from the small (200–300 mm) West Indian Ground Boas, *Tropidophis,* to the large (10–12 m) South American Anaconda, *Eunectes murinus.* They occur throughout the tropics and subtropics, and a few, such as the western United States Rubber Boa *(Charina bottae),* extend into temperate areas (Figure 16–11). Boas are typically stout-bodied and short-

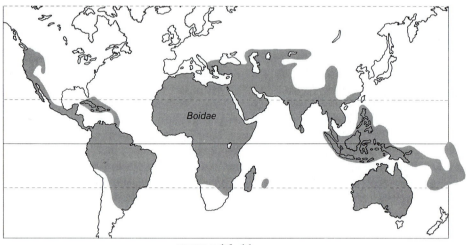

FIGURE 16–11
Distribution of the boids.

tailed. The palatomaxillary arch is movably attached to the rest of the skull. Teeth are present on the maxillary, palatine, pterygoid, and dentary bones, and sometimes on the premaxillary. The ventral scales form enlarged, transverse plates; the dorsal scales are small and sometimes iridescent. In all but one subfamily there are vestiges of the pelvis and hindlimbs (the latter terminate in clawlike spurs that are usually visible on either side of the vent, and are longer in the male than in the female).

Boids occupy a variety of habitats. Many of the small forms are burrowers in sandy soils. Some are arboreal, with short, more or less prehensile tails. The huge Anaconda of South America *(Eunectes murinus)* is largely aquatic and can remain submerged in water for a long time. A number of the large boids, such as the Indian Python *(Python molurus),* seem to be equally at home in trees or in water (Figure 16–12).

Boids feed largely on birds and mammals, and usually kill their prey by constriction. Contrary to popular opinion, they do not crush the bones of their victims. Two or three coils of the snake's body are wrapped around the upper trunk of the prey. These exert enough pressure to stop breathing, and the animal suffocates.

The vestigial hindlimbs of the boids are apparently functional structures used in courtship (Figure 16–13). The male Boa Constrictor vibrates them rapidly and rubs them on the back and flanks of his mate. By beating with his spurs, the male Anaconda stimulates the female to move her body into a

FIGURE 16–12
The largest Old World snake, the Reticulated Python, *Python reticulata*. It is exceeded in size only by the Anaconda of South America, which has a maximum recorded length of 1143 centimeters.

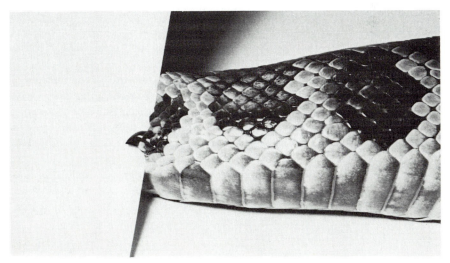

FIGURE 16–13
External portion of the hindlimb of the Cuban Boa, *Epicrates angulifer*. The spurs apparently function during courtship.

copulatory position. A male Python, after having arranged himself alongside the female with the anterior part of his body on her back, taps the region around her cloaca with his claws in a slow and rhythmic manner. This behavior may continue for as long as two hours, and stops only when the female inclines her anal region on the side and allows the male to insert his hemipenis into her cloaca.

The boids display both oviparity and ovoviviparity. The Reticulated Python lays eggs, the number of which depends partly on the size of the mother. A large, full-grown female may lay as many as 100 eggs, whereas small females have been known to lay as few as 15. The incubation period lasts 60 to 80 days, and the newly hatched young are 600 to 750 millimeters long. Pythons brood their eggs: the mother coils around them and provides extra heat through an elevation of her body temperature. The size of this temperature increase is still a matter of debate. Some authors have recorded the body temperature of the brooding animal to be no higher than that of the surrounding environment, whereas in other species, at other times, body temperatures for the female on the eggs have been reported to be 6° or 7°C warmer than for the nonbrooding male, and 12° to 15°C warmer than the air. From these divergent results, it seems that the body temperature of the female may increase during the brooding process, but that the amount of increase varies from one species to another. The temperature of the female is higher at the onset of brooding than toward the end of it. Many more observations of temperature and brooding are needed before the overall pattern can be determined.

There are also many ovoviviparous snakes in this family. The Anaconda (*Eunectes murinus*) of South America gives birth to between 4 and 39 young (each about 800 mm long) and the Rosy Boa *(Lichanura)* has been known to give birth to 6 young (each about 280 mm long).

The boids are divided into five subfamilies.

Subfamily Boinae.   The true boas of the boine subfamily are confined to the New World tropics with the exception of the Malagasian *Acrantophis* and *Sanzinia* and the South Pacific *Candoia*. Although commonly linked to the Old World pythons, to which they show many similarities in body form and habits, they differ from the latter in their ovoviviparous reproduction, absence of supraorbital bones, more flexible primary palate, and other characters. *Corallus,* the South American Tree Boas, are the smallest boines, averaging 500–600 millimeters; *Boa* and *Epicrates* are medium-sized snakes commonly 1 to 2 meters and occasionally 3 meters long (Figure 16–14). *Eunectes murinus,* the Anaconda, is the giant of the group.

Subfamily Bolyerinae.   The Round Island Boas *(Bolyeria* and *Casarea)* occur only on a small island off the coast of Mauritius in the Indian Ocean.

FIGURE 16-14
The Cuban Boa, *Epicrates angulifer,* in its characteristic feeding posture.

Their continued existence is definitely threatened, and they may already be extinct through predation by introduced pigs and rats. They are unique boids, apparently very early offshoots of the main boid lineage. Structurally, these small, semifossorial snakes are set off by a unique maxilla with a median joint dividing it into an anterior and a posterior bone. This character, along with their multifarious sharing of characters with other lineages of henophidian and colubrid snakes, leads some herpetologists to elevate the two genera to separate familial or superfamilial status, and this is not an unreasonable decision.

Subfamily Erycinae. Three genera are commonly grouped together in the erycine subfamily: *Charina* (Rubber Boa) and *Lichanura* (Rosy Boa) of western North America and *Eryx* (Sand Boa) of Africa and Asia. They are closely related to the true boas, but differ from them in their adaptations to burrowing habits. The skull is more compact, and the tail vertebrae are reduced in number but increased in size.

Subfamily Pythoninae. The pythonines are the boa counterparts of Africa, Asia, and Australia. All share oviparous reproduction and the presence of a supraorbital bone. *Python* is the most widespread genus, occurring on all three continents. It and *Liasis* of Australia are medium to large snakes living in a variety of habitats, from arid scrub to humid rain forest. *Chondropython* (the Green Tree Python of Australia and New Guinea) is a medium-sized, arboreal snake. *Aspidites* of Australia and *Calabaria* of West Africa are terrestrial snakes, distinct from the other pythons and from each other.

Subfamily Tropidophiinae. Four small ground boas *(Exiliboa, Trachyboa, Trophiophis,* and *Ungaliophis)* of Central America and the West Indies form the tropidophiine subfamily. In many ways, they seem intermediate between the boids and the colubrids. They possess boid characters, such as vestiges of pelvic girdles and paired common carotid arteries, and their similarities may be convergences resulting from an early radiation into a niche then lacking colubrids.

Family Uropeltidae

The uropeltids (roughtails) are secretive, frequently fossorial snakes with rigid, cylindrical bodies and very short tails. Most are less than 600 millimeters long. Many species are brightly marked with red, orange, or yellow; some are a shiny, iridescent black. The pupil of the eye is round. The ventral scales are little larger than the dorsals. The bones of the skull are more solidly united than in any other family of snakes. Burrowing snakes dig with their heads, and the remarkably solid skull of the roughtails is probably an adaptation to this mode of life. The maxilla has 6 to 8 teeth; the mandible, 8 to 10; and the palatine, 3 or 4 minute ones or none at all. The occipital condyle projects markedly beyond the back of the skull. There are no vestiges of hindlimbs or a pelvic girdle.

The most striking characteristic of the roughtails is the enlarged scale at the end of the tail. It is either very rugose, spiny, or reduced to two short points. The tail of a freshly caught specimen is often coated with mud. The purpose of this unique appendage has never been satisfactorily explained.

There are seven genera in the family. All are found in damp places in mountainous regions of southern India and Sri Lanka (see Figure 16–9). These snakes are quiet and inoffensive: they do not bite when handled, nor do they apparently show any fear. When picked up, they do not try to escape, but entwine themselves around the fingers or a stick and remain in that position for a long time. They are easily kept in captivity, and have been known to eat immediately after being caught. Like most small burrowing snakes, they feed on worms and soft-bodied larvae. So far as is known, all are ovoviviparous, producing from three to eight young at a time.

Family Xenopeltidae

The small xenopeltid family contains only one species, *Xenopeltis unicolor,* the Sunbeam Snake of southeastern Asia (see Figure 16–9). Its common name comes from the highly iridescent scales. As it crawls in sunlight it flashes with electric blue, emerald green, blood red, purple, and copper. This brilliant display is seldom seen, however, for these snakes are secretive and largely nocturnal. The body is cylindrical, the tail is short, and the ventral scales are enlarged to form transverse plates. The female may be 900 millimeters long or more, the male somewhat smaller. The bones of the skull are united. The teeth are small, close together, and sharply curved; there are four or five teeth on each premaxilla and 35 to 45 on each maxilla. The palatine, pterygoid, and dentary also bear teeth. Hypapophyses are absent on the posterior dorsal vertebrae. There is no trace of the pelvic girdle or hindlimbs. *Xenopeltis* has two well-developed lungs, the left being about half as large as the right.

Sunbeam Snakes are frequently found beneath logs and stones in rice fields and gardens near human habitations. They can burrow rapidly in soft earth, and those kept in captivity usually spend the day under cover and go forth only at night. They feed on other snakes, small rodents, and frogs. Apparently nothing has been recorded about their breeding habits.

INFRAORDER CAENOPHIDIA

The huge caenophidian infraorder includes most of the snakes of the world. Its members occupy a wide variety of habitats: some are terrestrial, some arboreal, some fossorial, some aquatic in fresh water, some marine. Paralleling this diversity of habitats is a great diversity of structure, so that it is difficult to find characters to define the infraorder. In addition, nothing is known of the internal anatomy of many of the genera. All forms that have been examined so far have only a single carotid artery, whereas members of the other infraorders have two. There are never any vestiges of the pelvic girdle. The facial bones are movable and loosely attached to the skull (see Figure 16–1). The optic foramen is usually bordered by the frontal, parietal, and parasphenoid bones. Caenophidians range in size from very small to very large, though none approaches the great boas in bulk. Most are harmless to man, but the infraorder also includes all the venomous snakes of the world.

Many attempts have been made to divide this huge and unwieldy assemblage of forms into families and subfamilies. We adopt a conservative position and recognize three families, each with two or more subfamilies.

Family Colubridae

The majority of snake are included in the colubrid family. Usually the belly scales are as wide as the body. Teeth are normally present on the maxillary, palatine, pterygoid, and dentary, but are never found on the premaxillary. Most species have solid teeth, without grooves (aglyphous) and unconnected with any venom glands. A few have several of the rear teeth grooved (opisthoglyphous). The supralabial gland above is specialized to produce a venom, which is channeled down the grooves. Such rear-fanged snakes do not inject venom by striking, but by chewing an object that has been taken into the mouth. The venom is thus used, not for capturing prey, but for quieting the struggles of the animal being swallowed, and for initiating digestion. Most rear-fanged snakes are small and harmless, but both the African Boomslang *(Dispholidus typus)* and the African Twig Snake *(Thelotornis kirtlandi)* have caused human fatalities, and several other species may also be dangerous (Figure 16–15).

FIGURE 16–15
The African Boomslang, *Dispholidus typus,* a snake whose bite is potentially fatal to humans.

As would be expected in such a large and varied family, colubrids show great diversity in feeding habits. Some will eat almost anything they are able to catch and engulf, whereas others, such as the egg-eating and snail-eating snakes, have specialized diets. The smaller forms eat worms and insects; many of the larger ones feed exclusively on birds and mammals, and usually kill their prey by constriction. Aquatic colubrids prey on fishes and amphibians. The King Snakes *(Lampropeltis)* seem to be especially fond of other snakes.

The colubrids include oviparous, ovoviviparous, and viviparous forms. The life history of the Racer *(Coluber constrictor)* of the eastern United States is typical of the oviparous members of the family. Mating has been observed in May, and the eggs are laid in decaying vegetable matter through

June and early July. There may be as many as 25 eggs in a clutch, but the average is around 12. Eggs range from 26.8 to 46.5 millimeters in length and are somewhat elongated and granular in texture. The young, which hatch in August, are between 200 and 300 millimeters long.

The breeding habits of the Northern Water Snake, *Natrix s. sipedon,* may be taken as normal for the ovoviviparous forms. Mating takes place in the early spring and the young are born in the late summer and early fall. The average number reported in eight broods was 31, with the number per brood ranging from 16 to 40—but much larger broods have been recorded, some snakes giving birth to more than 75 at one time. Length of the young at birth averages about 225 millimeters.

True viviparity, with placenta formation, occurs in the Garter Snake, *Thamnophis sirtalis.*

If any generalization can be made about the life history of the colubrids, it is that the more terrestrial forms lay eggs and the more aquatic ones give birth to living young. Perhaps this is because eggs laid in the places normally inhabited by aquatic snakes might be endangered by flooding. But even this generalization cannot be carried too far, for the highly aquatic Mud Snakes of the genus *Farancia* lay 50 to 100 eggs, which the female broods. There is variation even within a single genus: the young of American forms of *Natrix* are born alive, whereas females of the European *Natrix* (less aquatic than their American congeners) lay eggs. On the other hand, *Xenochrophis piscator,* which is as fully aquatic as any North American *Natrix,* lays eggs in the delta of the Indus in late March at the onset of the dry season. The very small young are first to be found when the rains moisten the soil in early July.

Such a large and unwieldy family as the Colubridae can be discussed much more easily if it is subdivided into subfamilies, but even this is very difficult to do. The colubrids are very successful, both in number of individuals (biomass) and in adaptation to a large number of niches. The latter results from a broad adaptive radiation in each of three tropical regions—America, Africa, and Asia—with a bewildering convergence of structure and habits among the three areas (Figure 16–16). This convergence confounds any attempt to establish a natural classification. Six subfamilies are recognized here in an attempt to divide these snakes into discrete geographic—and, perhaps, distinct evolutionary—units. Snake classification can be expected to change significantly in the next decade or so, as knowledge of snakes and methods of phylogenetic analysis improve.

Subfamily Colubrinae. The colubrine subfamily is the largest and most diverse group of colubrids. All colubrines have posterior body vertebrae with small or no hypapophyses, and a single undivided seminal groove on the hemipenis. The head is covered with large symmetrical scales, and the

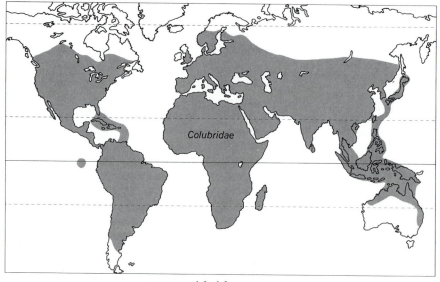

FIGURE 16–16
Distribution of the colubrids.

nostrils are lateral. The maxillary teeth may be aglyphous or opisthoglyphous. Most colubrines are oviparous. They are cosmopolitan in distribution, with representatives on all continents.

The majority of North American snakes are colubrines. They range from small, semifossorial, insectivorous genera such as *Diadophis* (Ringneck Snakes) and *Tantilla* (Black-headed Snakes) to large, terrestrial or semiarboreal, carnivorous genera such as *Elaphe* (Rat Snakes), *Lampropeltis* (King Snakes), and *Masticophis* (Whip Snakes). In the arid southwestern United States and northern Mexico, colubrines have become sand dwellers *(Sonora, Hypsiglena, Arizona).* Farther south, in the neotropics, there are burrowers *(Ficimia),* terrestrial racers *(Dendrophidion, Chironius)* and tree snakes *(Imantodes, Oxybelis)* (Figure 16–17).

Within the Old World tropics, the colubrines have formed a number of specialized groups. *Dispholidus,* the African Boomslang, is an arboreal snake with a venom that is extremely toxic to endothermic vertebrates. *Boiga* and its allies are another group of arboreal snakes with well-developed poison glands. This group is widespread in Africa and Indomalaya (Figure 16–18). The Earth Snakes *(Calamaria)* and their relatives, of the East Indies and Southeast Asia, are small, secretive, burrowing snakes with smooth, cylindrical bodies, blunt snouts, and short tails. This habitus with small eyes and a reduced number of head scales is typical of most burrowing snakes.

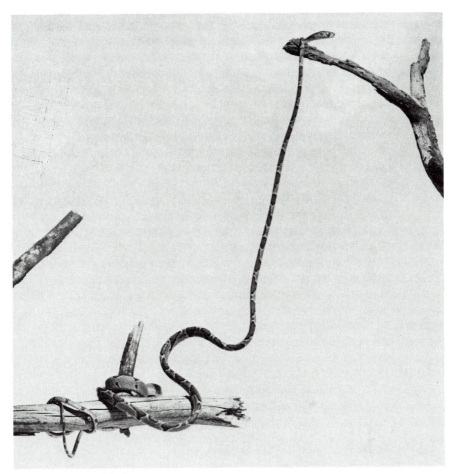

FIGURE 16–17
The Blunt-headed Tree Snake, *Imantodes cenchoa*, a neotropical species.

The most unusual colubrines are the Oriental snail-eating snakes, *Pareas* and *Aplopeltura* of Southeast Asia and the East Indies. They are slender little snakes with short, wide heads, slim necks, and big, childlike eyes. Hypapophyses are present only in the cervical region. The teeth are agly-phous. The mouth opening extends far behind the fringe of the buccal membrane. The nasal gland (one of the salivary glands) is enormous. The dentary bone is immovably fused to the surangular. The most striking fea-ture of this group is the arrangement of the scales under the lower jaw. In most snakes these chin shields are separated at the midline by a furrow, the mental groove, which is lined with distensible skin. This allows the two

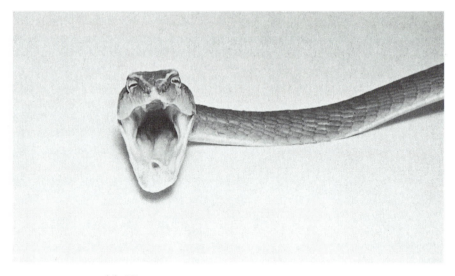

FIGURE 16–18
Dryophis nasuta, the Malayan Green Tree Snake, an Asian colubrine.

halves of the jaw to be spread widely for the swallowing of large prey. In *Pareas* and *Aplopeltura,* the chin shields of the two sides dovetail, and there is no mental groove. Consequently, their jaws cannot be spread widely, and their diet is restricted to such small items as snails, slugs, and grubs. They apparently cannot crush the shell of a snail, but use their sharp teeth to extract the body before swallowing it. There are about fifteen species, some terrestrial, some arboreal. They are quiet, inoffensive little snakes, mostly nocturnal. So far as is known, they are all oviparous.

Subfamily Natricinae. The natricines—water snakes and their allies—are cosmopolitan in distribution, although two centers of radiation (North America and Indomalaya) are apparent. Natricines are almost exclusively aquatic or terrestrial. All have well-developed hypapophyses throughout the vertebral column and symmetrical hemipenes. They range in size from small (*Virginia,* 250 mm) to medium (*Natrix,* 1 m). Many, but not all, are ovoviviparous. *Natrix* and *Thamnophis* are the most familiar representatives in North America (Figures 16–19, 16–20), *Amphiesma* and *Rhabdophis* in Asia.

Subfamily Homalopsinae. The rear-fanged water snakes occur throughout the Indoaustralian region, from India and China to northern Australia. These stout, cylindrical-bodied snakes are highly aquatic and have narrow ventral scales, valvular nostrils, and compressed tails. All have opisthoglyphous dentition and are ovoviviparous. *Enhydris* is the largest genus,

FIGURE 16–19
The Northern Water Snake of the eastern United States, *Natrix s. sipedon*. This snake is commonly called "Water Moccasin."

FIGURE 16–20
Natrix erythrogaster, the Red-bellied Water Snake of the eastern United States, in the act of shedding. The snake will ultimately crawl completely out of the shed skin, leaving it more or less entire and inside out.

with 21 species occurring throughout the entire range of the subfamily, in both fresh and salt water. Most homalopsines (*Cerberus, Erpeton*) eat frogs and fish, but *Fordonia* feeds primarily on crabs.

Subfamily Lycodontinae. The predominantly African lycodontine subfamily is composed largely of terrestrial and fossorial snakes. They are small to moderate-sized and are commonly opisthoglyphous. Most have short hypapophyses on all body vertebrae, and bilobed hemipenes with a bifur-cate sulcus. *Lycodon* and its allies (wolf snakes) are aggressive racerlike snakes feeding on birds and mammals. *Boaedon, Geodipsas* and their allies are generally small, semifossorial forms. *Aparallactus* and its relatives are fossorial.

The latter are of special interest because their lineage has apparently given rise to two highly specialized groups, the egg-eating snakes, *Dasypeltis* and *Elachistodon,* and the Mole Vipers, *Atractaspis.* For an animal not much thicker than a man's finger to engulf an object the size of a hen's egg is indeed an astounding feat. The egg-eating snakes accomplish this by a strik-ing series of adaptations. Although there is no mental groove, the skin along the angle of the mouth and cheek region is specially modified for expansion. The teeth are minute, reduced in number, and restricted to the posterior parts of the maxilla, palatine, and dentary, and although the upper-jaw elements are rigidly fused together, the bones of the lower jaw are very loosely connected. The most remarkable adaptation of all is the extension of some of the cervical hypapophyses downward to pierce the esophagus. The hypapophyses in *Dasypeltis* are highly modified: the ones in front have their ventral edges enlarged into sledlike runners, whereas those behind form elongate, forward-pointing spines. The egg, which is swallowed whole, glides down the runners and is forced against the sharp edges of the spikelike hypapophyses, thereby being cut open. The contents of the egg pass into the stomach, and the crushed shell is then regurgitated. The hypapophyses that pierce the esophagus in *Elachistodon* are less highly modified in shape: snakes of this genus have one or two enlarged and grooved teeth on the rear of the maxilla. It is possible that they may also feed on small birds and mammals.

There are about six species of *Dasypeltis,* widely distributed in tropical and southern Africa. *Elachistodon* is known only from a few specimens taken in northeastern India. Both genera are apparently oviparous.

The Mole Vipers *(Atractaspis)* of Africa and the Middle East are strongly fossorial. They have a narrow head, small eyes, a cylindrical body, and a short tail. This innocuous-looking body form, coupled with an ability to bite without gaping, has surprised many snake collectors. The fangs are very long. By shifting the lower jaw to one side, the Mole Viper can expose the fang on the opposite side and sink it by a twist of the head into the thumb or finger of the most expert collector.

Subfamily Xenodontinae. The xenodontine subfamily contains the major-ity of the neotropical snakes. Most members lack hypapophyses on the posterior vertebrae and possess a bifurcated sulcus on the hemipenis. Many have opisthoglyphous dentition. *Alsophis* and its relatives are racerlike and kill their prey by constriction. *Heterodon, Lystrophis,* and their allies are toad eaters with enlarged, but ungrooved, posterior maxillary fangs for deflating toads before swallowing. The slug-eating snakes, *Dipsas* and allies, are surprisingly similar to the Oriental snail-eating snakes, *Pareas*, in body form and behavior. The xenodontines exploit most niches open to snakes in the neotropics, and show numerous convergences with the snakes of the Old World tropics.

Family Elapidae

The elapids are the family of the extremely venomous coral snakes, cobras, mambas, kraits, and sea snakes. Like the viperids, they have venom fangs in the front part of the upper jaw. Snakes with fangs of this sort are called proteroglyphs. These two families have given all snakes a bad name, though the venomous snakes constitute but a small part of the snake fauna of the world. The fang of the elapids is a more or less enlarged, canaliculate tooth, which is held permanently in an erect position and fits into a pocket in the gum tissue, on the outside of the mandible but inside the lip when the jaw is closed. The canaliculate tooth has apparently evolved from a grooved tooth like those of the opisthoglyphous colubrids. The groove has sunk in to form a horseshoe-shaped cavity. In the elapids, the gap between the ends of the horseshoe is usually more or less filled in with calcium, but it still shows as a furrow on the front surface of the tooth. The duct from the venom gland is not attached directly to the fang, but expands into a small cavity in the gum above the opening of the tooth canal. Two fangs are normally present on each maxilla, lying side by side, though usually only one at a time is firmly attached and functional. Each is followed by a series of developing replacement fangs. Snake teeth are constantly being shed and replaced: this arrangement ensures that the snake is never without functional fangs. When the fang on the inner side of the maxilla drops off, the one on the outer side is either ready to be used or already in use. It serves while the next replacement fang on the inner side is growing into place. The maxillary bone is shortened, and probably represents only the hindpart of the maxilla found in the rear-fanged, opisthoglyphous snakes. It usually bears one or more small, solid or slightly grooved teeth behind the fang. Teeth are also present on the pterygoids, palatines, and dentaries. The facial bones are movable. Hypapophyses are developed throughout the vertebral column, and the pelvic girdle and left lung are lacking.

The distribution of the elapids is shown in Figure 16–21. There are three subfamilies.

Subfamily Elapinae.   Elapines are found throughout the tropical and subtropical regions of the world, but they are most numerous in Australia, where most snakes belong to this group. They are absent from Europe

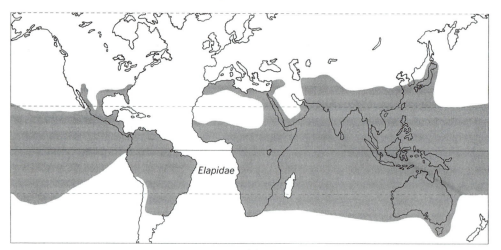

FIGURE 16–21
Distribution of the elapids.

today, but fossil forms have been found from the Miocene and Pliocene of France. About 30 genera are known, including the longest of all venomous snakes, the King Cobra *(Ophiophagus hannah)*, which may attain lengths of more than 540 centimeters.

The snake so often pictured with Indian snake charmers is an Indian Cobra in its defense position. It is not "charmed," but is reacting to the presence of a possible aggressor with a threat display—raising the forepart of its body and drawing up its long anterior ribs to spread the skin of its neck into a hood.

Many of the elapines are unaggressive and seem reluctant to bite, but their venom is highly toxic. Some of the cobras of Africa and Asia have the extremely unpleasant trait of "spitting" their venom into the eyes of their enemies. Their fangs are modified to permit the streams of venom to be ejected outward instead of downward. Spitting Cobras can spray their poison with great force for up to 180 centimeters, and seem to show a high degree of accuracy in aiming for the eyes. If the venom is washed away immediately no permanent damage results, but untreated victims become blind.

The Indian Cobra *(Naja naja)* mates during January and February, and the eggs are laid in May. Apparently the pair remain together from the time of mating until the young are hatched; the male may also share in guarding the eggs. Incubation takes between 69 and 84 days. The usual number of eggs is rather low, from one to two dozen, but as many as 45 have been recorded.

The elapines may have undergone three adaptive radiations. The coral snakes *(Micrurus, Micruroides)* of the New World are typically semifossorial snakes feeding on lizards. All have a slender, cylindrical body with a ringed pattern of bright warning colors (Figure 16–22). Their disjunct dis-

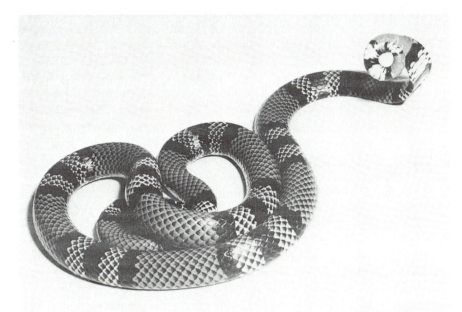

FIGURE 16–22
The defense attitude of the South American Coral Snake, *Micrurus frontalis*.

tribution far removed from other elapines suggest a separate origin from the colubrids. Kraits and cobras of Africa and Asia show more diversity in body form and habits: aquatic *Boulengerina* (Water Cobras), semifossorial *Callophis,* terrestrial *Naja* and *Bungarus,* arboreal *Dendroaspis*. The Australian elapines also show great diversity, although they have largely been confined to fossorial *(Apistocalamus, Toxicocalamus)* and terrestrial *(Demansia, Oxyuranus)* habitats.

Most elapines lay eggs, but the Ringhals or Spitting Cobras *(Hemachatus hemachatus)* of South Africa and some of the Australian forms are ovoviviparous, and *Denisonia* of Australia has been reported to be truly viviparous.

Subfamily Hydrophiinae.   The hydrophiines are the sea snakes, derived from a lineage of Australian elapines. All but one of the true sea snakes belong to this subfamily. The largest reach a length of 275 centimeters, and most species are only about one third as long, so they hardly qualify as the huge sea serpents of legend—but they may have provided the basis for many such tales. They differ from the elapines mainly in the characters by which they are adapted to life in the sea. The body is more or less laterally compressed posteriorly, and the tail is very compressed and paddle-shaped. The nostril opens on the upper side of the snout, and can be closed tightly by a valve. The tongue is short, so that only the cleft portion can be protruded.

Four genera—*Hydrelaps, Aipysurus, Ephalophis,* and *Emydocephalus*—are less specialized for marine life than are the other sea snakes. Their ventral shields are relatively large, one third to one half as wide as the body. They are able to move well on land and apparently spend a good bit of time there. At least two of the species, and possibly all of them, are oviparous, coming ashore to lay their eggs. They are never found far from land, but live in the shallow coastal waters and river estuaries of southeastern Asia, Australia, and the islands of Oceania. As would be expected, they feed on fish and are often caught in the nets of fishermen.

The majority of the sea snakes (about 15 genera) are completely aquatic. Their ventral scales are either very small or absent. They are graceful, competent swimmers, though, like the less specialized forms discussed above, they are seldom found far from shore. Some species bask on the surface of the water, and on days when the sea is calm they may be seen from the bows of steamers, sometimes by the hundreds. One naturalist reported seeing a mass of snakes forming a line across the surface of the sea about 300 centimeters wide and nearly 100 kilometers long. The snakes were so closely packed that the line could be seen from several kilometers away. He estimated that the column included millions of snakes—probably the largest congregation ever reported.

So far as is known, all these snakes are ovoviviparous. Although they are so highly aquatic, at least some species come ashore to bear their young. Female sea snakes of the Philippines have reportedly come onto the smaller islets to bear their young among the rocks and tidal pools. They are found in the Indian and Pacific Oceans, along the Asiatic coast, and throughout the Indoaustralian seas to the coast of tropical Australia and the oceanic islands of the South Pacific. One form *(Pelamis)* has extended its range eastward across the Pacific to the shores of tropical America and westward to Madagascar and Africa.

All the sea snakes are venomous. The venom of some does not appear to be strongly toxic to humans, but laboratory experiments have shown that the venom of others is even more powerful than that of the cobra. The deaths while bathing of at least four people have been attributed to sea-snake bites. A number of nonfatal bites have also been reported. What induced the snakes to bite is unknown, because in general hydrophiines can be stimulated to bite only after considerable provocation.

Subfamily Laticaudinae. *Laticaudus* is the only member of the laticaudine subfamily. It differs most noticeably from the other sea snakes in the typically elapine position of the nostrils on the anterolateral surface of the snout, the well-developed ventral scales, and the compressed tail lacking the support of enlarged neural spines. Other characteristics suggest that *Laticaudus* was independently derived from an Afroasian lineage of elapines.

Family Viperidae

In the viperids, the whole mechanism for the injection of venom reaches its highest development. The maxillary bone is very short but deep vertically, and is movably attached to the prefrontal and ectopterygoid bones. The large fang is on its posterior end, but because the bone is so shortened anteroposteriorly, the fang lies in the front part of the mouth (Figure 16–23). The canal for the transmission of venom is usually completely closed,

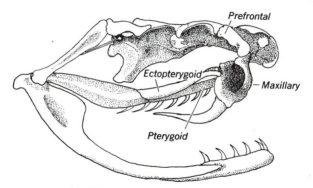

FIGURE 16–23
Skull of the Eastern Diamondback Rattlesnake, *Crotalus adamanteus.*

so that no external groove is visible (solenoglyphous). At rest, the tooth is folded to lie horizontally along the upper jaw. In striking, the fang is brought forward from the resting position by a movement of the bones forming the palatomaxillary arch, with the maxilla turning like a hinge on the anterior end of the prefrontal. Most elapids, with their smaller fixed fangs, tend to bite and hold on, but the viperids, with their large and powerful fangs, are able to inject a greater amount of venom at the instant of bite, and tend to strike and then draw back. Replacement of the fangs is the same as in the elapids, and two fangs are often present at the same time. There are no other maxillary teeth. In other characteristics, the viperids resemble the elapids.

Snake venoms are highly complex protein mixtures that vary in composition, and hence in effect, from species to species, In general, though, the venom of the elapids acts primarily on the nervous system (neurotoxic), and of the viperids acts on the blood (hemotoxic).

Figure 16–24 shows the distribution of the viperids. The family is divided into three subfamilies.

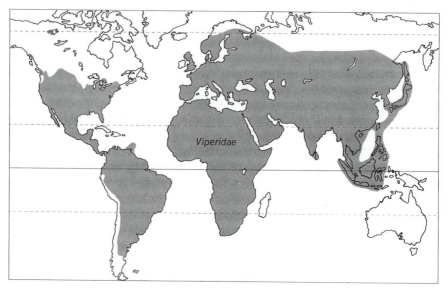

FIGURE 16–24
Distribution of the viperids.

Subfamily Azemiopinae. The rare Southeast Asian *Azemiops feae* is the only member of the azemiopine subfamily. This mountain viper possesses many primitive traits, and can be separated from the true vipers and pit vipers by internal characters such as the absence of the levator anguli oris muscle. It appears to be an early derivative of the viperid stock, with its closest affinities to the pit-viper lineage.

Subfamily Viperinae. The viperines, or true vipers, are usually stockily built, with short bodies and short tails. The large Gaboon Viper *(Bitis gabonica)* (Figure 16–25) may be nearly 180 centimeters long, but no viperid matches the big cobras in length. The maxillary bone is not hollowed, and there is no pit on the side of the face between the nostril and the eye. The eight genera of true vipers are found only in Eurasia and Africa. Some are among the deadliest snakes in the world.

A few of the vipers lay eggs, but in most the young are born alive. A chorioallantoic placenta forms in the European Adder *(Vipera berus)*. Mat-

FIGURE 16–25
An Old World viperid, *Bitis gabonica,* the Gaboon Viper of Africa.

ing takes place in the spring after the snakes have emerged from hibernation. Rivalry between males is keen during the mating season, and they sometimes give a spectacular performance called the "dance of the Adders." Although copulation may take place in April, ovulation does not occur until the end of May. The young are born during August and September. The eggs, which will produce the young of the next year, are formed in the ovaries following parturition. In the northern part of the range, the short summer is apparently insufficient for the complete intraovarian development of the eggs, and the snake is forced back into hibernation before the young are ready to be born. In consequence, the vipers in the northern half of Sweden and in Finland breed only every second year. Numbers of young range from 6 to 20, but there are usually about 10 to 14. They may be 130 to 180 millimeters long.

Subfamily Crotalinae. The crotalines are called pit vipers, because the maxillary bone is hollowed out above by a pit that opens between the eye and the nostril. The membrane in this pit is extremely sensitive to changes in temperature and serves to detect the presence of the warm-blooded animals on which the snake preys. Pit vipers occur from eastern Europe across Asia to Japan and in the Indoaustralian Archipelago, but the majority of the species are found in North, Central, and South America. Among the five genera are such dreaded forms as the Rattlesnakes *(Crotalus)* and the tropical American Bushmaster *(Lachesis muta)* and Fer-de-Lance *(Bothrops at-rox)*. The Bushmaster reaches a length of 360 centimeters and is truly one of the most formidable snakes in the world (Figure 16–26).

FIGURE 16–26
A New World pit viper, *Lachesis muta stenophrys,* the Bushmaster of South America.

Apparently all the New World crotalines except *Lachesis* are ovovivip-arous, but there are some egg-laying species of *Agkistrodon* and *Trimeresurus* in southeastern Asia. As in the viperids, there is an extensive premating dance by the males. The Cottonmouth *(Agkistrodon piscivorus)* mates in March, and the young are born in late August and early September. The number in a brood varies from 5 to 15 with an average of about 8. Length of the young ranges from 150 to more than 250 millimeters.

READINGS AND REFERENCES

Angel, F. *Vie et Moeurs des Serpentes*. Paris: Payot, 1950.

Boulenger, G. A. *Catalogue of the Snakes in the British Museum (Natural History)*, vols. 1–3. London: British Museum (Natural History), 1893–1896.

Bourgeois, M. "Contribution à la morphologie comparée du crâne des ophidiens de l'Afrique centrale." *Publications de l'Université officielle du Congo à Lubumbashi,* vol. 18, 1968.

Dowling, H. G. "A provisional classification of snakes." *In 1974 Yearbook of Herpetology*. New York: HISS Publications, 1975.

Gyi, K. K. "A revision of colubrid snakes of the subfamily Homalopsinae." *University of Kansas Publications, Museum of Natural History*, vol. 20, 1970.

Hoffstetter, R. "Review: A contribution to the classification of snakes." *Copeia,* no. 1, 1968.

Klauber, L. M. *Rattlesnakes.* Berkeley and Los Angeles: University of California Press, 1972.

Liem, K. F., H. Marx, and G. B. Rabb. "The viperid snake *Azemiops:* its comparative cephalic anatomy and phylogenetic position in relation to Viperinae and Crotalinae." *Fieldiana: Zoology,* vol. 59, 1971.

McDowell, S. B. "Notes on the Australian sea-snake *Ephalophis greyi* M. Smith (Serpentes: Elapidae, Hydrophiinae) and the origin and classification of sea-snakes." *Zoological Journal of the Linnean Society,* vol. 48, 1969.

————. "The genera of sea-snakes of the Hydrophis group (Serpentes: Elapidae)." *Transactions of the Zoological Society of London,* vol. 32, 1972.

————. "A catalogue of the snakes of New Guinea and the Solomons, with special reference to those in the Bernice P. Bishop Museum. Part I: Scolecophidia." *Journal of Herpetology,* vol. 8, 1974.

————. "A catalogue of the snakes of New Guinea and the Solomons, with special reference to those in the Bernice P. Bishop Museum. Part II: Anilioidea and Pythoninae." *Journal of Herpetology,* vol. 9, 1975.

Marx, H., and G. B. Rabb. "Relationships and zoogeography of the viperine snakes (Family Viperidae)." *Fieldiana: Zoology,* vol. 44, 1965.

Minton, S. A., and M. R. Minton. *Venomous Reptiles.* New York: Scribner's, 1969.

Oliver, J. A. *Snakes in Fact and Fiction.* New York: Doubleday, 1963.

Parker, H. W., and A. Grandison. *Snakes: A Natural History.* London: British Museum, 1977.

Rossman, D. A. (coord.). "Symposium on colubrid snake systematics." *Herpetologica,* vol. 23, 1967.

Smith, M. A. *Fauna of British India: Reptilia and Amphibia,* vol. III, *Serpentes.* London: Taylor and Francis, 1943.

Underwood, G. A. *A Contribution to the Classification of Snakes.* London: British Museum (Natural History), 1967.

Wright, A. H., and A. A. Wright. *Handbook of Snakes,* vols. 1 and 2. Ithaca, N.Y.: Comstock, 1957.

CHAPTER SEVENTEEN

RHYNCHOCEPHALIANS
AND CROCODILIANS

The two remaining orders of reptiles are both relics. The modern crocodilians are all that remain of the mighty archosaur stock that once gave rise to the Mesozoic dinosaurs, as well as to the modern birds. The Tuatara *(Sphenodon punctatus)* is the only surviving member of the order Rhynchocephalia, now placed in the subclass Lepidosauria, to which the Squamata also belong. No fossil remains of *Sphenodon* have ever been found, but all other members of the family are known only from the Mesozoic. In spite of this enormous time gap, *Sphenodon* seems to have changed little through the ages, and remains today a relatively unspecialized representative of the reptiles of the early Mesozoic. It has aptly been called a "living fossil."

Although they are placed in different subclasses, in one respect *Sphenodon* and the crocodilians resemble each other and differ from all other living reptiles: they have diapsid skulls, which have both dorsal and temporal fossae with their bounding arches. The turtles have anapsid skulls, without fossae; the lizards and snakes have lost one or both of the arches.

ORDER RHYNCHOCEPHALIA

The rhynchocephalians are primitive reptiles that look like lizards, but differ from them not only in having a two-arched skull, but in many other structural characters as well. Teeth are present on the premaxillary, maxillary, palatine, and dentary, and vestigially on the vomer. Well-developed gastralia are present. The male lacks a copulatory organ. The cloacal opening is a transverse slit. A nictitating membrane, or third eyelid, can be moved slowly across the eyeball from the inner corner of the eye outward while the upper and lower lids remain open. A well-developed parietal eye, with small lens and retina, is present on top of the head. In the young it can be seen clearly through the translucent covering scale, but in the adult the skin above thickens. A similar structure is present in many lizards.

Family Sphenodontidae

Sphenodon punctatus is scarcely known to most nonzoologists, but to the professional it is one of the most fascinating creatures alive. It shows us, in the flesh, what some of the early reptiles of the Mesozoic must have been like. One of the most perplexing things about it is its apparent failure either to evolve or to become extinct. If we could learn why it has remained virtually unchanged through such a long time, we might understand more of the forces that do bring about evolutionary change in most living things.

The adult Tuatara is about 500 to 800 millimeters long. It is brownish-olive, and has a small yellow spot in the center of each scale. Enlarged scales form a crest down the back and tail.

Tuataras are found only in New Zealand. Those on the main islands succumbed rapidly to the onslaught of the mammals introduced by European settlers, and the remainder now inhabit only the waterless offshore islands, where they are rigidly protected. They live in close association with vast colonies of nesting shearwaters, called Mutton Birds by the New Zealanders. These birds nest in underground burrows, which they share with the Tuataras. For the most part the association seems amicable, although the normally insectivorous Tuataras occasionally feed on eggs or nestling birds.

The Tuatara remains in its burrow during the day and prowls at night, when the temperature drops sharply and cold gusts of wind sweep over the islands. These animals are active at lower temperatures than other reptiles, and their body temperature tends to be lower than that of their surroundings. Body temperatures ranging from 6.2° to 13.3°C have been reported in nature.

Mating has been observed both in the field and in the laboratory. Insemination takes place by simple cloacal apposition. During the summer (which, south of the equator, is from November to January) the female lays about ten white, hard-shelled, elongate eggs about 28 millimeters long. They are usually deposited well away from the home burrow, in a shallow hole in sand where they can be warmed by the sun. By August, the embryos are nearly mature. However, the late-stage embryo apparently undergoes a sort of estivation over the second summer, and does not hatch until it is about 13 months old. During this estivation period, the nasal chambers become blocked with a proliferating epithelium that is resorbed shortly before hatching.

ORDER CROCODYLIA

Crocodilians are elongated reptiles with a muscular, laterally compressed tail, a more or less elongated snout, and two pairs of short legs. There are five toes on the front feet and four on the hindfeet. A nictitating membrane is present. The cloacal opening is a longitudinal slit, the penis is single, and there is no urinary bladder. The tongue is not protrusible. Gastralia and dorsal and ventral epidermal scales, reinforced by bony plates, are present. The skull is diapsid.

In some respects, crocodilians are the most advanced of all living reptiles. They possess a true cerebral cortex and a completely four-chambered heart. The teeth are thecodont (set in sockets in the jaw). The skin of the head is fused to the skull bones, and there are no fleshy lips to make a watertight closure of the mouth possible. The nostrils are far forward on top of the elongate snout. The maxillaries, palatines, and pterygoids meet in the midline of the roof of the mouth to form a secondary palate, which separates the nasal passages from the mouth. These air passages open into the throat behind a valve formed by a fleshy fold at the back of the tongue, which meets a similar fold on the palate. Water in the mouth is thus kept separate from the inspired air, and the crocodile is able to breathe while submerged, with only the tip of its snout protruding, or while holding prey in the water.

Whereas *Sphenodon* is of interest mainly to zoologists, crocodilians have a horrible fascination for almost everyone. They truly look like monsters out of the prehistoric past. The huge size attained by some—the maximum length recorded for an American Crocodile is 690 centimeters—and the man-eating proclivities of a few so color our concept of the whole group that we think of all crocodiles as ferocious monsters. Actually, some are dwarf forms quite harmless to man. The Congo Dwarf Crocodile is rarely more than a meter long.

Crocodiles are an old and diverse group. Today's members represent only one (suborder Eusuchia) of the five lineages that existed in the Mesozoic.

Only eight genera remain, and these are facing extinction through human exploitation and the destruction of habitat caused by human population growth.

Family Crocodylidae

The snout of the true crocodiles is not sharply set off from the posterior section of the skull. The maxillary bones do not meet dorsally to separate the nasals from the premaxillaries (Figure 17–1). There are from 14 to 24 teeth on each side of the lower jaw.

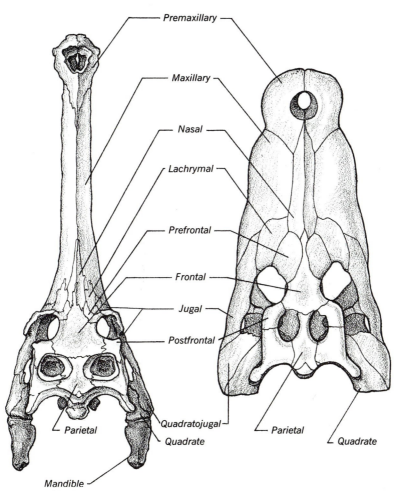

FIGURE 17–1
The skulls of *Gavialis gangeticus (left)* and *Crocodylus palustris (right)*, showing the disposition of the nasal bones (separated from the premaxillaries by the maxillaries in the former but not in the latter). [After M. A. Smith, 1931.]

All crocodilians are aquatic, living in rivers, marshes, and lakes, and some take freely to salt water. They frequently bask in the sun on shore, but most species seldom wander far from water. The long, strong tail is used for swimming and is also a powerful weapon of offense or defense. With it a victim can be literally swept off his feet, and is then seized and dragged into the water before he can recover and escape. If the prey is too large to be swallowed whole, the crocodile tears it limb from limb by the gruesomely effective method of rotating its whole body rapidly over and over in the water.

The sharp-pointed teeth are used for seizing, but not for chewing. The muscular stomach functions like the gizzard of a bird, and like a bird, a crocodilian swallows hard objects to aid in grinding its food. Sixty-nine pebbles were once found in the stomach of a Smooth-fronted Caiman (*Paleosuchus*). In the southeastern United States, where stones are scarce, alligator stomachs have been found to contain Coca-Cola bottles, bottle tops, the brass portions of shotgun shells, and "lightered knots," as the natives call hard knots of resinous pine.

A few lizards and turtles have voices, but for the most part reptiles are silent creatures. Crocodilians are an exception. The young of some species make a curious, high-pitched croak when disturbed. The voice of the adult male is something between a deep bark and a bellow. In the spring months, the bellowing of a bull alligator is perhaps the most impressive sound of the southeastern swamps, and is something that every herpetologist looks forward to hearing.

All crocodilians lay large, hard-shelled eggs, either in a shallow excavation on a sandy shore or in a nest piled by the mother. The female American Alligator scoops mud and vegetation with her jaws to build a mound about 90 centimeters high and 150 to 210 centimeters wide at the base. She deposits 20 to 70 eggs in a hollow in the center and covers them with material from the rim of the nest. She remains on guard nearby for 9 or 10 weeks, until she hears the young, now ready to emerge, beginning to peep loudly. She then tears open the nest and allows them to escape.

The Crocodylidae are commonly divided into three subfamilies. Because each of the subfamilies has been recognizable since the late Cretaceous, some believe they should be elevated to family status. The distribution of the crocodylids is shown in Figure 17–2.

Subfamily Alligatorinae.   The alligatorines are the broad-snouted and predominantly New World genera. They differ from the crocodilians in having the fourth mandibular tooth inside the mouth when it is closed; the tooth fits into a pit in the upper jaw. Also the fourth maxillary tooth is the largest tooth, and the upper teeth overlap the lower.

The Caimans (*Caiman, Melanosuchus,* and *Paleosuchus*) are neotropic crocodilians (Figure 17–3). They range in size from the small *Paleosuchus*

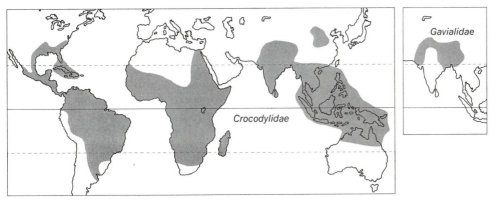

FIGURE 17–2
Distribution of the crocodylids and gavialids.

FIGURE 17–3
Young Spectacled Caimans, *Caiman sclerops,* of the Amazon and Orinoco Rivers of South America.

palpebrosus (average adult, 90–120 cm) to the large *Melanosuchus niger* (300–400 cm). Like the alligators, they are primarily fresh-water animals of rivers, lakes, and swamps. *Alligator* is the most northern in distribution of all crocodilians, being found in the southeastern United States *(A. mississippiensis)* and southern China *(A. sinensis).* Both species hibernate in semiaquatic burrows to avoid the cold winter temperatures.

Subfamily Crocodylinae. The crocodylines, or true crocodiles, range from narrow- to broad-headed species and are circumtropical in distribution.

When the mouth is closed, the fourth mandibular tooth is visible, fitting into a groove on the outside of the upper jaw, and the upper and lower teeth alternate. The fifth maxillary tooth is the largest.

The eleven species of *Crocodylus* are found in the tropics of five continents. Three species, *C. niloticus* (Nile Crocodile), *C. palustris* (Indian Mugger), and *C. porosus* (Salt-water Crocodile), are the ones that have given the crocodilians a bad image, as they find man an acceptable food item. *Crocodylus* has both brackish and fresh-water species. In many areas, one species occupies the brackish habitats and is replaced inland by a fresh-water species. Of this group, the African *Osteolaemus tetraspis* is the smallest (100–150 cm).

Subfamily Tomistominae. The False Gavial, *Tomistoma schlegeli,* is the only member of the tomistomine subfamily. It shares many characters with the crocodylines but differs in possessing an extremely long and narrow snout. Both the first and fourth mandibular teeth are exposed in grooves on the upper jaw when the mouth is closed. The long, thin teeth and elongate jaws are an adaptation for catching fish.

Family Gavialidae

The true Gavial, the sole member of the gavialid family, also has an extremely long and slender snout (Figure 17–4). When the mouth is closed,

FIGURE 17–4
A Gavial, *Gavialis gangeticus,* with a fish in its mouth. [Courtesy of C. A. Ross.]

the fourth mandibular tooth and all teeth anterior to it lie in grooves on the outside of the upper jaw, giving the tip of the snout a pincushion appearance. The lower-jaw symphysis extends almost to the end of the tooth row. Adults range from 350 to 450 centimeters; the largest reported specimen was 625 centimeters.

The single living species, *Gavialis gangeticus*, is found only in India and Burma. Its colloquial name in India is "gharial." The generic name was based on this, but through a clerical error was published as *Gavialis* rather than *Garialis*. Now the misformed scientific name has given rise to a new common name, Gavial, which has largely replaced the original Gharial.

The very narrow snout is an adaptation for eating fish, which the Gavial catches by a sudden, sideward sweep of the head through the water. A broad snout could not be moved so rapidly. The Gavial has been reported to catch birds and such fair-sized mammals as goats and dogs occasionally, but in spite of its large size it seldom if ever attacks man.

The female lays 40 or more eggs in a hole in a sand bank. These eggs are 85 to 90 millimeters long by 65 to 70 millimeters wide. The young, which appear in March and April, are about 375 millimeters long. Obviously they must have been very tightly coiled within the eggs.

READINGS AND REFERENCES

Brazaitis, P. "The identification of living crocodilians." *Zoologica*, vol. 58, no. 3–4, 1974.

Cope, E. D. "The crocodilians, lizards, and snakes of North America." *Report of the United States National Museum for 1898*, pt. 2, 1900.

Guggisberg, C. A. W. *Crocodiles: Their Natural History, Folklore and Conservation.* Harrisburg, Pa. Stackpole, 1972.

McIlhenny, E. A. *The Alligator's Life History.* Boston: Christopher Publishing House, 1935.

Neill, W. T. *The Last of the Ruling Reptiles: Alligators, Crocodiles and their Kin.* New York: Columbia University Press, 1971.

Reese, A. M. *The Alligator and its Allies.* New York: Putnam, 1915.

Ricciuti, E. R. *The American Alligator: Its Life in the Wild.* New York: Harper & Row, 1972.

Sharell, R. *The Tuatara, Lizards and Frogs of New Zealand.* London: Collins, 1966.

Wermuth, H., and R. Mertens. *Schildkröten, Krokodile, Brückeneschsen.* Jena: Gustav Fischer, 1961.

CLASSIFICATION
OF LIVING AMPHIBIANS
AND REPTILES

This table presents the classification of living amphibians and reptiles followed in this book. All groups are classified to the subfamilial level.

CLASS AMPHIBIA
    Superorder Lissamphibia
      Order Gymnophiona (Apoda)
          Family Caeciliidae
            Subfamily Caeciliinae
            Subfamily Dermophiinae
          Family Ichthyophiidae
          Family Scolecomorphidae
          Family Typhlonectidae
      Order Caudata (Urodela)
        Suborder Cryptobranchoidea
          Family Cryptobranchidae
          Family Hynobiidae

Suborder Salamandroidea
    Family Ambystomatidae
      Subfamily Ambystomatinae
      Subfamily Dicamptodontinae
      Subfamily Rhyacotritoninae
    Family Amphiumidae
    Family Plethodontidae
      Subfamily Desmognathinae
      Subfamily Plethodontinae
    Family Proteidae
    Family Salamandridae
Order Meantes (Trachystomata)
    Family Sirenidae
Order Salienta (Anura)
  Suborder Xenoanura
    Family Pipidae
      Subfamily Pipinae
      Subfamily Xenopinae
    Family Rhinophrynidae
  Suborder Scoptanura
    Family Microhylidae
      Subfamily Asterophryninae
      Subfamily Brevicipitinae
      Subfamily Cophylinae
      Subfamily Dyscophinae
      Subfamily Hoplophryninae
      Subfamily Microhylinae
      Subfamily Phrynomerinae
  Suborder Lemnanura
    Family Ascaphidae
    Family Discoglossidae
    Family Leiopelmatidae
  Suborder Acosmanura
    Family Pelobatidae
      Subfamily Megophryinae
      Subfamily Pelobatinae
      Subfamily Pelodytinae

Family Hyperoliidae
  Subfamily Arthroleptinae
  Subfamily Astylosterninae
  Subfamily Hyperoliinae
  Subfamily Scaphiophryninae
Family Ranidae
  Subfamily Hemisinae
  Subfamily Phrynobatrachinae
  Subfamily Platymantinae
  Subfamily Raninae
Family Rhacophoridae
  Subfamily Mantellinae
  Subfamily Rhacophorinae
Family Sooglossidae
Family Allophrynidae
Family Brachycephalidae
Family Bufonidae
Family Centrolenidae
Family Dendrobatidae
Family Hylidae
  Subfamily Amphignathodontinae
  Subfamily Hemiphractinae
  Subfamily Hylinae
  Subfamily Phyllomedusinae
Family Heleophrynidae
Family Leptodactylidae
  Subfamily Ceratophryinae
  Subfamily Hylodinae
  Subfamily Leptodactylinae
  Subfamily Telmatobiinae
Family Myobatrachidae
  Subfamily Limnodynastinae
  Subfamily Myobatrachinae
  Subfamily Rheobatrachinae
Family Pelodryadidae
Family Pseudidae
Family Rhinodermatidae

CLASS REPTILIA
  Subclass Anapsida
    Order Testudines (Testudinata)
     Suborder Cryptodira
       Superfamily Testudinoidea
         Family Chelydridae
           Subfamily Chelydrinae
           Subfamily Platysterninae
         Family Emydidae
           Subfamily Batagurinae
           Subfamily Emydinae
         Family Testudinidae
       Superfamily Trionychoidea
         Family Dermatemydidae
         Family Kinosternidae
           Subfamily Kinosterninae
           Subfamily Staurotypinae
         Family Carettochelyidae
         Family Trionychidae
           Subfamily Lissemyinae
           Subfamily Trionychinae
       Superfamily Chelonioidea
         Family Cheloniidae
           Subfamily Carettinae
           Subfamily Cheloniinae
         Family Dermochelyidae
     Suborder Pleurodira
         Family Pelomedusidae
         Family Chelidae
  Subclass Lepidosauria
    Order Rhynchocephalia
         Family Sphenodontidae
    Order Squamata
     Suborder Amphisbaenia
         Family Amphisbaenidae
         Family Trogonophidae
     Suborder Sauria (Lacertilia)

Infraorder Gekkota
    Family Gekkonidae
       Subfamily Diplodactylinae
       Subfamily Eublepharinae
       Subfamily Gekkoninae
       Subfamily Sphaerodactylinae
    Family Pygopodidae
    Family Xantusiidae
       Subfamily Cricosaurinae
       Subfamily Xantusiinae
Infraorder Iguania
    Family Agamidae
    Family Chamaeleonidae
    Family Iguanidae
Infraorder Anguinomorpha
   Superfamily Anguinoidea
    Family Anguinidae
       Subfamily Anguininae
       Subfamily Anniellinae
       Subfamily Diploglossinae
       Subfamily Gerrhonotinae
    Family Xenosauridae
       Subfamily Shinisaurinae
       Subfamily Xenosaurinae
   Superfamily Varanoidea
    Family Helodermatidae
    Family Lanthanotidae
    Family Varanidae
Infraorder Dibamia
    Family Dibamidae
Infraorder Scincomorpha
    Family Cordylidae
       Subfamily Cordylinae
       Subfamily Zonurinae
    Family Lacertidae
    Family Scincidae
       Subfamily Acontinae

Subfamily Feylininae
Subfamily Lygosominae
Subfamily Scincinae
Family Teiidae
Subfamily Gymnophthalminae
Subfamily Teiinae
Suborder Serpentes (Ophidia)
Infraorder Scolecophidia
Family Anomalepidae
Family Leptotyphlopidae
Family Typhlopidae
Infraorder Henophidia
Family Acrochordidae
Family Aniliidae
Family Boidae
Subfamily Boinae
Subfamily Bolyerinae
Subfamily Erycinae
Subfamily Pythoninae
Subfamily Tropidophiinae
Family Uropeltidae
Family Xenopeltidae
Infraorder Caenophidia
Family Colubridae
Subfamily Colubrinae
Subfamily Natricinae
Subfamily Homalopsinae
Subfamily Lycodontinae
Subfamily Xenodontinae
Family Elapidae
Subfamily Elapinae
Subfamily Hydrophiinae
Subfamily Laticaudinae
Family Viperidae
Subfamily Azemiopinae
Subfamily Crotalinae
Subfamily Viperinae

Subclass Archosauria
    Order Crocodylia
            Family Crocodylidae
                Subfamily Alligatorinae
                Subfamily Crocodylinae
                Subfamily Tomistominae
            Family Gavialidae

SOURCES
OF IDENTIFICATION

The following books and scientific articles are a small fraction of the published literature on the identification of amphibians and reptiles. This list does not attempt to be complete, but does attempt to provide worldwide coverage. For the reader who wishes to dig deeper, we recommend the *Zoological Record,* an excellent yearly summary of the herpetological literature.

NEARCTIC REGION

Conant, Roger. *A Field Guide to the Reptiles and Amphibians of Eastern and Central North America.* Boston: Houghton Mifflin, 1975.

Smith, Hobart, and Edward H. Taylor. "An annotated checklist and key to the snakes of Mexico." *United States National Museum Bulletin* (187), 1945.

———. "An annotated checklist and key to the Amphibia of Mexico."*U.S. Natl. Mus. Bull.* (194), 1948.

———. "An annotated checklist and key to the reptiles of Mexico exclusive of the snakes." *U.S. Natl. Mus. Bull.* (199), 1950.
These three volumes have been reprinted as *Herpetology of Mexico,* 1966, with an updated species list.

Stebbins, Robert. *A Field Guide to Western Reptiles and Amphibians.* Boston: Houghton Mifflin, 1966.

NEOTROPICAL REGION

Cochran, Doris M. "Frogs of southeastern Brazil." *U.S. Natl. Mus. Bull.* (206), 1954.

———, and Coleman J. Goin. "Frogs of Colombia." *U.S. Natl. Mus. Bull.* (288), 1970.

Duellman, William E. "The hylid frogs of Middle America," vol. 1 and 2. *University of Kansas Museum of Natural History,* monograph no. 1, 1970.

Meyer, John R., and Larry D. Wilson. "A distributional checklist of the amphibians of Honduras." *Los Angeles County Museum, Contributions in Science* (218), 1971.

———. "A distributional checklist of the turtles, crocodilians, and lizards of Honduras." *L.A. Co. Mus., Cont. Sci.* (244), 1973.

Peters, James A., *et al.* "Catalogue of the neotropical Squamata. Part I: Snakes. Part II. Lizards and amphisbaenians." *U.S. Natl. Mus. Bull.* (297), 1970.

Schwartz, Albert, and Richard Thomas. "A check-list of West Indian amphibians and reptiles." *Carnegie Museum of Natural History, Special Publication* (1), 1975.

Villa, Jaime. *Anfibios de Nicaragua.* Managua: Instituto Geográfico Nacional y Banco Central de Nicaragua, 1971.

PALEARCTIC REGION

Hellmich, W. *Reptiles and Amphibians of Europe.* London: Blandford, 1962.

Kahalaf, Kamel T. *Reptiles of Iraq with Some Notes on the Amphibians.* Baghdad: Ar-Rabitta Press, 1959.

Liu, Ch'eng-Chao. "Amphibians of western China." *Fieldiana: Zoology Memoirs* (2), 1950.

———, and Hu Shu-Qin. *The Anura of China.* Peking: Science Press, 1960. (In Chinese with Latin index and figure legends.)

Marx, H. *Checklist of the reptiles and amphibians of Egypt.* Cairo: Special Publication of U.S. Naval Medical Research Unit Number Three, 1968.

Mertens, Robert. *Die Amphibien und Reptilien Europas.* Frankfurt: Waldemar Kramer, 1960.

Okada, Yaichiro. *Fauna Japonica Anura (Amphibia)*. Tokyo: Tokyo Electrical Engineering College Press, 1966.

Pasteur, G., and J. Bons. "Les batraciens du Maroc." *Travaux de l'Institut Scientifique Chérifien, Série Zoologique* (17), 1959.

———. "Catalogue des reptiles actuels du Maroc." *Trav. Inst. Sci. Chér., Sér. Zool.* (21), 1960.

Pope, Clifford H. *The reptiles of China*. New York: American Museum of Natural History, 1935.

Stejneger, Leonhard. "Herpetology of Japan and adjacent territory." *U.S. Natl. Mus.* (58), 1907.

Terent'ev, P. V., and S. A. Chernov. *Key to Amphibians and Reptiles* [U.S.S.R.], translation. Springfield: Clearinghouse for Scientific and Technical Information, 1965.

Thorn, Robert. *Les Salamandres d'Europe, d'Asie, et d'Afrique du Nord*. Paris: Editions Paul Lechevalier, 1968.

ETHIOPIAN REGION

Broadley, Donald G. "The reptiles and amphibians of Zambia." *The Puku* (6), 1971.

Dunger, G. T. "The lizards and snakes of Nigeria." (11 parts). *The Nigerian Field,* 32–38, 1967–1973.

Fitzsimons, Vivian F. *The Lizards of South Africa*. Pretoria: Transvaal Museum, 1943.

———. *A Field Guide to the Snakes of Southern Africa*. London: Collins, 1970.

Loveridge, Arthur, and Ernst E. Williams. "Revision of the African tortoises and turtles of the suborder Cryptodira." *Bulletin of the Museum of Comparative Zoology,* 115(6), 1957.

Perret, Jean-Luc. "Les amphibiens du Cameroun." *Zoologische Jahrbücher Abt. Systematik, Ökologie und Geographie der Tiere,* 93, 1966.

Pitman, Charles R. S. *A Guide to the Snakes of Uganda*. Codicote: Wheldon & Wesley, 1974.

Poynton, J. C. "The Amphibia of southern Africa." *Annals of the Natal Museum,* 17, 1964.

Stewart, Margaret M. *Amphibians of Malawi*. Albany: State University of New York Press, 1967.

Villiers, A. *Tortues et crocodiles de l'Afrique Noire Française*. Dakar: Institut Français d'Afrique Noire, 1958.

———. *Les Serpents de l'Ouest Africain*. Dakar: Inst. Franc. d'Afr. N., 1963.

INDIAN REGION

Inger, Robert F. "Systematics and zoogeography of Philippine Amphibia." *Fieldiana: Zoology,* 33, 1954.

———. "The Amphibia of Borneo." *Fieldiana: Zool.,* 52, 1966.

Kampen, P. N. van. *The Amphibia of the Indo-Australian Archipelago.* Leiden: E. J. Brill, 1923.

Minton, Sherman A. "An annotated key to the amphibians and reptiles of Sind and Las Bela, West Pakistan." *American Museum Novitates* (2081), 1962.

Rooij, Nelly de. *The Reptiles of the Indo-Australian Archipelago.* I, *Lacertilia, Chelonia, Emydosauria,* and II, *Ophidia.* Leiden: E. J. Brill, 1915, 1917, respectively.

Smith, Malcolm A. *The Fauna of British India, including Ceylon and Burma,* vol. I, *Loricata, Testudines,* vol. II, *Sauria,* vol. III, *Serpentes.* London: Taylor and Francis, 1931, 1935, 1943, respectively.

Taylor, Edward H. *The Snakes of the Philippine Islands.* Manila: Bureau of Printing, 1922.

———. *The Lizards of the Philippine Islands.* Manila: Bureau of Printing, 1922.

———. "The amphibian fauna of Thailand." *University of Kansas Science Bulletin,* 63, 1962.

———. "The lizards of Thailand." *Univ. Kansas Sci. Bull.,* 64, 1963.

———. "The serpents of Thailand and adjacent waters." *Univ. Kansas Sci. Bull.,* 65, 1965.

AUSTRALIAN REGION AND OCEANIA

Cogger, H. *The Reptiles and Amphibians of Australia.* Sydney: Reed, 1975.

Loveridge, A. *The Reptiles of the Pacific World.* New York: Macmillan, 1945.

Menzies, J. *The Common Frogs of Papua New Guinea.* Honolulu: University of Hawaii Press, 1975.

Sharell, Richard. *The Tuatara, Lizards, and Frogs of New Zealand.* London: Collins, 1966.

Tyler, Michael J. "Papuan hylid frogs of the genus *Hyla.*" *Zoologische Verhandelingen* (96), 1968.

See also Kampen and Rooij, cited under Indian Region.

WORLDWIDE

Brazaitis, Peter. "The identification of living crocodilians." *Zoologica,* 58(¾), 1974.

Moore, Granville M. (ed.). *Poisonous Snakes of the World.* Washington: United States Government Printing Office, (NAVMED P–5099), 1966.

Pritchard, Peter C. H. *Living Turtles of the World.* Jersey City: T. F. H. Publ., 1967.

Taylor, Edward Harrison. *The Caecilians of the World.* Lawrence: University of Kansas Press, 1968.

Wermuth, H., and R. Mertens. *Schildkröten, Krokodile, Brückeneschsen.* Jena: Gustav Fischer, 1961.

SCIENTIFIC NAME INDEX

Page numbers in italic indicate the pages containing the characterization of the taxon. A page number followed by *ff* indicates that the taxon is used as an example over several pages.

SUBJECT INDEX